AF324023

Molecular Simulations and Biomembranes
From Biophysics to Function

Molecular Simulations and Biomembranes
From Biophysics to Function

Edited by

Mark S.P. Sansom and Philip C. Biggin
Department of Biochemistry, University of Oxford, Oxford, UK

RSCPublishing

RSC Biomolecular Sciences No. 20

ISBN: 978-0-85404-189-3
ISSN: 1757-7152

A catalogue record for this book is available from the British Library

© Royal Society of Chemistry 2010

Published by The Royal Society of Chemistry,
Thomas Graham House, Science Park, Milton Road,
Cambridge CB4 0WF, UK

Registered Charity Number 207890

For further information see our web site at www.rsc.org

Preface

The importance of membrane proteins is clear from their prevalence in the genome (*ca.* 25% of genes encode membrane proteins[1]). They also constitute ∼50% of current drug targets, and have key roles in a wide range of cellular functions, including transport, signalling and cell–cell interactions. Despite their physiological and pharmaceutical importance, membrane proteins constitute <1% of known protein structures. Thus, there are currently less than 200 distinct high-resolution membrane protein structures, of which just over half consist of bundles of hydrophobic transmembrane (TM) α-helices. For membrane proteins, interaction with lipids is essential for protein function; bilayer properties, such as hydrophobic thickness or lipid composition, can affect membrane protein activity.[2] Although often crystallized as membrane protein–detergent complexes, in most cases only a few tightly bound lipid molecules remain.[3] Thus, the crystal structure rarely contains explicit information on where the protein is located in the bilayer.

In addition to the proteins in membranes, the lipid bilayer environment is a complex two-dimensional liquid crystalline system. Furthermore, it is difficult to map details of protein–membrane interactions using experimental techniques. This makes biomembranes an ideal subject for computer simulations. However, due to the size of most membrane proteins and the simulation timescales involved, it is only recently that simulations have enabled prediction of biological properties to be made. Atomistic-detail (AT) approaches still generally fall short of timescales for, *e.g.*, helix aggregation/folding, but recent, more extended simulations have revealed aspects of functionally relevant local motions and of interactions with lipid molecules. New developments in coarse-grained (CG) simulations potentially can probe multi-microsecond dynamics of extremely large systems. Furthermore, the combination of CG and AT simulations in multi-scale approaches offers the possibility of combining efficient sampling with accurate modelling of key local interactions.

RSC Biomolecular Sciences No. 20
Molecular Simulations and Biomembranes: From Biophysics to Function
Edited by Mark S.P. Sansom and Philip C. Biggin
© Royal Society of Chemistry 2010
Published by the Royal Society of Chemistry, www.rsc.org

This book reviews recent progress in simulations of biomembranes, ranging from fundamental methodologies and advances in modelling of lipid bilayers through to applications to a number of transport and signalling proteins.

In Chapter 1, Peter Tieleman provides an introduction to the methodologies and parameters that are commonly used in molecular dynamics simulations of biomembranes. He begins by discussing force fields that are commonly used and how parameters are developed before going on to highlight some of the problems that exist in trying to simulate a greater variety of compounds and their interaction with the membrane. A central issue to performing simulations of membrane proteins embedded within a lipid bilayer is the initial setup. The various options for this are discussed, followed by what is probably the biggest problem not only for membrane simulations, but also for molecular dynamics (MD) simulations in general: the sampling problem. Although there are problems, we can also be reassured that there are several methods available that can improve sampling to the point that the results are both reliable and reproducible. The chapter then focuses on three main methodological areas that are of particular importance for membrane simulations: pressure coupling, electrostatics and periodicity. In fact, a lot of the parameter choices made for pressure coupling and electrostatics stem from the fact that the membrane is nearly always treated as a periodic system. Finally, the chapter ends with an indication of the future developments in this area, including continued work in force field development. For example, the inclusion of polarizable lipids is something that is going to appear with increasing frequency.

Chapter 2, by Ollila and Vattulainen, illustrates very nicely how detailed application of the principles laid out in Chapter 1 can provide detailed information, some of which is often not accessible very easily by experimental techniques. They focus in particular on the lateral pressure profiles that exist within biomembranes and discuss how they can be influenced by various factors, including lipid composition, the presence of sterols and anaesthetics. They also demonstrate that the lateral pressure profile exhibits large variations as one moves along the bilayer normal and that as a consequence it is not surprising that membrane proteins, such as the mechanosensitive channel (MscL), are able to exploit this to great mechanistic effect. Although much of the work on lateral pressure profiles has been performed with atomistic molecular dynamics, Ollila and Vattulainen also discuss contributions from coarse-grained molecular dynamics simulations which, provided that care is taken with respect to the parameterization, offer the advantage of larger systems being run for longer periods of time.

Coarse-grained simulations are also making an increasing impact in molecular dynamics simulations of membrane proteins and peptides. The focus of Chapter 3, by Rouse, Carpenter and Sansom, is to provide a brief introduction to the general methodology and to highlight some recent test cases. In particular, it addresses how CG simulations may be used to explore (i) the interactions of α-helices with a lipid bilayer and (ii) the interactions of transmembrane α-helices with one another within a lipid bilayer. The latter is of relevance both to modelling of membrane protein folding and to signalling across membranes by changes in helix oligomerization and/or packing.

In Chapter 4, Orsi and Essex provide an overview of how atomistic and coarse-grained simulations have contributed to our understanding of passive permeation of small molecules and drugs. Such information is of enormous benefit to the pharmaceutical industry, where advance knowledge of the permeation properties of a candidate drug molecule can potentially make huge savings by avoiding the synthesis of candidate molecules that have bad permeation characteristics. After describing some of the main experimental techniques and various solubility–diffusion models, the authors describe how the '*z*-constraint method' in particular has been used to good effect to calculate permeability coefficients of small molecules. The chapter also discusses a relatively new area of interest, that of nanomaterials, such as fullerenes, and their interaction with membranes, before concluding that computational power is now at a point where significant progress in this area can be expected.

In Chapter 5, Martin and Jakob Ulmschneider provide a comprehensive discussion of the progress made in using implicit membrane models to study peptide folding. Implicit models have the advantage that sampling times can be significantly expanded compared with atomistic representations. Indeed, for the well-studied model peptide WALP, they argue that by most commonly used metrics used to evaluate conformational exploration, the speed-up is somewhere between a factor of 50 and 100. Although there are certain limitations to the implicit model (for example, the lack of bilayer deformation), the authors demonstrate that these models can be extremely useful for addressing certain questions and can match experimentally derived data very well indeed. Although currently limited to small peptides, there is no doubt that future developments in this area will ultimately allow us to make progress in *ab initio* predictions of membrane proteins.

In Chapter 6, Yin, Arkhipov and Schulten provide an excellent example of the application of a multi-scale approach. Through a systematic study of the interaction of BAR domains with membranes, we are guided through four different levels of treatment, starting from atomistic molecular dynamics and all the way up to a continuum representation based on elastic theory. Some of their findings could only really have been obtained by using this approach due to the sheer size and timescale of these systems. For example, it appears that the lattice arrangements of these BAR domains is critical in determining the end result in terms of membrane curvature. The authors allude to the fact that multi-scale approaches will be an increasing necessity if we are to make progress properly in the growing field of systems biology.

Ion channels are an important class of membrane proteins, from a both a biological and a pharmacological perspective, and have been the focus of a wide range of computation studies. In Chapter 7, Robertson, Jogini and Roux describe how continuum electrostatics calculations can be used to provide valuable insights into the relationship between structure and biological function in potassium channels. These studies focus on a careful treatment of electrostatics both within the transmembrane region and also in those regions of some K^+ channel protein pores which project beyond the lipid bilayer. They also discuss how the voltage difference across a biological membrane results in a field which is focused on the voltage-sensing elements of Kv channels.

Apparently related to potassium channels, at least in the transmembrane domain, are the ionotropic glutamate receptors. This family of receptors mediates nearly all of the fast synaptic neurotransmission in the brain and consequently is of interest in terms of both understanding memory and learning, and also the pathology of many neurological disorders and conditions. In Chapter 8, Vijayan, Iorga and Biggin discuss how computational methods have contributed to our understanding of this important class of proteins. In addition to summarizing the work performed so far, they indicate the areas for future study. In particular, the recent tetrameric structure opens up the field to a whole range of questions.

In Chapter 9, we move to the outer membrane of Gram negative bacteria as Khalid and Baaden describe how molecular dynamics simulations have contributed to our understanding of beta barrel proteins. Even in the context of so-called simple barrels, MD studies (on OmpA) have shown that underlying motions may play a prominent role in controlling the nature of the pattern of water flux through these channels *via* modulation of conformation of the salt bridges. This once again highlights the intimate nature of the relationship between structure, dynamics and function, as inferred by Ollila and Vattulainen in Chapter 2. Khalid and Baaden also draw out attention to the fact that despite possessing relatively simple architecture, some of these proteins appear to be more than just holes in the membrane. Autotransporters exploit a barrel (formed by a C-terminal domain within the autotransporter itself) to secrete themselves out of the cell. The precise mechanism of this process is still unclear and it is likely that MD will help to improve models of secretion for these proteins. It also appears that the barrel proteins serve as suitable scaffolds for common enzymatic classes such as phospholipases (OMPLAA) and proteases (OmpT). Often the key residues for these reactions are located on loops which are presumably highly flexible and particularly prone to lattice forces in X-ray crystallography. Thus MD can offer a complementary approach to the study of these proteins. The chapter ends with a discussion of how barrel proteins are also being increasingly exploited in technological applications such as bio-sensors. One approach is to recognize analytes by the 'signature' single-channel signal they give. MD simulations can be used to study the key interactions and to predict, *in silico*, mutations that might lead to better discriminatory power.

Chapter 10 by Emad Tajkhorshid and co-workers illustrates how large increases in computer power and recent progress in membrane protein X-ray crystallography can give insight into the mechanisms of active transport. Using large-scale molecular dynamics simulations, they have shown that although simulation of the complete cycle of a membrane transporter is still some way off, these calculations can be used to capture successfully many of the key steps involved in active transport. This approach is particularly complementary to the study of these systems where experiments are still difficult to design in the absence of structural knowledge of the full pathway of translocation.

Simulations can now be used to explore the possible exploitation of biomembranes in bionanotechnology. One such application is described in Chapter 11, where Wallace and Sansom review the use of both atomistic and coarse-grained simulations to explore the interactions between carbon nanotubes (CNTs) and model biomembranes. Issues of parameterization of CNTs for simulations are of special importance, and are likely to be an area of future refinement of methodologies. Simulations have been used to characterize the interactions of CNTs with detergent and lipid molecules, and also with simple lipid bilayers. Once embedded within a bilayer, CNTs may form transbilayer pores. Simulations have been used to explore the behaviour of water and ions in such pores, and to explore their potential as 'nanosyringes' for injection across cell membranes.

In summary, the studies reviewed in this book show that MD simulations and related computational studies are now a mainstream component of studying the biophysics and function of membranes and their proteins. The evidence for this is provided in part by the increasing use of simulation studies to aid the functional analysis of membrane protein structures, *e.g.* in characterizing the specificity mechanisms of transmembrane pores.[4]

From a simulation perspective, it is evident that multi-scale simulation approaches will become increasingly important, carefully matching the granularity of the simulation to that of the biological problem being examined. We will see the continued importance of methodological developments for membrane simulations, and of especial importance will be careful parameterization of a wider range of lipids. Given these developments, simulation studies in the future will address a number of key problems in membrane biology (*e.g.* transport, signalling, fusion) with increasing emphasis on improved biological realism and complexity. In this fashion, membrane simulations will have a major impact on fundamental biomedical science (*i.e.* mechanistic understanding of membrane cell biology), and on more 'applied' areas such as pharmacology and bionanotechnology of membranes.

References

1. J. Nilsson, B. Persson and G. von Heijne, *Proteins Struct. Funct. Bioinf.*, 2005, **60**, 606.
2. C. Hunte, *Biochem. Soc. Trans.*, 2005, **33**, 938.
3. H. Palsdottir and C. Hunte, *Biochim. Biophys. Acta*, 2004, **1666**, 2.
4. J. S. Hub and B. L. de Groot, *Proc. Natl. Acad. Sci. USA*, 2008, **105**, 1198.

Philip C. Biggin
Mark S.P. Sansom

Contents

RSC Biomolecular Sciences No. 20
Molecular Simulations and Biomembranes: From Biophysics to Function
Edited by Mark S.P. Sansom and Philip C. Biggin
© Royal Society of Chemistry 2010
Published by the Royal Society of Chemistry, www.rsc.org

Chapter 3 Coarse-grained Molecular Dynamics Simulations of Membrane Proteins **56**
Sarah Rouse, Timothy Carpenter and Mark S. P. Sansom

Methods and Parameters for Membrane Simulations

D. PETER TIELEMAN

Department of Biological Sciences, University of Calgary, 2500 University Drive NW, Calgary, Alberta, T2N 1N4, Canada

1.1 Introduction

Computer simulation is a powerful approach to studying the properties of models of biological membranes. Because lipids have a certain degree of intrinsic disorder in biologically relevant states, direct structural and spectroscopic experimental methods necessarily average over a large number of different lipid conformations. Experimental methods to study membrane proteins are also complicated by their lipid environment. Although much progress has been made, experimental structural, dynamic and functional data on membrane proteins lag behind water-soluble proteins. In principle, simulation can be used to track the behaviour of individual atoms, with the potential for very high-resolution information, provided that the simulation models are sufficiently accurate for the properties of interest.

Biomolecular simulation, initially focused on proteins, has matured into a widely used method over the past 30 years, while more recently simulations of lipids have made significant progress. Figure 1.1 gives a graphical view of a typical simulation setup for simulating a bilayer, in this case a mixture of several lipids with a peptide. Review articles in 1994 and 1997 could still be comprehensive,[1,2] but after that the field became too large and even reviews of subtopics such as membrane protein simulations are now rarely comprehensive.[3]

RSC Biomolecular Sciences No. 20
Molecular Simulations and Biomembranes: From Biophysics to Function
Edited by Mark S.P. Sansom and Philip C. Biggin
© Royal Society of Chemistry 2010
Published by the Royal Society of Chemistry, www.rsc.org

A recent book volume gives a good overview of the state of the field,[4] with a broad range of chapters. Other recent reviews include reviews on membrane protein simulations,[3,5–8] simulations including cholesterol[9–11] and simulations involving significant changes in the basic bilayer structure, including lipid defects, pores, domain formation, phase transitions and curvature.[12]

Biomolecular simulation in general is a combination of an interaction model to describe the interactions between atoms or molecules and a sampling algorithm to explore this model numerically, using the methods of statistical mechanics. Molecular dynamics (MD) simulation is the most commonly used method, but not the only one. Because most other methods are important primarily to explore interaction models that contain less detail (coarse-grained models), this chapter will focus exclusively on MD simulation. In molecular dynamics, the force field consists of the equations chosen to model the potential energy and their associated parameters, while the standard sampling algorithm is a numerical solution of classical equations of motion based on the forces given by the energy function. The general form of the potential energy is a sum of terms similar to

$$V(\boldsymbol{r}) = \sum_{\text{bonds}} k_b(b - b_0)^2 + \sum_{\text{angles}} k_\theta(\theta - \theta_0)^2 + \sum_{\text{dihedrals}} k_\phi[1 + \cos(n\phi - \phi_0)]$$

$$+ \sum_{\text{impropers}} k_\psi(\psi - \psi_0)^2 + \sum_{\substack{\text{non-bonded} \\ \text{pairs}(i,j)}} 4\varepsilon_{ij}\left[\left(\frac{\sigma_{ij}}{r_{ij}}\right)^{12} - \left(\frac{\sigma_{ij}}{r_{ij}}\right)^{6}\right]$$

$$+ \sum_{\substack{\text{non-bonded} \\ \text{pairs}(i,j)}} \frac{q_i q_j}{\varepsilon_D r_{ij}}$$

$$(1.1)$$

The potential function V depends on $\boldsymbol{r}$, the position vector of particles in the system, expressed in terms of distances (r_{ij} and $b - b_0$) and angles θ, ϕ and ψ between atoms. Despite occasional extra terms in some force fields, the functional form of the potential function is essentially the same in all common force fields, so that they are not fundamentally different. Indeed, the potential function embodies the most serious assumptions and will provide an upper limit to the accuracy of any parameter set to describe the interactions between atoms.

In practice, the parameters in equation (1.1) are generally stored in data files with lists corresponding to each term in the sum. The force field includes a list of particle types corresponding to the different types of atoms that occur in a system of interest. For example, the carbon atom in a methyl group CH_3 and in a carbonyl group $C{=}O$ can be described by two different particle types with two different charges [q in equation (1.1)] and potentially different Lennard-Jones parameters (ε, σ), and also specific bonds, angles and dihedral parameters. The Lennard-Jones parameters may be different for each pair of different atom types or they may be systematically derived from parameters attached to a single type of atom. The force field also includes a list of parameters to describe bonds (force constant k_b, equilibrium bond length b_0), angles

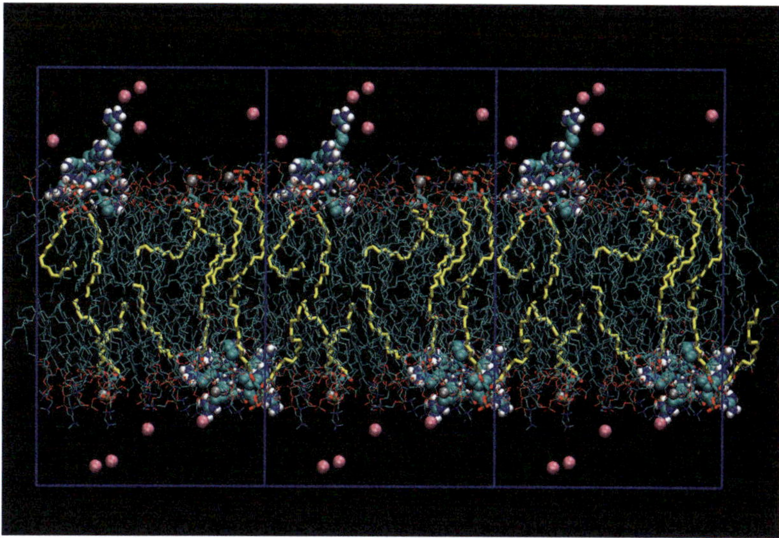

Figure 1.1 Snapshot of a typical simulation system with a lipid bilayer consisting of DOPC and DOPA (yellow) lipids and a peptide, showing the central simulation box and two of the periodic images. The peptide is $(Arg)_8$. Ions are shown; water has been omitted for clarity. Details of the simulation are given in ref. 91. Figure made with VMD.[92]

(force constant k_θ, equilibrium angle θ_0), dihedrals (force constant k_θ, phase angle θ_0, multiplicity n) to describe rotation around a central bond and improper dihedrals (force constant k_ψ, equilibrium angle ψ_0) to enforce certain specific geometries such as planar or tetrahedral groups. Combined, these parameters describe all combinations of different particle types necessary to model a particular molecule or class of molecules, *e.g.* proteins, nucleic acids or lipids, or in several modern force fields nearly every molecule one can think of.

The scale of the systems that have been studied continues to increase with increasing computer power and with more efficient simulation software. The range of applications has become extremely broad, but generally a reasonable limit on simulations at the moment is 10^6 particles, corresponding to *ca.* 5000 all-atom lipids or *ca.* 50 000 lipids in coarse-grained models. Simulation times of hundreds of nanoseconds have become routine in simple bilayer simulations, while microsecond simulations for all-atom and millisecond simulations at the coarse-grained level are pushing current limits. In terms of both time and length scales, although more prominently in time scale (which increases linearly with computer power), simulations can now probe into the microscopic regime, allowing a direct bridge to new classes of experiments such as vesicle aspiration, fluorescence imaging and atomic force microscopy (AFM) measurements.

This chapter briefly reviews the main technical choices that have to be made in simulations of lipids and membrane proteins. Most of this material is based on a number of publications from the author's group and collaborators: on electrostatics and algorithmic choices,[13,14] on force field issues for lipids and membrane proteins[15,16] and on creating starting structures and a number of other issues.[17] The goal is not to provide a comprehensive literature review, but to indicate which technical choices are considered important. For an informed decision on any of these, a broader consultation of the literature is essential.

It is assumed that simulation is a sensible approach for a particular problem, but this is not a given. In fact, the most important decision in a simulation project is likely whether a given problem is suitable for computer simulation at all. The answer to this question will depend on the problem itself – are models available that are sufficiently accurate to address the problem? Is there a possible link to experimental data? Is it likely enough sampling can be obtained with a sufficiently detailed model, given the available computational resources? Are there experimental approaches that are likely to give more insight or that are easier routes to answering a particular question? Once the decision has been made to attempt simulations, the remaining choices are important but of a more technical nature.

1.2 Force Fields/Descriptions of Interactions

The force field is the description of interactions between atoms in the simulation system. Equation (1.1) gives a typical classical potential function which combines with the (many) parameters for each type of interaction in the equation to form the force field. Strictly, there are additional modifiers, which affect the pairs of particles for which interactions are actually calculated (modulated by neighbour searching details, cut-offs, lattice sums, periodic boundary conditions and long-range corrections), the form of the interaction potential (shift or switch functions) and external influences including temperature and pressure control algorithms and potentially additional terms such as external electric fields. Several of these will be discussed in more detail below.

Different force fields exist that are fundamentally similar, but have their own strengths and weaknesses. Because the parameters are empirical, there is no unique solution for an optimal set of parameters. Hence there is a certain degree of experience and judgement involved in picking force fields and interpreting results from simulations.

Most current force fields contain terms for all atoms, while there continues to be a subset that uses 'united' atoms, in which non-polar hydrogens are grouped together with a carbon atom to make effective 'CH_1', 'CH_2' or 'CH_3' atoms. In biomolecular simulation in general, other levels of detail exist, ranging from *ab initio* quantum mechanics to very coarse-grained models that represent, for example, an entire amino acid by a single interaction site. For lipid simulations, two intermediate levels are likely to be important. Towards more detail, modifications of classical force fields to include explicitly electronic polarizability have seen significant progress in several groups and the first simulations of lipid bilayers using (partially) polarizable force fields have begun to appear in the literature.[18,19]

Towards less detail and longer time and length scales, there has been significant progress in developing coarse-grained models that average over a number of atoms, interacting through effective potentials that reproduce the average interactions of the details that have been left out. An example is the MARTINI model of Marrink and co-workers, in which on average four non-hydrogen atoms are replaced by an effective particle type that is parameterized on the properties of a library of small organic molecules.[20,21] A range of other methods exist, some of which are much coarser (and computationally cheaper), but the rest of this chapter is focused on classical atomistic simulations only.

1.2.1 Current Atomistic Force Fields

Major force fields commonly used in MD simulations of biomolecules include AMBER,[22,23] CHARMM,[24] GROMOS[25,26] and OPLS.[27] Parameters in each force field are supposed to be internally consistent, but this is not necessarily true between different force fields and may not be true between different force field versions from the same family. Most major force fields have developed a specific set of parameters for apolar solvents, common organic molecules and common phospholipids (typically at least DPPC, POPC and DOPC). However, only two phospholipid force fields are in common use today: an all-atom force field that is part of the official CHARMM distribution[24,28,29] and a united-atom force field by Berger *et al.*,[30] created by combining parameters taken from united-atom versions of OPLS and AMBER with some modifications to charges and lipid chain parameters and bonded parameters based on an older version of GROMOS. Both CHARMM-based lipids and Berger lipids reproduce the available experimental information on the structure and dynamics of phospholipid bilayers reasonably well, particularly for the experimentally well-studied phosphatidylcholine lipids, and in our opinion there is no compelling experimental information that indicates that either force field is substantially better, although their errors differ.[31] In addition to these two force fields, there has been recent progress in developing GROMOS96-based lipids,[32] AMBER parameters to describe lipids[33,34] and modifications of CHARMM parameters,[35–37] and there are several modifications of older force fields.

Only CHARMM currently provides a full set of parameters for all types of biomolecules, but recent GROMOS96 and AMBER developments are moving in that direction. Other choices of lipid parameters require combining these parameters with other parameter sets, unless one is interested in pure lipids. However, even in that case it is a problem to develop new lipids – phosphatidylcholine lipids have been the traditional 'test' lipid, but other headgroups such as phosphatidylserine and phosphatidylglycerol require specific groups that look like a peptide or carbohydrate force field, respectively, while modifying lipid chains requires parameters for (poly-)unsaturated bonds and common types of other lipids such as sphingomyelin, ceramide, cardiolipin and sterols require a broad range of chemical groups.

1.2.2   Development of Force Field Parameters

The parameterization of all lipid force fields is typically based on *ab initio* and other calculations on molecules mimicking lipid fragments, such as dimethyl phosphate and liquid alkanes,[29] combined with parameterization based on density and thermodynamic parameters for long-chain alkanes. Lipids are complex molecules but share a limited number of functional groups, so that a building block approach similar to proteins seems sensible. Detailed structural and thermodynamic information on complete phospholipids is scarce, but more readily available for well-chosen model compounds. High-level quantum mechanics calculations have a high computational cost and are limited in size and cannot deal accurately with solvation phenomena. This general approach assumes that bond, angle and dihedral parameters and also charge distributions in model compounds such as dimethyl phosphate and butane are reasonably close to those of phospholipids.

In practice, lipid force fields continue to have issues when parameters that work well for the model compounds are combined for use in full bilayers. *Ad hoc* adjustments can improve the results for a single lipid or class of lipids (phosphatidylcholine), but this often runs counter to chemical intuition and in practice such adjustments have often turned out not to be transferable to other lipids, such as the closely related phosphatidylethanolamine and phosphatidylserine lipids. For example, Chiu *et al.* have reparameterized GROMOS-based lipids to give the right area for PC[38] and Sonne *et al.* have shown that is possible partially to solve the area per lipid problem of CHARMM27 by treating DPPC molecules as a whole and not as a sum of building blocks.[35] A key problem appears to be that lipid bilayers are extremely sensitive to an accurate balance between water–water, water–lipid and lipid–lipid interactions, at different levels along the bilayer normal.

A significant additional problem is that it is difficult to test a lipid model. Although it is fairly easy to make sure a particular lipid model produces, for example, the area or molecular volume of DMPC, this does not mean that 'trivial' modifications such as adding or removing two carbons on the alkane chains also give accurate results; area and volume contain relatively little information and may be reproduced even if the actual distribution of atoms in the system is very different from experiment; and more important chemical changes such as changing the CH_3 groups of the PC headgroup into H-atoms to make a phosphatidylethanolamine headgroup are rarely accurate based on a model optimized for DMPC.

A number of different issues complicate the systematic development of lipid models. First, equilibration times for small one-component bilayers are tens of nanoseconds and to obtain statistically accurate areas or other properties may take several times that. This makes it hard to test, modify, etc., parameters. A second problem is that there fore few experimental data on most lipids that are useful for testing parameters. Areas per lipid, chain order parameters and density profiles (or form factors) are useful and essential, but are only available for a handful of lipids. Lipid properties change with temperature and hydration level, both of which are difficult to obtain accurately experimentally and expensive to test across a wide range of lipids. Phase diagrams can be measured experimentally and are obviously biologically relevant, but it is a major challenge to use these in testing simulation parameters. As simulations become more sophisticated or longer, more properties, such as mechanical properties, come within reach, but it remains to be seen how critical these are as test cases.

1.2.3 Issues with Combining Force Fields

As outlined above, it is often necessary or tempting to combine force fields to describe systems which contain lipids and a broad range of other molecules, ranging from small drug-like molecules to crystals and polynucleotides. CHARMM has a full set of parameters but in that case it may still be desirable to use united-atom lipids with other molecules, and atomistic CHARMM lipid parameters historically have required a constant area or poorly known surface tension to represent a biologically relevant phase. In practice, combinations are very often used, but unexpected consequences occur and care has to be taken in interpreting the results. We have previously compared a number of different ways of combining the GROMOS-based ffgmx force field that was for years the default in the GROMACS software with the Berger *et al.* parameters.[16] As one of the test systems, we chose the hydrophobic peptide alamethicin in a hydrophobic solvent or in a bilayer. One sensitive parameter in these simulations was the partial volume of the peptide, which depended strongly on the specific combination rules used.[16] This dependence is disturbing, as it will affect peptide–lipid simulations, but possibly also simulations with bilayers containing, *e.g.*, cholesterol that is based on a different force field.

Another approach is to test solvation free energies of small molecules in different environments, which has become an important tool in testing force fields.[39–42] Interestingly, these solvation free energies appear relatively insensitive to force field changes, although they appear to depend more strongly on water models.[43]

There may also be practical problems with implementations of force fields in different simulation programs. Although these can, of course, be prevented by careful testing, differences in units, dihedral conventions, combination rules, neighbour searching groups, different scaling rules for one to four interactions in different force fields and special terms such as the CMAP backbone potential of CHARMM[44] require careful attention.

1.3 Starting Structures

Molecular dynamics simulations sample only a very small part of phase space, usually near the initial conformation of the system. When simulations are short relative to the slowest relaxation times in the system of interest, starting structures are very important. Historically, this was true even for the simplest systems, such as liquid argon or water, but with modern computers starting structures for such systems are no longer critical as equilibration occurs on time scales that are easily accessible. If sampling would not be an issue, starting structures would be irrelevant: starting from any conformation, no matter how unrealistic, the composition, temperature and pressure or geometric constraints would automatically lead to the equilibrium conformation. Unfortunately, the slowest relaxation times in lipid bilayers can be very slow: perhaps of the order of 10–100 ns for a small bilayer in the liquid crystalline state for pure lipids, hundreds of nanoseconds to microseconds for small mixtures and in many cases longer for larger or more complex systems or for systems that contain gel-like parts.

The importance of starting structures is, of course, undesirable, but has been an unfortunate reality in many studies. Many early bilayer simulations gave good results based on 100 ps simulations because they were prepared in a very realistic state, not because 100 ps is sufficient to sample even moderately complex motions such as rotation of a single lipid around its long axis.[45] With the current computer power this is less critical: if one starts from a random 'solution' of lipids in water, given a particular box geometry, the lipids will self-assemble into bilayers on a 50–100 ns time scale[46] and even self-assembly into a vesicle is possible.[47] Below a number of different systems are considered.

1.3.1 Bilayers

Bilayers consisting of a single type of lipid appear relatively straightforward and can be made in a number of ways, including a rigid-body packing procedure implemented in CHARMM,[48] random placement on a grid[49] and self-assembly from a random mixture in solution.[46] Water can be added by geometric criteria (avoiding overlap with lipid atoms), although this will often place water in the centre of the bilayer where there is a significant amount of free volume. Such water molecules can simply be deleted, as they may take a long simulation time to move. If lipids are placed systematically, as identical copies, long-range correlations will be introduced that can persist for hundreds of nanoseconds, which is a serious problem. Correlation times for rotation around a lipid long axis can be tens of nanoseconds, which is also a concern.

Mixtures can be made in the same way, but have an additional problem: sampling over lateral diffusion is not routinely possible at present and the initial distribution of lipid components will persist throughout a simulation. This is undesirable and a real problem in many cases. Much longer simulations than currently feasible, resolution exchange (see below) and smarter sampling algorithms[50] may be approaches forward.

Asymmetric bilayers are biologically of great interest, but introduce an additional problem. Because simulation systems are by necessity extremely small compared with macroscopic bilayers, the number of lipids in each membrane leaflet has a significant effect on the bilayer if there is an imbalance.

The leaflets couple both directly and through the enforced periodic boundaries, which can introduce strong artificial curvature effects. Perhaps one way around this is to simulate two symmetric bilayers consisting of two leaflets of each of the leaflets of the asymmetric bilayer of interest to make sure their equilibrium areas are exactly the same. This remains a relatively unexplored area for simulations and other issues may arise.

1.3.2 Membrane Proteins

Simulations of membrane proteins are a booming field, with obvious appeal.[3,5,6] They are subject to the same considerations as soluble proteins, but membrane proteins appear to be somewhat harder. Many structures are very low resolution, which is a problem for reliable MD simulations. Often, parts of the structures are missing and may require modelling (inaccurate and/or unreliable) or leaving gaps (undesirable), side chains may not be resolved and proteins may have unrealistic conformations due to the crystallization conditions that cannot mimic the native environment of membrane proteins. In many cases, crystal structures are of prokaryotic homologues of eukaryotic proteins of interest, which makes it tempting to construct homology models. However, simulations of homology models in well-defined tests for soluble proteins rarely improve homology models and, given the very low degrees of similarity/identity in many membrane protein homology models, such simulations should be viewed with some scepticism. Drug binding and other applications of homology models that are known to require very accurate homology models in soluble proteins are routinely used in membrane proteins, with much less consideration of their limitations than may be desirable. Membrane proteins may also be more complicated to simulate than soluble proteins because critical titratable residues in the transmembrane regions may have significant pK_a shifts that are difficult to calculate, they may have complex redox factors or may require tight interactions with specific lipids that are not resolved in crystal structures.

1.3.3 Embedding Proteins in Bilayers

If a membrane protein structure is of sufficient quality to warrant simulations, the technical challenge of embedding the protein in a suitable bilayer remains. The naive approach of simply putting the protein in a pre-equilibrated bilayer and deleting all lipids that overlap with the protein causes a large 'moat' around the protein that takes a long time to disappear (Figure 1.2A). Several better methods have been proposed and used. One option is to use a cylindrical repulsive potential (Figure 1.2B) or a repulsive potential perpendicular to a molecular surface (Figure 1.2C) to move lipids out of the volume occupied by the protein, which has been implemented in a version of GROMACS.[51] Another option is to pack lipids one by one around the protein, analogous to how one could create a bilayer (Figure 1.2D).[48] This has recently been

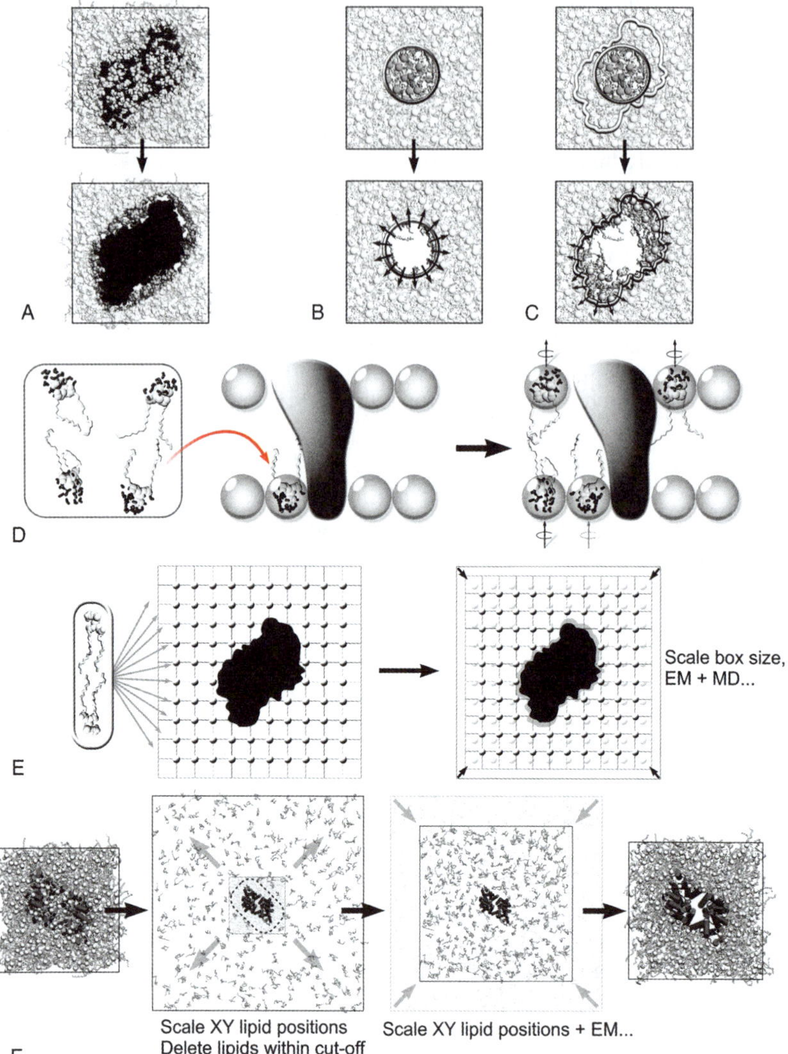

Figure 1.2 Methods to incorporate a protein in a bilayer (A–F) or to make a bilayer (D, E). (A) Deleting lipids within the range of a simple protein–lipid distance cut-off; (B) applying a repulsive potential to drive out lipid and water from the area to be occupied by the protein as in the cylinder approach, or with (C) mdrun_hole; (D) constructing the membrane lipid by lipid using structures from a pool of hydrated lipid conformations; (E) using scaling of a grid of lipids: starting by placing copies of lipids on a 2D grid and then scaling down the box size stepwise with energy minimization and short molecular dynamics steps in between or (F) starting from a pre-equilibrated bilayer, which is expanded within the xy plane. Lipids within a given cut-off distance are removed and the system is brought back to natural dimension by a series of scaling steps of the lipid xy positions and energy minimizations. Reproduced with permission from reference 49.

automated by Im's group.[52] We recently proposed two methods (depending on whether a pre-equilibrated bilayer is available or not) to use a series of steps that can be automated to incorporate a protein in any bilayer while maintaining much of the original bilayer structure (Figure 1.2E and F).[17]

Both methods rely on placing lipids and the protein(s) on a widely spaced grid and then 'shrinking' the grid until the bilayer with the protein has the desired density, with lipids neatly packed around the protein. When starting from a grid based on a single lipid structure or several potentially different lipid structures (method 1), the bilayer will start well packed but requires more equilibration. When starting from a pre-equilibrated bilayer, either pure or mixed, most of the structure of the bilayer stays intact, reducing equilibration time (method 2). The main advantages of these methods are that they minimize equilibration time and can be almost completely automated, nearly eliminating one time-consuming step in MD simulations of membrane proteins.

Another method is to self-assemble a bilayer around a protein. Although this probably works for smaller proteins at the atomistic level, it is clear that it does not work for large membrane proteins at the atomistic level. Sansom's group has established a database of all membrane protein structures embedded in lipids using the coarse-grained MARTINI model.[53] This is an important resource to investigate the orientation of membrane proteins (which is not always obvious from the crystal structure) in a lipid environment but also demonstrates that this is a feasible method for making starting structures provided that the coarse-grained models can be translated back to atomistic models. This is only a practical problem, however, for which there are some solutions already that are bound to become available in a user-friendly way.

1.4 Sampling

If MD simulations are run sufficiently long, the system will explore all its relevant states with the appropriate weight. In practice, such an exhaustive exploration of all possible states is not possible since the calculation time is limited by the computer hardware and straightforward MD simulation is unable to cross high energetic barriers. Practical limits currently are on the order of 100 ns for typical membrane systems, which is short compared with several relaxation times in the membrane, even in the absence of obvious high energy barriers. *Sampling is often by far the greatest source of errors in MD simulations of lipids and typically much more limiting than relatively modest differences caused by force field choice or choice of simulation algorithms.*

Although it is not possible in general to state how long a given simulation has to be in order to obtain good averaging, and statistical analyses are essential, there are several studies in the literature that help to provide some guidelines for common cases.

In an extensive set of comparative simulations of the same DPPC bilayer, Anézo *et al.* have estimated that *ca.* 10–20 ns is required for small well-tested DPPC bilayers to equilibrate properly,[13] while slow fluctuations in area or, for constant area simulations, pressure require sampling during several times that period to obtain reasonable averages. Figure 1.3 illustrates this. If lateral diffusion is of interest, longer simulations are required.[54] Interestingly, the time scale required for a simulation depends on the size of the system and increases steeply with system size. Larger bilayers develop slow undulatory motions that are a feature of real bilayers as well, but are very difficult to sample. Ironically, the 10–20 ns simulations that give good averages for small DPPC bilayers do so by virtue of simulating a system that suppresses larger scale slow fluctuations, essentially an artefact of a small system. In this study, different areas per lipid were found when averaged over 10 ns blocks, all of which may appear converged just by looking at the area. In fact, Figure 1.3A–C shows three different DPPC simulations that initially are not at equilibrium because some conditions were changed. Detailed are described in Anézo *et al.*,[13] from which the figure is taken. It is clear that based on 10 or 20 ns stretches of simulation, wildly different conclusions can be reached about average area, stability of the bilayer and perhaps quality of the force field, based on areas alone. This is similar to protein simulations that may appear converged on a time scale of 100 ps, 1 ns or 10 ns, but actually show major changes on longer time scales. From peptide-folding simulations, it is known that microsecond simulations may sample only a few transitions and for some peptides, particularly beta-sheet peptides, no transitions at all. It is also noteworthy that less commonly checked properties may equilibrate much more slowly also, such as collective orientations of headgroups. These may affect, *e.g.*, electrostatic potential values calculated from a bilayer (unpublished work).

A second striking example of an at-first-sight simple process that takes surprisingly long to equilibrate is ion binding to a bilayer. Two interesting papers by Bockmann and co-workers show that equilibration of sodium occurs on a tens of nanoseconds time scale,[55] whereas binding of calcium requires equilibration times of over 100 ns.[56] Although not directly relevant for lipid–protein interactions, these sobering conclusions on simpler ions suggest that extreme caution is necessary in the interpretation of simulations of, *e.g.*, peptide binding to a membrane on a shorter time scale.

Simulations of simple pentapeptides Ac-W-L-X-L-L in POPC bilayers illustrate this point.[57,58] These peptides are largely hydrophobic and rapidly 'bind' to a POPC bilayer on a 10 ns simulation time scale, most of the time. However, the motions of these peptides are so slow that the distribution of different side chains does not accurately equilibrate on a 50 ns time scale.[58] When two of the same peptides are present in one simulation, one bound to each side of the membrane, the distribution of side chains on both sides is similar, especially at a coarse level like the centre of mass of the peptide relative to the centre of mass of the membrane. However, detailed distributions of individual side chains do not average accurately on a 50 ns time scale. This makes it currently very difficult to try to reproduce directly the Wimley–White

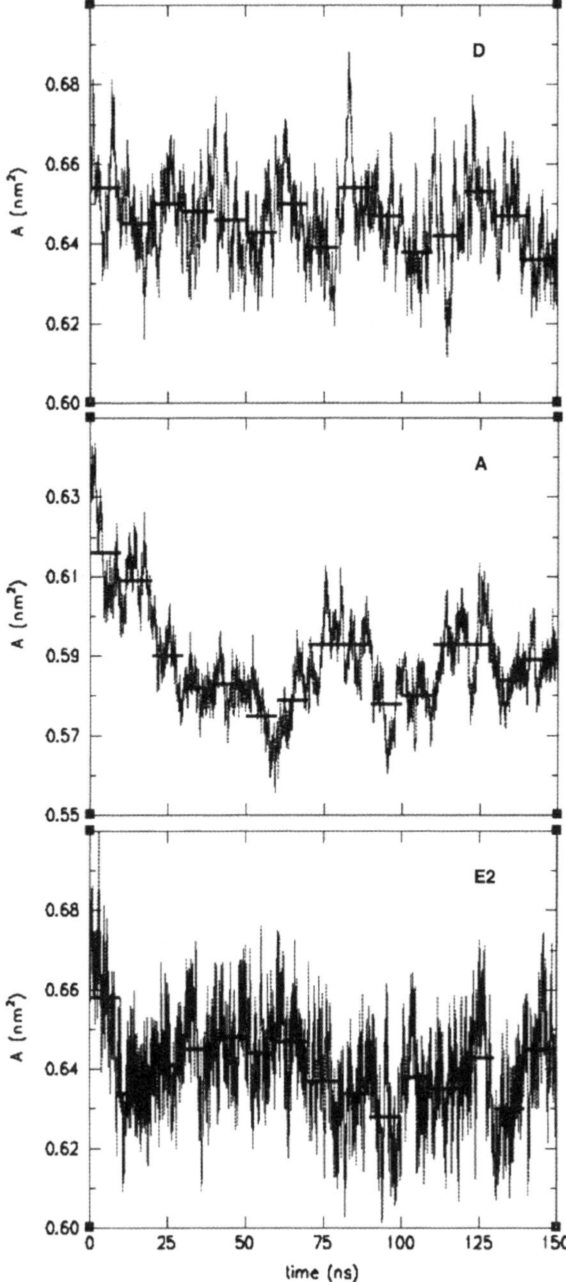

Figure 1.3 Area fluctuations of a DPPC bilayer over 150 ns. The three panels are the same bilayer, but with minor changes in simulations conditions explained in reference 13. Black bars indicate averages over 10 ns. Reproduced with permission from reference 13.

hydrophobicity scale that is based on binding of these peptides to vesicles,[59] although this would be extremely useful for validation and improvement of force fields.

In simulations of integral membrane proteins, the property of interest might not be directly associated with lipids, but in all cases lipids provide the environment for such proteins and in several cases lipid and protein interact in a functionally important way.[60] The G-protein coupled receptor rhodopsin is an excellent example of a protein that shows both interesting internal phenomena and important interactions with lipids. Based on an interesting study of rhodopsin at an unprecedented degree of sampling,[61] Grossfield *et al.* have shown that 26 independent simulations of rhodopsin run over 100 ns each were not sufficient to sample the structure correctly, especially the loops between transmembrane segments,[62] although averaging over all 26 simulations gave good agreement with available experimental data.

1.4.1   Improving Sampling

Despite the perhaps somewhat gloomy picture painted above, there are several useful approaches to improve sampling. In most of the examples above, sufficient sampling can be obtained, although at the time of those studies the simulation lengths required were at the edge of what is practical. This will improve with faster computers and better software. Multiple simulations and advanced sampling methods such as replica exchange may improve sampling, as do free energy and related methods for cases where a reaction coordinate can be identified. Free energy methods have been widely applied to calculating the distribution of small molecules, and more recently larger molecules up to the size of lipids, in membranes. We recently reviewed a large fraction of the literature on this topic.[63] As in most free energy methods, the goal is to apply a bias to the potential function to sample otherwise rare events. On a larger scale, similar methods have been applied to calculate interactions between helices in a membrane,[64,65] toxin distribution,[66] changes in pK_a values in ionizable residues that partition into the membrane[67,68] and the free energy (probability) of pore formation in a bilayer.[69] Figure 1.4 shows the free energy profile for moving a DPPC lipid within a DPPC bilayer, which involves high energies and would be impossible to sample accurately by equilibrium simulations. However, even the distribution of hexane cannot be accurately sampled by equilibrium simulations on a 100 ns time scale as the probability of finding hexane in the headgroup or water regions is just too low.[70]

1.4.2   Coarse Graining

In some cases, it may also be possible to use coarse-grained simulations to explore large parts of conformational space. In principle, it is possible to move between atomistic and coarse-grained levels of detail. To study the interactions between lipids and antimicrobial peptides, for instance, one could start with coarse-grained simulations to equilibrate mixtures of lipids and peptides and

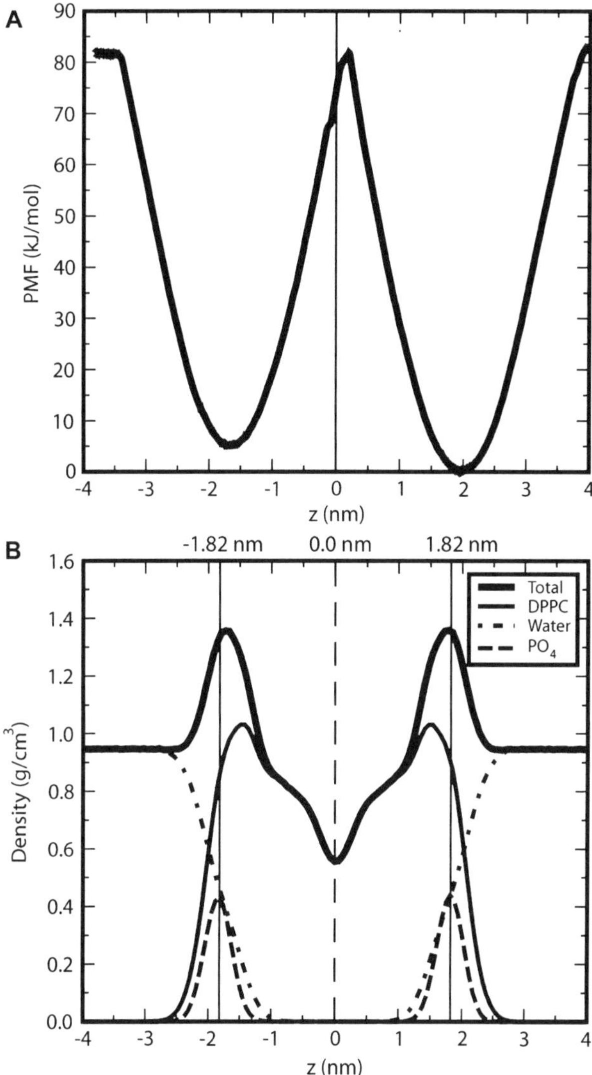

Figure 1.4 (A) The potential of mean force (free energy profile) for a DPPC lipid in a DPPC bilayer. The water phase is at 4 nm or − 4 nm, and 0 nm is the centre of the membrane. The wells correspond to the equilibrium position of a DPPC lipid. The centre is the barrier for translocation of a lipid and in this case corresponds to the free energy cost of pore formation. (B) Density profiles for the main components of the bilayer–water system. Reproduced with permission from reference 69.

then translate the structures back into atomistic structures for further simulations to explore the details of interactions in the system. A recent example of this is a paper on the aggregation behaviour of the antimicrobial peptide alamethicin.[71]

1.5   Pressure Coupling

Most simulations are done at constant temperature and pressure, although simulations at constant energy and volume are possible. In membrane simulations, temperature control is not substantially different from other simulations,[72] but pressure coupling is significantly more complicated. The presence of an interface gives an additional thermodynamic variable that specifies the state of the system (this has been reviewed[73]). In addition to the correct averages for the correct thermodynamic variables, the fluctuations in area are also important as they are linked to, *e.g.*, the area compressibility modulus of a bilayer, which can be measured.

A detailed comparison of several algorithms, some more rigorous than others, for pressure coupling and also some pragmatic approaches showed no substantial difference in tests of pure lipids on a 150 ns time scale,[13] neither in averages nor in fluctuations. Different methods are expected to give the same averages and differences in fluctuations, but the latter are difficult to sample accurately in the noise of simulations of small bilayers. However, more recent simulations on coarse-grained models that allow orders of magnitude more sampling do show that fluctuations in membrane area are different for different algorithms.[74] This is expected based on the properties of common algorithms: Berendsen weak coupling does not generate correct fluctuations, whereas Parinello–Rahman pressure coupling[72] and the Langevin piston method of Feller *et al.*[75] do. The last two or similar extended Hamiltonian methods are recommended, although for many current applications the differences between these and weak coupling are likely negligible.

Proteins in solution are often simulated under constant volume or constant total pressure conditions, averaging over the different directions in the box (isotropic pressure). Although simulations of membranes are possible under both conditions, they are ill-defined because there are implicit variables involving the area and height of the system or the lateral and normal pressure. Constant-volume simulations will potentially result in large surface tensions and unphysical pressures in lateral and normal directions. Constant total pressure has the same problem and is for practical purposes almost the same as constant volume: fluctuations in volume are restricted to those corresponding to overall density fluctuations in the lipid–water system, but these are negligible compared with physically important fluctuations in membrane area that do not involve changes in total density (Figure 1.5A). Although both of these boundary conditions are sometimes used in the literature, this is a serious error.

We either have to specify the area of the interface and the normal pressure or the surface tension and the normal pressure. Unfortunately, knowledge of the right area per lipid is very difficult to obtain, depends on the force field and possibly other system parameters such as system size, and is actually not possible to obtain experimentally for arbitrary systems that contain proteins or lipid mixtures or undergo, *e.g.*, insertion of peptides. Furthermore, this approach will not easily allow the reproduction of effects such as temperature-dependent swelling of the bilayer. However, constant area simulations

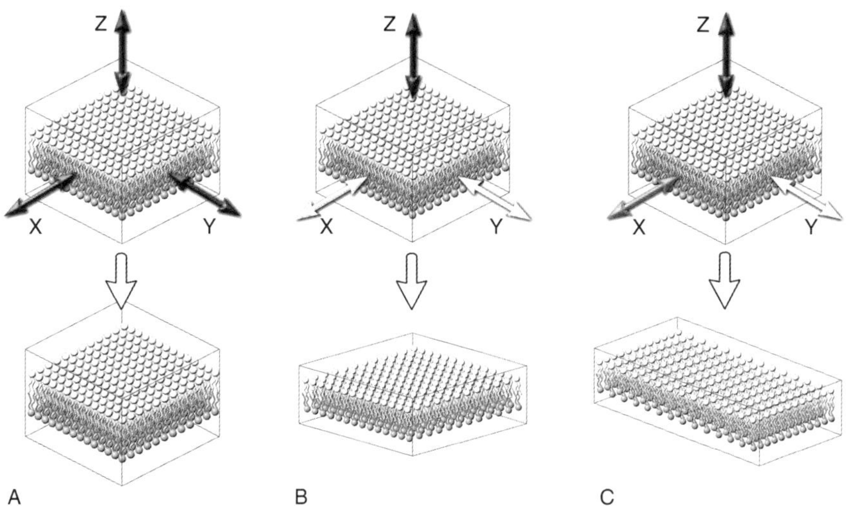

Figure 1.5 Pressure coupling schemes. (A) Isotropic pressure coupling (not recommended for membrane simulations); (B) semi-isotropic or constant surface tension pressure coupling (recommended); (C) anisotropic pressure coupling. Reproduced with permission from reference 49.

are widely used in simulations of membrane proteins, where the details of area fluctuations or surface tension may not have much effect on, *e.g.*, ion conduction.

Semi-isotropic pressure coupling and surface tension coupling (Figure 1.5B) do allow area fluctuations. In this case, only the pressure contributions in the x and y directions are coupled together, but normal pressure (in the z direction) is not. This is the recommended type of pressure coupling for membrane protein simulations. With the united-atom parameters in GROMACS, a reference pressure of 1 bar in both xy and z is appropriate and corresponds to a realistic pressure.[76] Simulations using CHARMM parameters sometimes use different values in xy to impose a non-zero surface tension. This corresponds to a membrane under tension but is related to force field issues that are still under investigation. Strictly, such simulations should not specify lateral and normal pressure but rather surface tension and normal pressure. This is implemented in GROMACS but gives essentially the same results as specifying corresponding lateral and normal pressures.

A third option is what could be called anisotropic coupling: the dimensions x, y and z are coupled separately to reference pressures. This is theoretically not correct, as there is no known ensemble that allows these three components to be specified separately, but in practice the combined xy pressure can be set in this way, making it similar to semi-isotropic or surface tension coupling. However, the method is also not recommended on practical grounds, because it allows large deformations of the membrane without a change in area that cause problems in long simulations (Figure 1.5C).

1.6   Electrostatics

A key issue in MD simulations is the treatment of electrostatic interactions. This especially concerns lipid bilayer systems, as phospholipids have a very high interfacial charge density and regions of low dielectric constant where electrostatic interactions are largely unshielded. The difficulty is to calculate interactions energies that formally decay as 1/distance, a common problem in physics. Many numerical methods exist to treat such interactions, some of which have been adopted for use in MD simulations. The most common ones are cut-off based techniques, lattice sum techniques, of which particle mesh Ewald (PME) is by far the most popular,[77,78] and reaction field (RF) approaches.[79]

Cut-off approaches neglect long-range Coulombic effects by truncating the electrostatic forces at a specific cut-off distance, which is typically between 1.0 and 2.0 nm, giving considerable savings in computational cost. However, such approximations are rather drastic and have a significant effect on the system's properties.[13,80] As an improvement over straight cut-off methods, shifting functions can be applied where artificial correlation effects are partly removed by smoothing the interaction energy to zero either within the full cut-off range or over a limited region.[13]

Lattice sum methods treat the simulation cell as one repeating unit of a crystal and calculate interactions between all the images in the crystal using Fourier methods. The basis of PME, the most commonly used method, is an interpolation of the reciprocal-space Ewald sum. All of the interactions in the periodically replicated system are summed, thus including all long-range electrostatic interactions. A potential concern with this method is that due to its periodicity, PME may induce an artificial ordering that enhances the system's stability. Detailed tests of what should be an extreme case – a system consisting of an alamethicin a helix bundle with six parallel helices, all aligned with their helical dipoles in the same direction – did not show a strong effect, however,[81] but corrections due to PME for the distribution of ions in an ion channel are non-negligible.[82]

Reaction field approaches are another technique that considers the effect of long-range electrostatic interactions. They add a correction term to the cut-off result based on a continuum electrostatics description of the solvent outside the cut-off sphere. This method was originally developed for the use in homogeneous systems such as liquid simulations. In simulations of DPPC, similar results to PME can be obtained,[13] although the water orientation in alamethicin channels is different.[14] Isotropic reaction fields clearly do not match the electrostatic environment of a lipid bilayer, however.

None of these methods is perfect for the simulation of interfacial systems: cut-off methods induce artificial orderings, PME enhances periodicity and causes spurious interactions between charges in low dielectric environments and their periodic images and RF approaches ignore the heterogeneous nature of the membrane.[13] However, against this background, it seems that today PME would be the recommended choice bearing the least drawbacks. The choice is not arbitrary: a particular force field really should be developed for use with one of these methods, as the accuracy of the force field is not maintained when switching between different methods to calculate electrostatic interactions. As force fields move towards full treatment of long-range interactions, parameters should become less dependent on the details of the methods. In this context, it is also worth mentioning the treatment of long-range Lennard-Jones interactions. Although these decay faster (as $1/r^6$), they are always attractive and have a significant effect on, *e.g.*, pressure. This becomes critical for the accurate and reproducible calculation of surface tension-related effects. Several methods exist, analogous to the options for Coulomb interactions, but these are not in common use yet and require testing and probably reparameterization of at least parts of most current force fields.

1.7 Periodicity

Molecular dynamics simulations are commonly done using periodic boundary conditions to mimic bulk conditions. The major alternative is some form of mean field treatment beyond a solvation shell, but for membranes periodic boundaries at least in the plane of the membrane are a natural extension from a small simulation box to an infinite extended sheet. In practice, periodicity in the third dimension – normal to the membrane – means that an infinite stack of membranes is simulated, with perfect correlations between the different layers. In simulations in solution, some artefacts due to periodicity have been identified in certain pathological cases, such as a helix in an elongated box with increased interaction with its images due to the dipole interactions[83] or difficulties with free energy calculations using lattice sum electrostatics.[84] A study of helix bundles in a membrane found little evidence of at least the first type of artefacts (see Section 1.6), but there are a number of important other considerations that do not arise in simulations of soluble proteins.

First, membranes show significant fluctuations activated by thermal energy. This is a natural feature of soft condensed matter, but such fluctuations are suppressed by small simulation boxes. In addition, the spectrum of wavelengths necessarily is different, because waves must fit inside the simulation box. Thus simulations aimed at studying undulations or similar large-scale phenomena are not easy.[12]

Second, perturbations of membranes are difficult to simulate, because they interact with periodicity. If one adds a number of peptides to a membrane in one leaflet, strong perturbations are often observed but it is difficult to separate the effect of the peptides and the effects due to the requirement of a continuous membrane due to periodicity. One approach is to add peptides symmetrically,[57,58,85] but this is often not the biological situation.

Third, the requirement for periodicity complicates simulating asymmetric membranes, *e.g.* ones mimicking bacterial plasma membranes with an asymmetric distribution of lipids. Since no equilibration occurs between both leaflets, the initial choice of the number of lipids of each type has an effect on the result. If the result is, *e.g.*, bending of the membrane, care must be taken to distinguish between actual effects due to the specific lipid distribution and effects due to a poor choice of initial system setup. Self-assembling a membrane might be one way to approach this, although this process may still occur too fast to give equilibrated results. Other options include systematic scanning of a range of possibilities, around some sensible guess based on lipid volumes and areas and simulation of vesicles with a trick to equilibrate lipids between leaflets.[86] Dolan *et al.* suggested modifying the symmetry of a simulation cell to allow diffusion from lipids from one leaflet into the other leaflet.[87] This seems an interesting idea but has not been used much yet in practice.

Fourth, in many biological problems an asymmetric ion solution is of interest. In a typical simulation setup there is one bilayer and one water phase, although graphically it often appears there are two water phases. With only one water phase there can be only one concentration of a particular ion. Work-arounds include simulating two bilayers and two water phases in a single system[88] or including two water–vacuum interfaces.[89] The latter approach yields a smaller simulation system, but introduces the additional complexity of two extra interfaces with a specific surface tension and water that behaves differently from both bulk and water near the membrane.

Periodicity is not necessarily a problem and indeed helps to model a membrane, but it does introduce complications. Simulations of entire vesicles are one approach that will become increasingly feasible,[86] although in that case one has to worry about the internal pressure and high intrinsic curvature of very small vesicles. Simulations of bilayer discs[90] or similar structures may also become attractive options in some cases.

1.8 Future Developments

Significant progress in the past few years has brought a rapidly growing number of interesting scientific problems within reach of simulations. In many cases simulations have also become one of many techniques in interdisciplinary studies. Although many state-of-the-art simulation studies require the largest computers available and a high degree of technical expertise, many scientific problems have aspects that can benefit from simulations on modest computers or modest allocations on shared computers, combined with modern software with a reasonable degree of user friendliness. In the author's opinion this is one of the greatest advances in recent years, as it has greatly expanded the field of simulation and its usefulness in a broad range of studies.

Clearly, there remains significant room for improvement. As outlined above, a number of technical choices involve tradeoffs and compromises. Improvements can be expected in several areas. It is likely that computer power will continue to increase, probably in a combination of faster processors, cheaper processors and increasing use of large-scale parallel machines with heterogeneous connections between processors. This poses an engineering problem for simulation software, but progress in this direction is being made. It is also likely that software will continue to become easier to use.

On a more technical level, force fields still have significant weaknesses. Classical force fields can probably be further improved. Consistency between lipid parameters and parameters for other biomolecules is important. Also, accurate united-atom parameters for lipids combined with all-atom parameters are of interest and initial steps in this direction require further work. It was recognized decades ago that explicit electronic polarizability would be desirable in classical simulations, but progress in this area has been slow until recently. However, in the past few years several groups have pursued different approaches towards including polarizability, with significant success. The first lipid simulations with polarizable chains have been reported.[18]

A major problem with lipids and especially lipid–protein systems is the lack of accurate experimental data on which to test lipids. Much of the discussion on different lipid force fields in the 1990s was somewhat confused because the nature and accuracy of experimental properties were not clearly understood. Finite size effects in lipid bilayers are significant, as a typical bilayer model is only a relatively small number of lipids, sometimes as low as 6×6 for one leaflet of a bilayer. The relative scarcity of critical experimental data continues to be a problem, although a tighter connection between experiments and simulation has brought significant progress in testing pure lipids.

In addition to concerns about parameters and finite-size effects, the long relaxation times involved in any lipid system cause serious sampling problems. Although simulations of the order of 100–250 ns may seem 'long' or perhaps 'standard' compared with the most recent literature (2008), these times may be very short compared with averaging times in experiments or physiological phenomena. The use of sampling-enhancing methods, including free energy calculations or even as simple as running multiple copies of a simulation, is for many problems essential. An interesting recent development in this respect is the use of coarser models that retain atomistic details in some way. One potential way of using such models is to increase sampling of lipid mixtures or lipid–membrane protein systems by switching between atomistic and coarse-grained levels of description.

Acknowledgements

The author thanks former and present group members and collaborators for their original research and review contributions on which much of this chapter is based. The author is a Senior Scholar of the Alberta Heritage Foundation for Medical Research and Canadian Institutes for Health Research (CIHR) New Investigator. Work in his group is supported by CIHR and the Natural Science and Engineering Research Council (NSERC).

References

1. R. W. Pastor, *Curr. Opin. Struct. Biol.*, 1994, **4**, 486–492.
2. D. P. Tieleman, S. J. Marrink and H. J. Berendsen, *Biochim. Biophys. Acta*, 1997, **1331**, 235–270.

3. W. L. Ash, M. R. Zlomislic, E. O. Oloo and D. P. Tieleman, *Biochim. Biophys. Acta*, 2004, **1666**, 158–189.
4. S. E. Feller (ed.), *Computational Modeling of Membrane Bilayers,* Elsevier, Amsterdam, 2008.
5. E. Lindahl and M. S. Sansom, *Curr. Opin. Struct. Biol.*, 2008, **18**, 425–431.
6. J. Gumbart, Y. Wang, A. Aksimentiev, E. Tajkhorshid and K. Schulten, *Curr. Opin. Struct. Biol.*, 2005, **15**, 423–431.
7. B. Roux and K. Schulten, *Structure*, 2004, **12**, 1343–1351.
8. P. J. Bond and M. S. P. Sansom, *Mol. Membr. Biol.*, 2004, **21**, 151–161.
9. T. Rog, M. Pasenkiewicz-Gierula, I. Vattulainen and M. Karttunen, *Biochim. Biophys. Acta*, 2009, **1788**, 97–121.
10. P. S. Niemela, M. T. Hyvonen and I. Vattulainen, *Biochim. Biophys. Acta*, 2009, **1788**, 122–135.
11. M. L. Berkowitz, *Biochim. Biophys. Acta*, 2009, **1788**, 86–96.
12. S. J. Marrink, A. H. de Vries and D. P. Tieleman, *Biochim. Biophys. Acta*, 2009, **1788**, 149–168.
13. C. Anézo, A. H. de Vries, H. D. Höltje, D. P. Tieleman and S. J. Marrink, *J. Phys. Chem. B*, 2003, **107**, 9424–9433.
14. D. P. Tieleman, B. Hess and M. S. P. Sansom, *Biophys. J.*, 2002, **83**, 2393–2407.
15. N. Sapay and D. P. Tieleman, in *Computational Modeling of Membrane Bilayers*, 2008, Vol. 60, pp. 111–130.
16. D. P. Tieleman, J. L. MacCallum, W. L. Ash, C. Kandt, Z. Xu and L. Monticelli, *J. Phys. Condens. Matter*, 2006, **18**, S1221–S1234.
17. C. Kandt, W. L. Ash and D. P. Tieleman, *Methods*, 2007, **41**, 475–488.
18. I. Vorobyov, L. B. Li and T. W. Allen, *J. Phys. Chem. B*, 2008, **112**, 9588–9602.
19. J. E. Davis, O. Rahaman and S. Patel, *Biophys. J.*, 2008, **96**, 385–402.
20. S. J. Marrink, H. J. Risselada, S. Yefimov, D. P. Tieleman and A. H. de Vries, *J. Phys. Chem. B*, 2007, **111**, 7812–7824.
21. L. Monticelli, S. K. Kandasamy, X. Periole, R. G. Larson, D. P. Tieleman and S. J. Marrink, *J. Chem. Theory Comp.*, 2008, **4**, 819–834.
22. D. A. Case, T. E. Cheatham, T. Darden, H. Gohlke, R. Luo, K. M. Merz, A. Onufriev, C. Simmerling, B. Wang and R. J. Woods, *J. Computat. Chem.*, 2005, **26**, 1668–1688.
23. J. W. Ponder and D. A. Case, *Protein Simul.*, 2003, **66**, 27–85.
24. A. D. MacKerell Jr, D. Bashford, D. Bellott, R. L. Dunbrack Jr, J. D. Evanseck, M. J. Field, S. Fischer, J. Gao, H. Guo, S. Ha, D. Joseph-McCarthy, L. Kuchnir, K. Kuczera, F. T. K. Lau, C. Mattos, S. Michnick, T. Ngo, D. T. Nguyen, B. Prodhom, W. E. Reiher III, B. Roux, M. Schlenkrich, J. C. Smith, R. Stote, J. Straub, M. Watanabe, J. Wiorkiewicz-Kuczera, D. Yin and M. Karplus, *J. Phys. Chem. B*, 1998, **102**, 3586–3616.
25. W. R. P. Scott, P. H. Hünenberger, I. G. Tironi, A. E. Mark, S. R. Billeter, J. Fennen, A. E. Torda, T. Huber, P. Krüger and W. F. van Gunsteren, *J. Phys. Chem. A*, 1999, **103**, 3596–3607.

26. W. F. van Gunsteren, S. R. Billeter, A. A. Eising, P. H. Hünenberger, P. Krüger, A. E. Mark, W. R. P. Scott and I. G. Tironi, *Biomolecular Simulation: the GROMOS96 Manual and User Guide*, Vdf Hochschulverlag, ETH Zürich, Zürich, 1996.
27. W. L. Jorgensen, D. S. Maxwell and J. Tirado-Rives, *J. Am. Chem. Soc.*, 1996, **118**, 11225–11236.
28. S. E. Feller and A. D. MacKerell Jr, *J. Phys. Chem. B*, 2000, **104**, 7510–7515.
29. J. B. Klauda, R. M. Venable, A. D. MacKerell and R. W. Pastor, in *Computational Modeling of Membrane Bilayers*, 2008, Vol. 60, pp. 1–48.
30. O. Berger, O. Edholm and F. Jahnig, *Biophys. J.*, 1997, **72**, 2002–2013.
31. R. W. Benz, F. Castro-Roman, D. J. Tobias and S. H. White, *Biophys. J.*, 2005, **88**, 805–817.
32. I. Chandrasekhar, M. Kastenholz, R. D. Lins, C. Oostenbrink, L. D. Schuler, D. P. Tieleman and W. F. van Gunsteren, *Eur. Biophys. J.*, 2003, **32**, 67–77.
33. A. Pertsin, D. Platonov and M. Grunze, *Langmuir*, 2007, **23**, 1388–1393.
34. L. Rosso and I. R. Gould, *J. Comp. Chem.*, 2008, **29**, 24–37.
35. J. Sonne, M. O. Jensen, F. Y. Hansen, L. Hemmingsen and G. H. Peters, *Biophys. J.*, 2007, **92**, 4157–4167.
36. J. Henin, W. Shinoda and M. L. Klein, *J. Phys. Chem. B*, 2008, **112**, 7008–7015.
37. C. J. Hogberg, A. M. Nikitin and A. P. Lyubartsev, *J. Comp. Chem.*, 2008, **29**, 2359–2369.
38. S. W. Chiu, M. M. Clark, E. Jakobsson, S. Subramaniam and H. L. Scott, *J. Phys. Chem. B*, 1999, **103**, 6323–6327.
39. A. Villa and A. E. Mark, *J. Comp. Chem.*, 2002, **23**, 548–553.
40. C. Oostenbrink, A. Villa, A. E. Mark and W. F. van Gunsteren, *J. Comp. Chem.*, 2004, **25**, 1656–1676.
41. J. L. MacCallum and D. P. Tieleman, *J. Comp. Chem.*, 2003, **24**, 1930–1935.
42. M. R. Shirts, J. W. Pitera, W. C. Swope and V. S. Pande, *J. Chem. Phys.*, 2003, **119**, 5740–5761.
43. B. Hess and N. F. A. van der Vegt, *J. Phys. Chem. B*, 2006, **110**, 17616–17626.
44. A. D. MacKerell Jr, M. Feig and C. L. Brooks III, *J. Am. Chem. Soc.*, 2004, **126**, 698–699.
45. R. W. Pastor, R. M. Venable and S. E. Feller, *Acc. Chem. Res.*, 2002, **35**, 438–446.
46. S. J. Marrink, E. Lindahl, O. Edholm and A. E. Mark, *J. Am. Chem. Soc.*, 2001, **123**, 8638–8639.
47. A. H. de Vries, A. E. Mark and S. J. Marrink, *J. Am. Chem. Soc.*, 2004, **126**, 4488–4489.
48. T. B. Woolf and B. Roux, *Proc. Natl. Acad. Sci. USA*, 1994, **91**, 11631–11635.
49. C. Kandt, W. L. Ash and D. P. Tieleman, *Methods*, 2007, **41**, 475–488.

50. J. de Joannis, Y. Jiang, F. Yin and J. T. Kindt, *J. Phys. Chem. B*, 2006, **110**, 25875–25882.

51. J. D. Faraldo-Gomez, G. R. Smith and M. S. Sansom, *Eur. Biophys. J.*, 2002, **31**, 217–227.

52. S. Jo, T. Kim and W. Im, *PLoS ONE*, 2007, **2**, e880.

53. A. P. Chetwynd, K. A. Scott, Y. Mokrab and M. S. Sansom, *Mol. Membr. Biol.*, 2008, **25**, 662–669.

54. J. Wohlert, W. K. den Otter, O. Edholm and W. J. Briels, *J. Chem. Phys.*, 2006, **124**.

55. R. A. Bockmann, A. Hac, T. Heimburg and H. Grubmuller, *Biophys. J.*, 2003, **85**, 1647–1655.

56. R. A. Bockmann and H. Grubmuller, *Angew. Chem. Int. Ed.*, 2004, **43**, 1021–1024.

57. M. P. Aliste, J. L. MacCallum and D. P. Tieleman, *Biochemistry*, 2003, **42**, 8976–8987.

58. M. P. Aliste and D. P. Tieleman, *BMC Biochem.*, 2005, **6**, 30.

59. W. C. Wimley and S. H. White, *Nat. Struct. Biol.*, 1996, **3**, 842–848.

60. A. G. Lee, *Biochim. Biophys. Acta*, 2004, **1666**, 62–87.

61. A. Grossfield, S. E. Feller and M. C. Pitman, *Proc. Natl. Acad. Sci. USA*, 2006, **103**, 4888–4893.

62. A. Grossfield, S. E. Feller and M. C. Pitman, *Proteins*, 2007, **67**, 31–40.

63. J. L. MacCallum and D. P. Tieleman, in *Computational Modeling of Membrane Bilayers*, 2008, Vol. 60, pp. 227–256.

64. D. S. An, H. G. Lee, W. T. Im, Q. M. Liu and S. T. Lee, *Int. J. Syst. Evol. Microbiol.*, 2007, **57**, 1828–1833.

65. J. Henin, A. Pohorille and C. Chipot, *J. Am. Chem. Soc.*, 2005, **127**, 8478–8484.

66. C. L. Wee, D. Gavaghan and M. S. Sansom, *Biophys. J.*, 2008, **95**, 3816–3826.

67. J. L. MacCallum, W. F. Bennett and D. P. Tieleman, *J. Gen. Physiol.*, 2007, **129**, 371–377.

68. J. L. MacCallum, W. F. D. Bennett and D. P. Tieleman, *Biophys. J.*, 2008, **94**, 3393–3404.

69. D. P. Tieleman and S. J. Marrink, *J. Am. Chem. Soc.*, 2006, **128**, 12462–12467.

70. J. L. MacCallum and D. P. Tieleman, *J. Am. Chem. Soc.*, 2006, **128**, 125–130.

71. L. Thogersen, B. Schiott, T. Vosegaard, N. C. Nielsen and E. Tajkhorshid, *Biophys. J.*, 2008, **95**, 4337–4347.

72. D. v. d. Spoel, E. Lindahl, B. Hess, et al., *GROMACS User Manual Version 4.0*, 2008, www.gromacs.org.

73. Y. H. Zhang, S. E. Feller, B. R. Brooks and R. W. Pastor, *J. Chem. Phys.*, 1995, **103**, 10252–10266.

74. J. Wong-Ekkabut, S. Baoukina, W. Triampo, I. M. Tang, D. P. Tieleman and L. Monticelli, *Nat. Nanotechnol.*, 2008, **3**, 363–368.

75. S. E. Feller, R. Zhang and R. W. Pastor, *J. Chem. Phys.*, 1995, **103**, 4613–4621.
76. S. J. Marrink and A. E. Mark, *J. Phys. Chem. B*, 2001, **105**, 6122–6127.
77. T. Darden, D. York and L. Pedersen, *J. Chem. Phys.*, 1993, **98**, 10089–10092.
78. U. Essmann, L. Perera, M. L. Berkowitz, T. Darden, H. Lee and L. G. Pedersen, *J. Chem. Phys.*, 1995, **103**, 8577–8593.
79. I. G. Tironi, R. Sperb, P. E. Smith and W. F. Vangunsteren, *J. Chem. Phys.*, 1995, **102**, 5451–5459.
80. M. Patra, M. Karttunen, M. T. Hyvonen, E. Falck, P. Lindqvist and I. Vattulainen, *Biophys. J.*, 2003, **84**, 3636–3645.
81. D. P. Tieleman, B. Hess and M. S. Sansom, *Biophys. J.*, 2002, **83**, 2393–2407.
82. T. W. Allen, O. S. Andersen and B. Roux, *Proc. Natl. Acad. Sci. USA*, 2004, **101**, 117–122.
83. P. H. Hunenberger and J. A. McCammon, *Biophys. Chem.*, 1999, **78**, 69–88.
84. A. Warshel, P. K. Sharma, M. Kato and W. W. Parson, *Biochim. Biophys. Acta*, 2006, **1764**, 1647–1676.
85. C. M. Shepherd, H. J. Vogel and D. P. Tieleman, *Biochem. J.*, 2003, **370**, 233–243.
86. H. J. Risselada and S. J. Marrink, *Proc. Natl. Acad. Sci. USA*, 2008, **105**, 17367–17372.
87. E. A. Dolan, R. M. Venable, R. W. Pastor and B. R. Brooks, *Biophys. J.*, 2002, **82**, 2317–2325.
88. J. N. Sachs, P. S. Crozier and T. B. Woolf, *J. Chem. Phys.*, 2004, **121**, 10847–10851.
89. L. Delemotte, F. Dehez, W. Treptow and M. Tarek, *J. Phys. Chem. B*, 2008, **112**, 5547–5550.
90. A. Y. Shih, P. L. Freddolino, A. Arkhipov, S. G. Sligar and K. Schulten, in *Computational Modeling of Membrane Bilayers*, 2008, Vol. 60, pp. 313–342.

Lateral Pressure Profiles in Lipid Membranes: Dependence on Molecular Composition

O. H. SAMULI OLLILA[a] AND ILPO VATTULAINEN[b]

[a] Department of Physics, Tampere University of Technology, P.O. Box 692, FI-33101, Tampere, Finland; [b] Department of Applied Physics, Helsinki University of Technology, P.O. Box 1100, FI-02015 HUT, Finland, and Memphys – Center for Biomembrane Physics, Physics Department, University of Southern Denmark, Campusvej 55, DK-5230, Odense M, Denmark

2.1 Introduction

The deepest part of the Earth's oceans is the Mariana Trench located in the Pacific Ocean. It is about 11 km below sea level and the pressure down there is about 1100 bar. Would you consider feeling comfortable under these conditions? Probably not.

Next imagine a tiny molecule inside a cell membrane. The pressure felt by this molecule is occasionally about 1000–2000 bar.[1] Does this in turn sound reasonable? Probably not. Yet the tiny molecule feels just fine.

The above example illustrates one of the major differences between macroscopic and nanoscale worlds. Under nanoscopic scales, pressures of the order of 1000 bar are not necessarily exceptional, since the related energies are of the order of $1–10k_\mathrm{B}T$,[1,2] and compete with thermal fluctuations characterized by the thermal energy $k_\mathrm{B}T$. Nonetheless, whereas the absolute value of the pressure may not be a decisive factor, changes in pressure may play a major role in

RSC Biomolecular Sciences No. 20

Molecular Simulations and Biomembranes: From Biophysics to Function

Edited by Mark S.P. Sansom and Philip C. Biggin

© Royal Society of Chemistry 2010

Published by the Royal Society of Chemistry, www.rsc.org

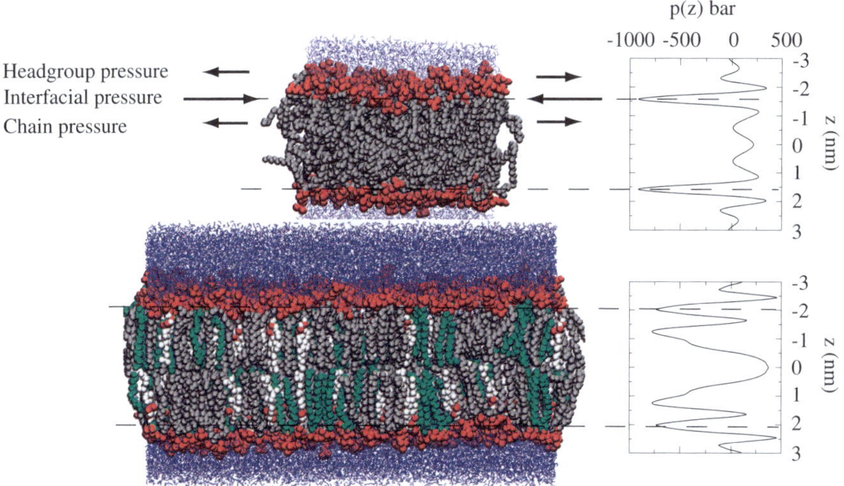

Figure 2.1   Description of the lateral pressure profiles in different lipid environments. Top: a saturated one-component model membrane comprised of DPPC lipids.[12] Bottom: raft-like model membrane comprised of POPC, sphingomyelin and cholesterol.[1] The molar concentrations of the three lipid components are equal here. Attention should be paid in both cases to the maximum values and gradients of the pressure. The results are shown along the membrane normal direction z (the case $z = 0$ corresponding to membrane center).

processes such as membrane protein activation, if pressure changes take place over a length scale that is considerably smaller than the size of the protein. This is precisely the case in cell membranes.

Figure 2.1 illustrates a typical situation in a lipid membrane. The many different interactions give rise to an inhomogeneous pressure distribution across a membrane – one talks about the *lateral pressure profile*.[3–8] For simplicity, here we assume that the pressure profile computed along the membrane normal direction can be averaged over the membrane. This assumption is obviously correct for a homogeneous membrane or for a membrane domain whose molecular composition is homogeneously distributed within the domain. The more general case for heterogeneous membranes is discussed in a recent paper.[9] As for interactions giving rise to the lateral pressure profile, there are several interaction types that result in essentially three regions (see Figure 2.1) characterized by different behavior of the pressure profile. First, close to the membrane–water interface there is an attractive pressure contribution due to the interfacial energy between the water and the hydrocarbon phases, trying to minimize the area of the membrane. Second, next to this interface the hydrophilic headgroups are connected to water, creating a repulsive component due to electrostatic and steric interactions, together with repulsion due to hydration. Finally, in the middle part of the membrane one finds a repulsive contribution due to steric interactions between hydrocarbon chains.[8,10,11]

In practice, it has been found that the pressure in the lipid headgroup region is about $+500$ bar. As one goes slightly deeper into the membrane and considers the region right below lipid headgroups, the pressure typically decreases to about -1000 bar. Finally, in the hydrocarbon region in the middle of a membrane, the pressure is repulsive and about $+200$ bar.

The change in pressure close to the membrane–water interface from the positive (repulsive) peak to the negative (attractive) peak takes place in about 1 nm, meaning that the pressure gradient is about 1500 bar nm^{-1}, which is about 1.5×10^{12} bar m^{-1}. For comparison, if one dives in a sea, the corresponding pressure gradient would be about 0.1 bar m^{-1}. One is tempted to think that such huge pressure gradients may have a role to play, *e.g.* in the activation of membrane proteins.

The connection between lateral pressure profiles and membrane protein functionality has been discussed rather extensively.[8] In the early 1990s, Gibson and Brown[13,14] discussed the role of curvature stress on rhodopsin activity. In the late 1990s, Cantor presented related views in a more concrete framework by formulating a theoretical connection between membrane protein functionality and the lateral pressure profile.[6,7,15,16] His analytical calculations for model systems demonstrated that the contribution of the lateral pressure profile to the total free energy barrier of protein (de)activation can be significant. Many experimental studies have used these ideas to interpret their findings. For example, Perozo[17] found that the large mechanosensitive channel (MscL) can be opened without external pressure by adding lysophosphatidylcholine molecules asymmetrically in only one of the two leaflets in a bilayer (see Figure 2.2). Meanwhile, symmetric addition of lysophosphatidylcholines was not found to open the channel. Related studies by other workers are also in favor of an idea that the lateral pressure profile may be involved in the opening of MscL.[18,19] More recently, it has been found[20,21] that addition of short-chain alcohols to a membrane resulted in dissociation of a membrane protein complex (KcsA) and the results were interpreted in terms of changes in the pressure profile induced by the alcohol. Findings of a similar nature, favoring a view that changes in lateral pressure profile have a role to play in protein functionality, have been made for the activation of rhodopsin[14,22] and the modulation of CTP:phosphocholine cytidylyltransferase.[23,24]

The general significance of the lateral pressure profile is further stressed by the fact that elastic coefficients for bending (and Gaussian bending) of membranes are connected to the pressure profile (see Section 2.2.3).[25,26] For the same reason, the topological phase behavior of lipid systems emerges in part from the pressure profile.[27] Thus, if the lateral pressure profile were important for membrane protein activation, then some proteins should likely also be sensitive to membrane elasticity. It has turned out that this is the case. There are membrane protein classes such as the mechanosensitive channels whose activation depends on the elasticity of a membrane in which the proteins are embedded.[28]

Considering the above examples, it would be of profound interest to understand the relation between the lateral pressure profile and the molecular

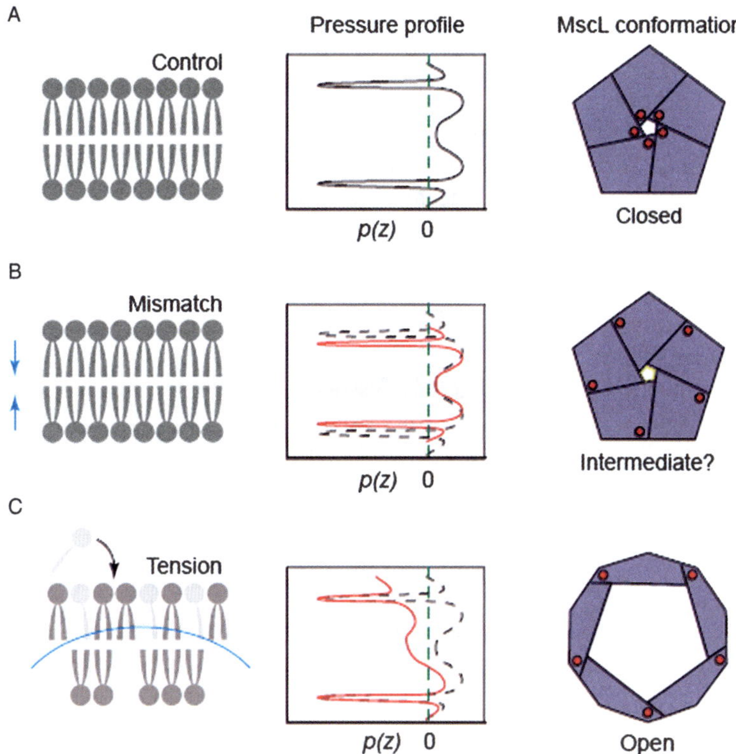

Figure 2.2 Schematic view of how an asymmetric distribution of lipids in the two leaflets of a membrane can result in lateral stress, which may be decisive for the opening of MscL. Shown here are plots for the type of membrane perturbation (left), the estimated transmembrane pressure profile (middle) and the corresponding state of MscL regarding its functionality. Of interest here is case (C) shown at the bottom, describing the possibility that asymmetric distribution of cone-shaped lipids would alter the pressure profile, favoring the open state of the channel. Adapted with permission from reference 17. Copyright 2003 Elsevier.

composition of a membrane. Unfortunately, doing this is not a simple task. Experimental determination of the lateral pressure profile is exceptionally difficult, because one should measure local pressure differences inside a membrane in a scale that is less than 1 nm. Pyrene probes with different lengths have been used to gauge the pressure in different parts of the acyl chain region (see Figure 2.3).[29,30] However, this technique allows one to measure only relative pressures. What is even more problematic with the use of probes, however, is that their effects on membrane structure and the resulting pressure profile are unknown. Atomistic simulations have shown[31] that pyrenes disturb the packing and dynamics of lipids in the vicinity of the probe, the perturbations being significant within a few nanometers around the probe, but their effect on the pressure profile has not yet been elucidated.

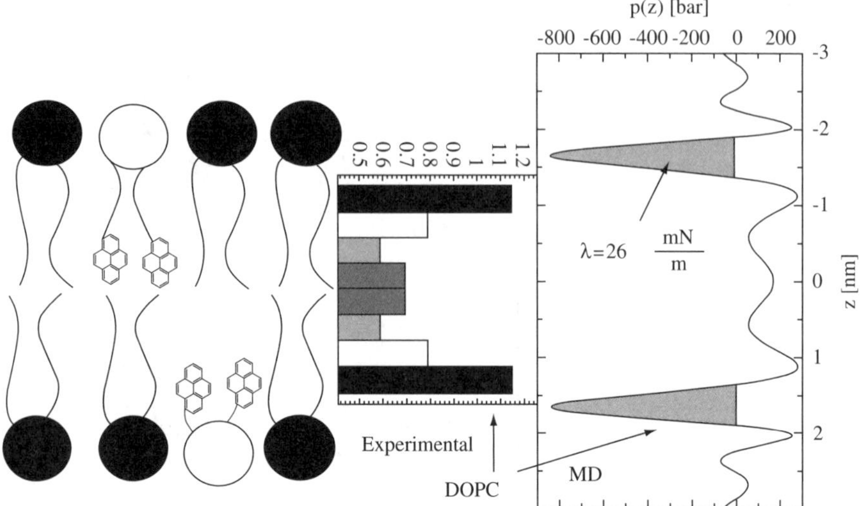

Figure 2.3 Illustration of the pressure distribution for DOPC bilayer measured by pyrene probes[29] (middle) and calculated from atomistic simulation[2] (right). The drawing showing the use of pyrene probes has been adapted from reference 30 (left). Experimental and simulated profiles are in agreement in the acyl chain region (see further discussion in the text). The interfacial energy density was calculated integrating over the shaded area in the profile.

As an alternative means, different approaches have been used to characterize the negative peak at the interface of polar and non-polar regions, the peak arising from the hydrophobic free energy density γ_{hpb}. Using thermodynamic measurements, it has been approximated that $\gamma_{hpb} \approx 30\text{–}35\,\mathrm{mN\,m^{-1}}$.[8] The relation between surface pressure in monolayer experiments and the hydrophobic free energy is still under discussion,[8,32–34] but it is often concluded that $\gamma_{hpb} \approx \Pi(A_0)$, where A_0 is the equilibrium area of the bilayer. This would allow comparison between the lateral pressure profile and the surface pressure determined from Langmuir monolayer experiments.

Given the difficulties in measuring pressure profiles through experiments, computational and theoretical techniques have been applied to complement experiments. In fact, since atomistic and coarse-grained (CG) model simulations are an exceptionally useful approach to deal with atomistic and molecular aspects of membrane systems in (sub)nanoscopic detail, most of the lateral pressure profile studies reported by far have been based on atomistic and CG simulations. Pressure distributions in the acyl chain region have been determined using mean field theory, statistical thermodynamics calculations and Monte Carlo methods.[15,35–37] The pressure profiles in the headgroup region have been studied using dissipative particle dynamics simulations or other CG approaches[38,39] and also atom-scale and CG molecular dynamics simulations.[1,2,12,40–48] Recent studies have concentrated on the dependence of pressure distribution on lipid composition[1,2,12,40,45] and anesthetics embedded in membranes,[46,49] and the connection of the pressure profile with elastic constants.[2,47,48] The possible interplay between protein structure and the lateral pressure profile[1,2] has also been received attention.

In this chapter, we discuss the concept of the lateral pressure profile starting from its definition. We focus on the dependence of the pressure distribution on the molecular composition in a membrane and in this context discuss the roles of various small molecules, including sterols and anesthetics. Particular attention is paid to elucidating the role of the pressure profile on membrane protein activation. Most of the results presented and discussed here are based on simulation studies, but comparison with the few related experimental studies is made where appropriate.

2.2   Theoretical Concepts

2.2.1   Lateral Pressure Profile

The pressure for an inhomogeneous system is represented by a tensor $\mathbf{P}(\mathbf{r})$ that depends on the location $\mathbf{r}$. This can be written as a sum of two components

$$\mathbf{P}(\mathbf{r}) = -\sigma_K(\mathbf{r}) - \sigma_C(\mathbf{r}) \tag{2.1}$$

The kinetic contribution of the local stress tensor δ_K is defined as[50]

$$\sigma_K^{\alpha\beta}(\mathbf{r}, t) = -\sum_i m_i v_i^\alpha v_i^\beta \delta(\mathbf{r} - \mathbf{r}_i) \tag{2.2}$$

where m_i, v_i and $\mathbf{r}_i$ refer to the mass, velocity and location of atom i, respectively. The second component in equation (2.1), that is, the configurational part of the local stress tensor σ_C, is defined as[38,50]

$$\sigma_C^{\alpha\beta}(\mathbf{r}, t) = \sum_i \sum_{j<i} \mathbf{F}_{ij} \times \oint_{C_{ij}} dl^\beta \delta(\mathbf{r} - \mathbf{l}) \tag{2.3}$$

where C_{ij} is a contour from the particle i to the particle j and $\mathbf{F}_{ij}$ is the force exerted by the particle j on i. This definition is in line with the definition of virial for total pressure,[50,51] except that a chosen contour is used to divide the space between particles that interact through pairwise forces. Note that in an inhomogeneous medium, the pressure distribution depends on the contour used, whose correct definition under these conditions is currently under discussion;[50–57] see below.

The pressure for any volume V can be calculated from the continuous pressure distribution, defined by equations (2.1)–(2.3) by taking an average over the volume $\mathbf{P}_V(t) = \int_V d\mathbf{R}\, \mathbf{P}(\mathbf{R}, t)/V$.

Although there has been a great deal of discussion of the choice of the contour, no consensus has yet been reached. Traditionally two different contours have been suggested: the Irving–Kirkwood contour, which is a straight line between particles,[58] and the Harasima contour, which goes along the coordinate axes.[59] There have been views in favor of the Irving–Kirkwood contour,[52,55] but these views have also been contested.[51,56] Sonne *et al.*[41] showed that long-range interactions through Ewald summation techniques can be included in the local pressure calculation using the Harasima contour. They also showed that in simulations of planar lipid bilayers, both contours yielded practically identical results. However, some authors have reported that in spherical systems the Harasima contour gives unphysical results,[53] unlike the Irving–Kirkwood contour. In all molecular dynamics (MD) studies of lipid membranes conducted so far, except for the studies by Sonne *et al.*[41] and Orsi *et al.*,[48] the Irving–Kirkwood contour has been the method of choice. Traditionally the pressure distribution in lipid bilayers is described using the lateral pressure profile, defined as

$$p(z) = p_L(z) - p_{zz}(z) \tag{2.4}$$

where $p_L(z) = [p_{xx}(z) + p_{yy}(z)]/2$,[8,25,60] and the coordinate z is along the membrane normal. This choice is favored because then the moments are related to elastic properties of the bilayer.[25]

It follows from planar symmetry and the condition for mechanical equilibrium $\nabla \cdot \mathbf{P} = 0$ that $p_L(z) = p_{xx}(z) = p_{yy}(z)$ depends only on z and that p_{zz} is constant. Furthermore, the surface tension of a layer between z_1 and z_2 is given by[60] $\gamma = -\int_{z_1}^{z_2} \mathrm{d}z\, p(z)$. In this chapter, we consider only lipid bilayers having planar symmetry. At present, the full pressure field in three dimensions without assumptions about planar symmetry has been determined in only one study.[9]

2.2.2 Calculation of Lateral Pressure Profile from Simulation

The lateral pressure profile calculations for planar lipid bilayers are essentially always implemented by dividing the membrane system into thin slices perpendicular to the membrane normal and calculating the local pressure tensor in each slice.[12,38,41,42,44] To this end, equations (2.2) and (2.3) are modified such that the Dirac delta function in equation (2.2) and the contour in equation. (2.3) are discretized. The forces between particles are calculated using potentials for m-body interactions (*i.e.* pair interactions, three-body interactions, *etc.*) and summation over pairs is made over clusters formed by the many types of interactions[12,38,42] as follows:

$$\mathbf{P}^{\alpha\beta}(\mathbf{R}, t) = \sum_{i \in \text{slice}} m_i v_i^\alpha v_i^\beta + \sum_m \frac{1}{mV_{\delta z}} \sum_{\langle j \rangle} \sum_{\langle k,l \rangle} \left(\nabla_{j_k}^\alpha U^m - \nabla_{j_l}^\alpha U^m \right) r_{j_k j_l}^\beta f(z_{j_k}, z_{j_l}, z_s) \tag{2.5}$$

where $f(z_{j_k}, z_{j_l}, z_s) = 0$ if particles are on the same side of the slice, $f(z_{j_k}, z_{j_l}, z_s) = 1$ if both particles are inside the slice, $f(z_{j_k}, z_{j_l}, z_s) = \delta_z/|z_{j_k} - z_{j_l}|$ if particles are on different sides of the slice and $f(z_{j_k}, z_{j_l}, z_s) = d_z/|z_{j_k} - z_{j_l}|$ if only one of the particles is inside the slice. The term d_z is the distance from a given particle inside the slice to the wall of the slice separating the two particles and δ_z and $V_{\delta z}$ are the thickness and the volume of the slice, respectively. The notion $\langle j \rangle$ stands for summation over all m-clusters in the system and $\langle k, l \rangle$ describes summation over all possible pairs of particles within a given m-cluster.

In published studies,[1,2,12,40–48] the bilayer is usually divided into slabs approximately 0.1 nm thick. The implementation of the pressure profile calculation is largely similar in all publications, although some details differ. The most evident difference concerns treatment of long-range electrostatic interactions. Using truncation for electrostatic interactions causes severe artifacts in lipid bilayer simulations,[61–63] which can be corrected using, for example, Ewald summation methods. However, in Ewald summation techniques, forces cannot be divided into pairwise forces and therefore equation (2.5) can no longer be used. This problem is usually handled by running the actual simulations using Particle-Mesh Ewald summation (PME) for Coulombic interactions and calculating the pressure profile from the obtained simulation results (trajectories) by using a long cut-off distance (typically about 2 nm) for long-range interactions. The error in this case is systematic, not cumulative as it would be if also the simulation were done with truncation. This is a reasonably good approximation for comparing pressure profiles of different systems, since the systematic error is then almost identical in all cases.

Sonne *et al.*[41] introduced a method to include Ewald summation in pressure distribution calculations using the Harasima contour. They showed that PME and truncation yield qualitatively similar results for pressure profiles if the truncation distance is larger than about 2.0 nm.

2.2.3 Elastic Properties

Surface curvature elastic energy per unit area can be expressed using the Helfrich approach:[25]

$$g(c_1, c_2) = \frac{1}{2}\kappa(c_1 + c_2 - c_0)^2 + \bar{\kappa}c_1 c_2 \tag{2.6}$$

where κ is the bending modulus, c_0 is the spontaneous curvature, c_1 and c_2 are local principal curvatures and $\bar{\kappa}$ is the Gaussian bending modulus. Here we concentrate on curvature energy and omit the area compressiblity modulus K_A. The constants in equation (2.6) are connected to the lateral pressure profile through the equations[25]

$$\kappa_m c_0^m = \int_0^h \mathrm{d}z(z - \delta)p(z) = \tau_m^{(1)} \tag{2.7}$$

and

$$\bar{\kappa}_m = \int_0^h \mathrm{d}z (z-\delta)^2 p(z) = \tau_m^{(2)} \tag{2.8}$$

where z is the normal coordinate of the layer and δ is the position of the mono-layer neutral plane. Here we define $z = 0$ to be the center of the bilayer and h is the thickness of the monolayer. c_0^m, κ_m and $\bar{\kappa}$ refer to the properties of the monolayer. Thus, equations (2.7) and (2.8) give elastic properties of a monolayer in the bilayer. The first and second moments are denoted by $\tau_m^{(1)}$ and $\tau_m^{(2)}$, respectively.

The moments and elastic properties of lipid monolayers described in these equations have been studied earlier using analytical theory,[64] statistical thermodynamics calculations,[16] coarse-grained molecular dynamics simulations[47,48,65] and atomistic molecular dynamics simulations.[2] The results are discussed in Section 2.4.5.

2.2.4 Interplay of Pressure Profile and Membrane Protein Activation

The influence of the lateral pressure profile on protein configurational energies can be approximated by calculating the mechanical work needed to create a cavity for the protein inclusion in the membrane:[16]

$$W = -\int_{-h}^h \mathrm{d}z p(z) A(z) \tag{2.9}$$

where $A(z)$ is the area of the protein; z and h are as in the previous section, hence integration now goes over the bilayer. The energy difference between two different conformational states of the protein is then

$$\Delta W = \int_{-h}^h \mathrm{d}z p(z) \, \Delta A(z) \tag{2.10}$$

where $\Delta A(z)$ is the difference in area between different states. The work can be connected to measurable quantities by expanding the protein cross-sectional area profile $A(z)$ using the Taylor expansion:[16,66]

$$A(z) = A(0) + a_1 z + a_2 z^2 + \dots \tag{2.11}$$

If we ignore higher order terms and substitute this in equation (2.9), we obtain

$$W = -a_1 \int_{-h}^h \mathrm{d}z z p(z) - a_2 \int_{-h}^h \mathrm{d}z z^2 p(z) \tag{2.12}$$

The moments of the pressure profile are connected to elastic constants through equations (2.7) and (2.8), hence we obtain[66]

$$W = (a_1 + 2a_2\delta)\kappa_b c_0^b - a_2\bar{\kappa}_b \qquad (2.13)$$

where c_0^b, κ_b and $\bar{\kappa}_b$ refer to the properties of a bilayer and the neutral surface of the bilayer is assumed to be in the center of the membrane ($\delta = 0$). From equation (2.13), one can see that the effect due to the pressure profile on changes in membrane protein shape can be written using elastic constants for bending and Gaussian bending. This means that the dependence of membrane protein functionality on membrane elasticity and the changes in pressure profile are related.

Comparing these energy differences for different lipid environments, one obtains estimates for the free energy difference between the states in each environment.[1,2,16,44,67] Unfortunately, conformations of membrane proteins in different states are not very well known, thus $\Delta A(z)$ is difficult to determine. Another general drawback of this method is that the effect of the protein on the lateral pressure profile is not accounted for, unless the pressure profile has been determined (from simulations) while the protein is present in a system. Nonetheless, the results using the above approach are discussed in Section 2.4.5.

2.3 Gauging Pressure Profile

As mentioned in the Introduction, several theoretical and computational methods have been used to calculate the lateral pressure profile.[1,2,12,15,35–48] By far the most successful technique has been atomistic molecular dynamics, since it allows the pressure to be calculated in atomic detail, providing at least semiquantitative insight into the pressure distribution inside membranes.[1,2,12,40–48] Below we compare lateral pressure profiles obtained from atomistic and in part also coarse-grained molecular dynamics simulations applied to a variety of different models. These results are compared with experiments, when appropriate.

Let us first discuss how reliably one can determine the pressure profiles on the whole, since a variety of different force fields are available for any lipid system and it is likely that the pressure profile depends at least to some extent on the force field used. Figure 2.4A depicts a comparison of pressure profiles for a DPPC bilayer found in four different atomistic models.[12,41,44,45] All profiles portray similar features, although some details are different. The differences likely arise from the differences in the force fields and also from differences in the implementation of the lateral pressure profile calculation. In the studies by Ollila *et al.*[12] and Patra,[45] the model and simulation protocols (using GRO-MACS[68,69]) are essentially identical but the implementation of the lateral pressure profile calculation is different. However, the only difference between the profiles found in these studies is the height of the interfacial and headgroup peaks, which likely results from the different averaging schemes used. The

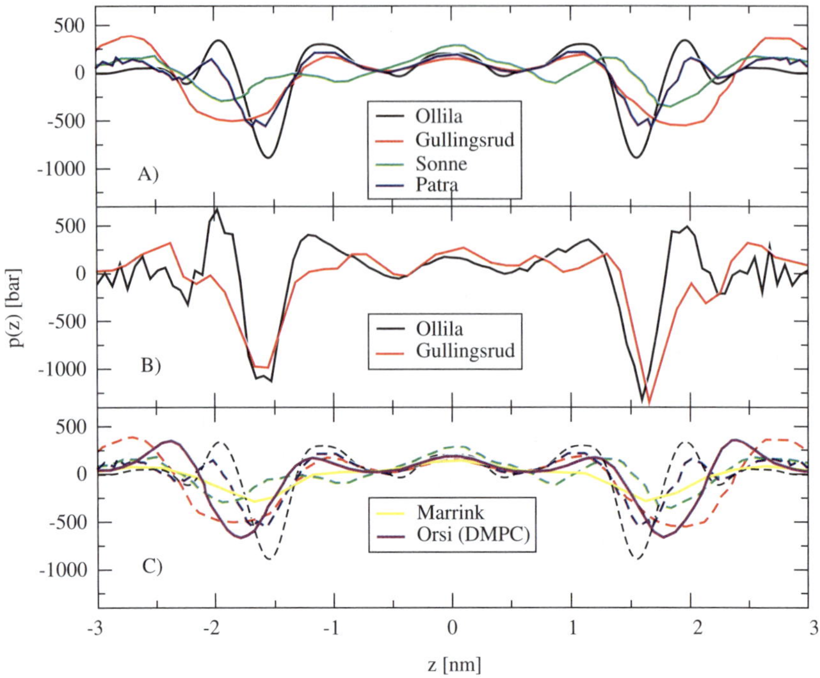

Figure 2.4 Comparison between lateral pressure profiles found by a number of different force fields for bilayer models. (A) Atomistic simulations for a DPPC bilayer; (B) atomistic simulations for a POPC bilayer; (C) coarse-grained simulations for DPPC and DMPC (atomistic data shown with dashed lines for comparison).

profiles found by Gullingsrud and Schulten[44] and Sonne *et al.*[41] were calculated with the same simulation protocol (using NAMD/CHARMM) but with different force field parameters for the DPPC bilayer. The profiles calculated using NAMD/CHARMM[41,44] seem to be broader than the results calculated by GROMACS and related force fields.[12,45] This might be due to the averaging schemes or to the differences in the models.

Figure 2.4B compares lateral pressure profiles for a POPC bilayer calculated using the NAMD/CHARMM[44] and GROMACS protocols.[12] In these profiles, the interfacial peaks are in good agreement, which likely results from the fact that here both profiles are pure data, not smoothed by running averages or related schemes. However, the positive peaks in the acyl chain region, close to the interface and headgroup regions, are in somewhat different locations.

In Figure 2.4C, we compare the pressure profiles found through two coarse-grained models: the MARTINI model[47,70] and the model of Orsi *et al.*[48] Generally, the features of both models are similar to those found in atomistic models that are also shown in the same figure. In the MARTINI model,[47] the heights of the interfacial and headgroup peaks seem to be small, but actually they are comparable to the profiles given by the atomistic study of Sonne *et al.*[41]

In general, although some details of the pressure profiles (such as the height of the peaks) may vary considerably, it is important to keep in mind that physical properties arising from the pressure profile are dictated by its moments (see Sections 2.2.3 and 2.2.4), hence the moments are likely more important than the actual form of the profile. We will discuss this issue in more detail in Section 2.4.5.

Experimental determination of the lateral pressure profile is very difficult, because one should measure local pressure differences on the nanometer scale. Templer *et al.*[29] used pyrene probes to measure pressure at different depths inside a DOPC bilayer using four probes with different chain lengths. Kamo *et al.*[30] also used a similar technique. The results of Templer *et al.* are shown and compared with computational results[2] in Figure 2.3. It is evident that experiments and simulations yield the same general form for the pressure profile in the acyl chain region: pressure decreases as one moves ahead from the interfacial region towards the membrane center, but then there is a peak in the center of the bilayer. This peak was not found in earlier models that were based on simplified descriptions of lipids,[15,35,36,38,39,64] but it is evident in all atomistic simulations,[1,2,12,40–46] recent coarse-grained simulations[47,48] and also theoretical studies.[37] However, it is important to note that the technique based on probes to measure pressure allows only relative pressures to be measured, in addition to which it is challenged by the fact that the effects of probes on the pressure profile are not understood. In simulations of membranes, pyrene and other probes have been found to perturb the membrane structure and dynamics significantly in the vicinity of the probe.[31,71,72]

The origin of the pressure profile peak in the membrane center has been discussed extensively.[12,21,37,48] The most obvious explanations would be the higher density in the membrane center compared with the acyl chain region or interdigitation of the leaflets.[21] However, in computer simulations where the peak in the pressure profile has been observed, there is no pronounced density in the middle of a membrane.[12] Also, it has been found that the peak in the pressure profile disappears with increasing level of unsaturation, while interdigitation remains essentially unchanged.[12] Mukhin and Baoukina suggested that the origin of the peak is related to increasing configurational freedom of the chains in the center of the bilayer, which hints at the possibility of an entropic origin.[37] This idea was used to explain the disappearance of the pressure profile peak in the membrane center due to an increasing number of double bonds in acyl chains.[12]

The above indicates that, despite their limitations, experiments based on the use of fluorescent probes seem to be consistent with atomistic simulations and to yield some insight into the pressure distribution inside membranes. This view is also supported by data for the hydrophobic free energy density (see Figure 2.3). This can be computed from the pressure profile as an integral over the negative peak in the interfacial region, in experiments found from thermodynamic and Langmuir monolayer studies with some degree of approximation. While simulations give $\gamma_{hpb} \approx 26\,\mathrm{mN\,m^{-1}}$, experimental values for γ_{hpb} range over 30–$35\,\mathrm{mN\,m^{-1}}$, in reasonably good agreement with simulations. While the relation between the surface pressure in monolayer experiments $\Pi(A)$ and the hydrophobic free energy is under discussion,[8,32–34] it is fairly widely accepted that γ_{hpb} and $\Pi(A_0)$ are related (A_0 being the equilibrium area of the bilayer). This suggests that qualitative comparison between lateral pressure profiles and surface pressures measured through Langmuir monolayer film experiments is possible. However, this has not yet been tested.

Acyl chain order parameters, S_{CD}, determined through deuterium ^{2}H NMR measurements, have also been suggested to be related to the lateral pressure profile.[73] The idea is that small order parameters would relate to large repulsive pressure. As S_{CD} often relates to the average area per lipid in a membrane, a more complex relation to describe the dependence of the lateral pressure profile on the area per molecule has also been suggested.[12,48] These propositions are essentially similar to the suggestion that pressure profiles depend on bending rigidity and occupied area of acyl chains.[37] Further, properties of the inverted hexagonal phase have also been used to approximate the intensities of different components in the lateral pressure profile.[74] Although none of the above suggestions have been systemically tested, the recent progress in computational resources together with validated force fields hints that it will soon be possible.

2.4 Dependence of Pressure Profiles on Molecular Composition

2.4.1 Dependence on Unsaturation Level

It has been suggested that inclusion of polyunsaturated lipids in lipid membranes could affect protein activity through changes in lateral pressure profile or consequent changes in elastic properties.[14–16,22,73,75] Especially an increase in the fraction of the MII state of rhodopsin has been suggested to follow from the altered pressure profile.[13,14] To corroborate these views, it is necessary to understand the dependence of the lateral pressure profile on the level of unsaturation.

The dependence of the lateral pressure profile on the unsaturation level has been studied through theoretical thermodynamics considerations[15] and atomistic molecular dynamics simulations.[12,40] The analytical studies of Cantor suggest that pressure in the acyl chain region is shifted towards the interfacial region with increasing number of double bonds.[15] Carillo-Tripp and Feller[40] studied pressure profiles in one-component membranes comprised of polyunsaturated lipids (with DPA or DHA chains) using atomistic simulations (see Figure 2.5). They showed that even though DPA and DHA differ only by one double bond, there is a clear change in the lateral pressure profile, in line with the prediction of Cantor.[15]

Ollila *et al.*[12] studied the effect of double bonds on the lateral pressure profile using atom-scale molecular dynamics simulations by comparing membranes where the number of double bonds in a lipid hydrocarbon chain varied from zero to six (see Figure 2.5). The most pronounced effect that they found was the

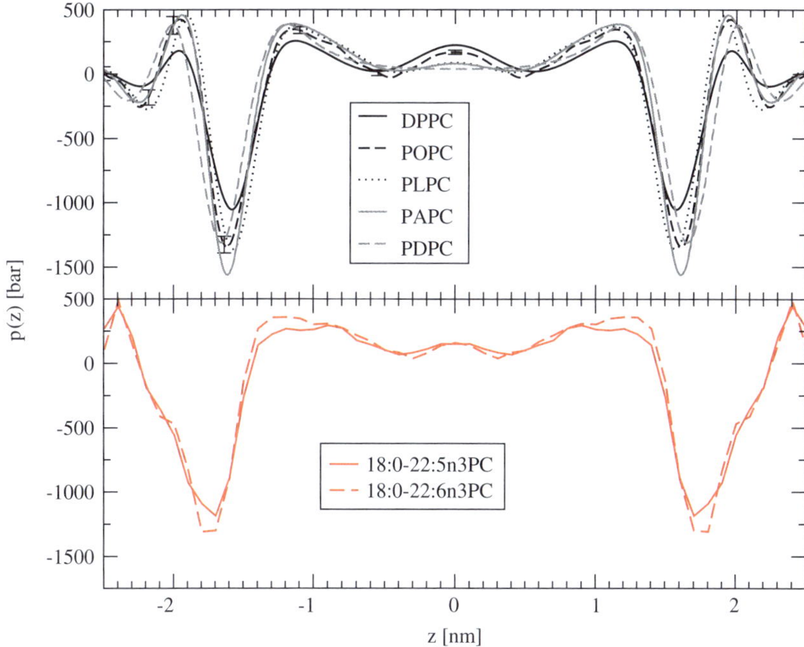

Figure 2.5 Top: dependence of the lateral pressure profile on the number of double bonds:[12] DPPC, POPC, PLPC, PAPC and PDPC bilayers, having zero, one, two, four and six double bonds, respectively. Bottom: influence of adding one extra double bond (from five to six) according to Tripp *et al.*[40]

reduction of the central peak with increasing number of double bonds. The reduction in pressure in the membrane center was found to be accompanied by an increase in pressure in the headgroup and interfacial regions.

Both simulation studies therefore support the prediction by Cantor that double bonds shift repulsive pressure from the acyl chain region towards the interfacial area. In addition, they suggested a slight increase in surface free energy and in the headgroup peaks of the pressure profile.

Further analysis and possible simulations with rhodopsin embedded in a membrane are needed to establish whether the reported changes in the pressure profile could explain the dependence of rhodopsin functionality on the level of unsaturation in a membrane.

2.4.2 Effects of Different Sterols in Two-component Membranes

Cholesterol (CHOL) is one of the most important lipid components of biological membranes, hence its effect on the lateral pressure profile is also of profound interest. Interestingly, concerning the above-mentioned dependence of rhodopsin activity on lipid composition, the fraction that rhodopsin spends in the MII state decreases with increasing cholesterol concentration,[76,77] in contrast to the trend observed with polyunsaturated lipids.[75] One way to explain this difference is to consider the lateral pressure profiles in the two cases.

The effect of cholesterol on the lateral pressure profile in two-component membranes has been studied using theoretical thermodynamic calculations[15] and also through atomistic molecular dynamics simulations.[2,40,45] Also, the effect of cholesterol as a function of its concentration[45] and the effects of other sterols (such as desmosterol, 7-dehydrocholesterol and ketosterol) in varying lipid matrices[2] have been elucidated through atom-scale simulations. All these investigations support the view that inclusion of cholesterol and other sterols in a membrane is associated with a significant change in the lateral pressure profile.

Simulation studies indicate that additional peaks emerge in the bilayer interior region due to cholesterol (see Figure 2.6). The relative significance of additional peaks increases for increasing cholesterol concentration at least until 50 mol%.[45] Studies of saturated (DPPC) and diunsaturated (DOPC) membranes have shown that peaks associated with headgroup repulsion and interfacial energy increase substantially due to cholesterol.[2,45] In polyunsaturated membranes with cholesterol, a small increase has been found in the headgroup peak, accompanied by a small decrease in the interfacial peak,[40]

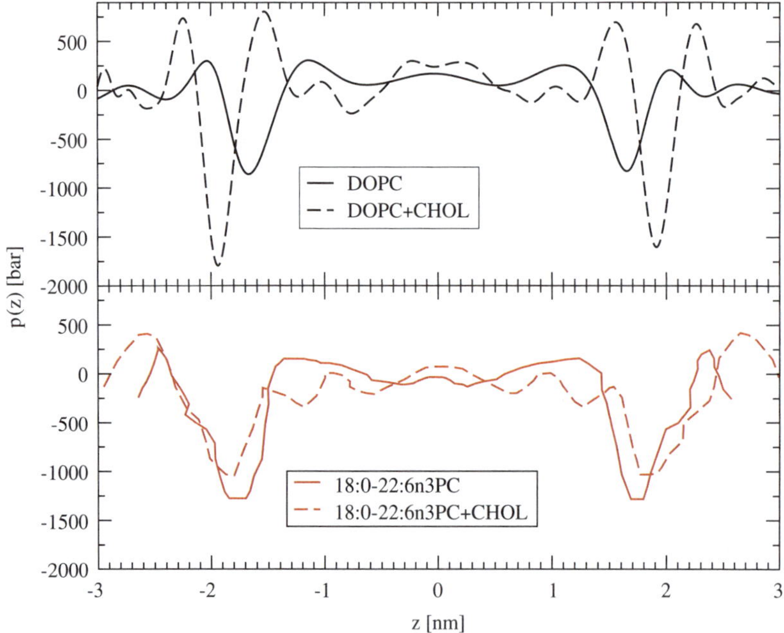

Figure 2.6   Effect of cholesterol on the lateral pressure profile in unsaturated membranes. Top: DOPC and DOPC–CHOL.[2] Bottom: SDPC and SDPC–CHOL.[40]

opposite to conclusions reported in other simulation studies for saturated or diunsaturated membranes (see Figure 2.6). Although the difference could be due to different force fields or different simulation protocols (GROMACS with constant pressure[2] and NAMD/CHARMM with constant area[40]), it could also be due to double bonds that are known to play an important role in membranes. In polyunsaturated membranes, in particular, the effect of cholesterol on membrane structure and ordering has been shown to be distinctly different compared with saturated lipid bilayers.[78,79]

Interestingly, peaks in the bilayer interior region have been found in all simulation studies of lipid bilayers with cholesterol,[2,40,45] for many different sterols[2] and also with different lipid matrices.[2,40] Hence it seems evident that these peaks are characteristic to all sterols. The origin of these peaks is likely related to the ordering effect of the rigid ring structure in sterols. This explanation is supported by similar peaks found in the pressure profiles associated with gel-phase membranes.[9] The fine structures slightly differentiating the pressure profiles of the different sterols probably arise from the chemical details such as the number and the locations of methyl groups and double bonds in the sterol structure.

The existing data support a view that the effect of sterols on the lateral pressure profile is the strongest in saturated membranes[2] (see Figure 2.7). In a similar fashion, the differences in pressure profiles between the many sterols have also been shown to be the strongest in saturated bilayers.[2] With increasing level of unsaturation, the role of sterols in the pressure profile becomes weaker.

There is ample evidence showing that the biological functions of the many sterols are different.[80–83] For example, it has been shown that if cholesterol is replaced by desmosterol, the structure of which differs from that of cholesterol only by a single double bond, the raft-dependent signaling via the insulin receptor is impaired.[80] In the framework of the lipid raft model,[84,85] some sterols such as cholesterol and ergosterol are known to be 'raft lipids' in terms of promoting the formation of strongly ordered membrane domains[86,87] (coexistence of sterol-rich and sterol-poor domains), whereas for sterols such as lanosterol this capability seems to be considerably weaker or even absent.[86] Whether the differences between the biological functions of the many sterols are mainly due to specific factors such as sterol–lipid or sterol–protein interactions or to more generic reasons through changes in, *e.g.*, membrane elasticity, remain to be clarified. Nonetheless, it is clear that the activation of some membrane proteins such as mechanosensitive channels depends on membrane elasticity[8,28,88,89] and the related elasticity coefficients are determined by the lateral pressure profile. Hence it is evident that the lateral pressure profile plays a role also with regard to the effects of sterols on membrane proteins.

2.4.3 Pressure Profiles in Three-component Bilayers

Biological membranes typically consist of hundreds of different lipid types, hence it is crucial to understand how membrane properties emerge from their

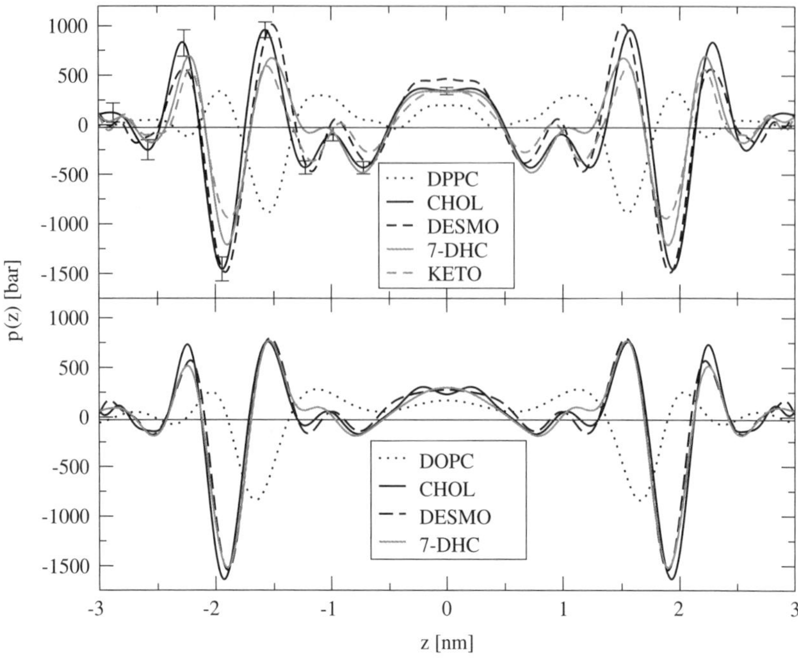

Figure 2.7 Top: lateral pressure profiles of DPPC, DPPC–CHOL, DPPC–DESMO, DPPC–7-DHC and DPPC–KETO bilayers. Statistical errors are presented for one leaflet of the DPPC–CHOL system, in which the fluctuations were the largest. Error bars in other systems were smaller. Bottom: lateral pressure profiles of DOPC, DOPC–CHOL, DOPC–DESMO and DOPC–7-DHC bilayers. Statistical errors are presented for one leaflet of the DOPC–CHOL system. Error bars in other systems were smaller. Adapted from reference 2.

molecular composition in many-component systems. On a general level, the nature of fluid-like membranes can be classified into two types: strongly disordered membranes comprised of polyunsaturated lipids and raft-like membrane domains with pronounced order characteristic of the liquid-ordered phase. The lateral pressure profiles for the latter, raft-like membrane domains have recently been studied by Niemelä *et al.*,[1] who considered three-component membrane systems containing POPC, cholesterol and palmitoylsphingomyelin (PSM) with two different concentration ratios for POPC:CHOL:PSM (1:1:1 and 2:1:1).

The pressure profiles calculated from these simulations are presented in Figure 2.8. For comparison, the pressure profiles of single-component POPC and PSM bilayers and of two-component DPPC–CHOL systems are also presented. Comparing the lateral pressure profiles of three- and two-component systems reveals that the general features remain the same: all peaks common to the one-component cases remain but become higher, although additional peaks are found in the acyl chain region. For the three-component systems, the

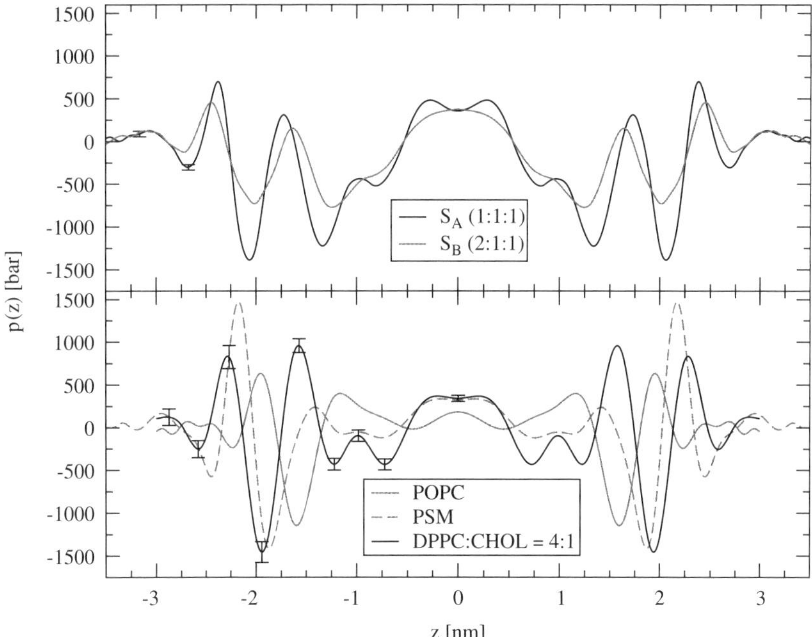

Figure 2.8 Top: lateral pressure profiles of the raft-like systems POPC:CHOL: PSM = 1:1:1 (SA) and 2:1:1 (SB). Bottom: pressure profiles for pure POPC and PSM single-component systems and a binary DPPC–CHOL membrane. The center of the membrane is at $z = 0$. Adapted from reference 1.

additional negative peaks in the acyl chain region are even more pronounced. This is logical if we look at the lateral pressure profile of a pure PSM bilayer, which also has lower values in the acyl chain region compared with the POPC bilayer. Thus, the high peaks in the acyl chain region in the three-component system could be understood in a generic sense to arise from the ordering of the bilayer rather than any specific property of cholesterol.

2.4.4 Implications of Anesthetics on Pressure Profile

One of the greatest mysteries in medical science is general anesthetics. Despite the fact that anesthetics have been used successfully for more than 150 years, the molecular mechanisms of how anesthetic molecules bring about their effects are not understood. On the one hand, it is possible that the mechanism, or at least one of the mechanisms, is associated with specific interactions and specific binding of an anesthetic molecule with the target molecule such as the GABA receptor channel complex. On the other hand, if this were the case, then there would be only a limited number of binding sites, which would imply that there is a limit to the action of anesthetics. However, many of those who have enjoyed ethanol – one of the anesthetics – know that the more one drinks, the more pronounced are the effects, thus giving one an impression that there is no limit to the action of ethanol. Therefore, the question is, could there also be an alternative and more generic mechanism?

In the late 1990s, Cantor suggested one plausible idea: he proposed that general anesthetics might affect membrane protein functionality in an indirect manner by changing the lateral pressure profile of a membrane where the proteins are embedded.[6,7,90] This idea seems reasonable since it is well known that the action of an anesthetic correlates with its partitioning between oil (membrane) and water: the larger is the partitioning coefficient for the membrane, the stronger are the effects of the given anesthetic. Although experimental verification of this view as regards the role of lateral pressure profile in anesthetics is difficult to achieve, there are some experimental data in favor of Cantor's proposition. For example, van den Brink-van der Laan and co-workers interpreted their KcsA dissociation measurements using this idea.[20,21] They proposed that during the dissociation process the shape of KcsA transforms from hourglass to cylindrical and that this process becomes easier when the compressing pressure is applied to the interfacial region. They also suggested that this kind of pressure can be induced using short-chain alkanols.

Considering the experimental difficulties in measuring lateral pressure profiles and especially changes in the profile due to anesthetics, it is a positive feature that theoretical and computational approaches have been able to shed some light on this issue. Regarding theoretical studies, the effect of alkanols with different chain lengths on the lateral pressure profile has been studied by Cantor.[6] He found that short-chain alkanols change the pressure profile close to the interfacial region and long-chain alkanols in the membrane center. Frischknecht and Frink[91] studied the pressure profiles as a function of ethanol, butanol and hexanol concentrations using coarse-grained molecular dynamics simulations and found that the peaks in the pressure profile decrease with increasing alcohol concentration. Terämä *et al.* studied the effect of ethanol on the pressure profile of DPPC and PDPC bilayers using atomistic molecular dynamics simulations[46] and found that ethanol reduces the peaks in the headgroup and interfacial regions (see Figure 2.9). Recent data by Griepernau and Böckmann[49] are in favor of these views.

Summarizing the present theoretical and simulation studies, they predict that the peaks in the pressure profile in the headgroup and interfacial regions are reduced due to alcohol, whereas changes in the acyl chain region are weaker. The decrease in the interfacial region is in agreement with conclusions drawn from micropipette aspiration studies.[92,93] The present understanding of anesthetics on membranes seems to indicate that ethanol and other short-chain alcohols decrease the hydrophobic free energy density γ_{hpb}. This may have implications on membrane properties. Interestingly, ethanol and methanol are often used as spreading solvents in Langmuir experiments, in favor of a view that ethanol reduces the surface pressure because $\Pi \approx \gamma_{hpb}$.[8,32–34]

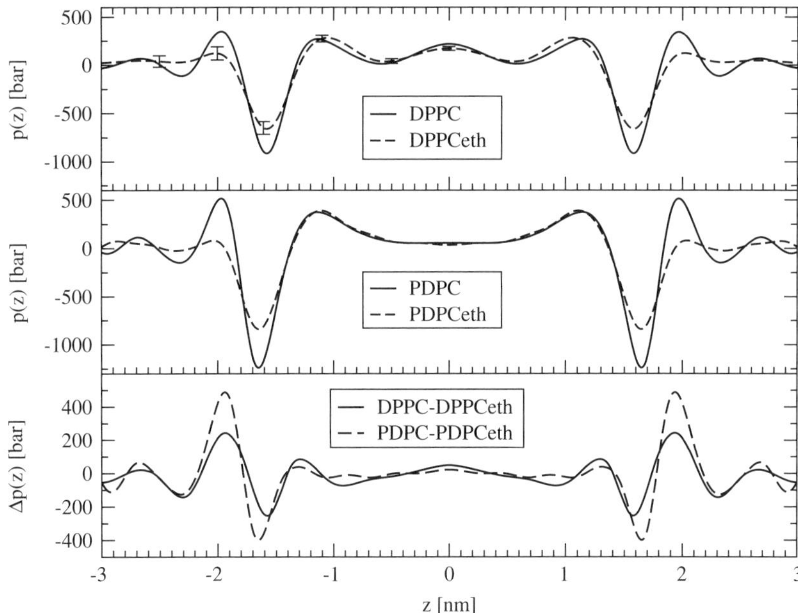

Figure 2.9 Lateral pressure profiles across a membrane for pure DPPC and PDPC bilayers and for the same bilayer under the influence of ethanol (DPPCeth and PDPCeth). Adapted from reference 46.

2.4.5 Elastic Properties Calculated from Lateral Pressure Profile

As discussed in Section 2.2.3, the Gaussian bending modulus κ and the product of the bending modulus κ and the spontaneous curvature c_0 of a monolayer are connected to the lateral pressure profile through equations (2.7) and (2.8). Further, the spontaneous curvature and also the bending and Gaussian bending moduli of a monolayer are related, *e.g.*, to topological phase behavior,[27] stalk formation[27,94,95] and membrane pore formation.[96] The product κc_0 is also interesting as such because it has been connected to the CTP activity[23,24] and membrane protein activity [see equation (2.13)].[66]

Experimental values for the bending modulus for different bilayers have been published,[33] but experimental or theoretical estimates for spontaneous curvatures and Gaussian bending moduli are rare.[33,94,95,97] Consequently, it is interesting to calculate the spontaneous curvatures using equation (2.7) and the Gaussian bending moduli using equation (2.8) (see below).

Recently, spontaneous curvatures for monolayers have been determined by calculating the first moment of the pressure profile, thus finding the bending modulus with another approach and calculating the spontaneous curvature from equation (2.7), $c_0^m = \tau_m^{(1)}/\kappa_m$.[47,48] Marrink *et al.*[47] used experimental estimates for the bending modulus of a bilayer, $\kappa_b \approx (6\text{--}14)\times10^{-20}\,\text{J}$, and Orsi *et al.*[48] calculated the bending modulus from simulations and obtained

$\kappa_b = (9.2 \pm 0.8) \times 10^{-20}$ J. The bending moduli for monolayers are then obtained by dividing these by two, that is, $\kappa_m = \kappa_b/2$.[33]

In this chapter, we gather together the published elastic constants calculated from the pressure profiles obtained through atomistic and coarse-grained molecular dynamics simulations. We also calculate spontaneous curvatures and Gaussian bending moduli from the published pressure profiles from our group.[2,12,46] For the spontaneous curvature we use the same approach as Marrink *et al.*,[47] that is, we use the experimental approximation for the bending modulus. The Gaussian bending modulus is calculated directly from equation (2.8). The location of the neutral surface, *i.e.* the surface with constant area when bending the monolayer, has been suggested to reside close to the polar–apolar interface, although its exact location is not known.[97,98] Orsi *et al.*[48] defined the neutral surface to be located at the interfacial minima of the pressure profile. On the other hand, Templer *et al.*[97] defined its location as residing close to the interface in the acyl chain region as the first maximum of the pressure profile after the interfacial peak. Our results were calculated using the surface in the middle of these two locations and the error bars were determined using the locations employed in previous studies.

Table 2.1 presents the moments and elastic constants calculated from pressure profiles for bilayers with several different compositions. The data for one-component bilayers are from coarse-grained studies by Marrink *et al.*[47] and Orsi *et al.*[48] and from atomistic simulations by Ollila and co-workers.[2,12] Results from atomistic simulations with different sterols[2] and bilayers with ethanol included[46] are also shown. Although moments of the pressure profile and their dependence on lipid content have also been studied by Cantor,[16,67] these results are difficult to compare with simulations because Cantor's statistical thermodynamic calculations included only the acyl chain region explicitly, so we concentrate here on molecular dynamics simulation results presented in Table 2.1.

Comparison between calculated and experimental values for spontaneous curvature shows that the coarse-grained simulations by Marrink *et al.*[47] are in good agreement with experimental values for DOPC, whereas atomistic simulations yield somewhat larger values. Meanwhile, for a DOPE bilayer the coarse-grained simulations give smaller values than experimental studies. For a DOPC–CHOL mixture, atomistic values are in good agreement with experiments.

Even though quantitative data for spontaneous curvatures of different monolayers are sparse, a qualitative dependence is better known based on topological phase transitions.[73] Double bonds, small headgroups such as PE and cholesterol have been found to promote inverted phases.[73] Thus, inclusion of these lipids increases the spontaneous curvature, whereas the effect of ethanol has been found to be the opposite.[73] From Table 2.1, one can see that the spontaneous curvatures are larger for polyunsaturated bilayers (about -0.3 nm^{-1} for PLPC, PAPC and PDPC) than for saturated ones (-0.22 nm^{-1} for DPPC), taken from the same study.[12] Comparing the spontaneous curvatures determined from the coarse-grained study by Marrink *et al.*[47] for DOPC

Table 2.1 Elastic coefficients calculated from lateral pressure profiles for several fluid-like pure and planar bilayers, PC–sterol binary systems and the influence of ethanol[a].

Lipid	Ref.	$\tau^{(1)} = \kappa c_0/10^{-12}\,J\,m^{-1}$	$\kappa_m^{expt}/10^{-20}\,J$	c_0/nm^{-1}	c_0/nm^{-1} (expt)	$\tau^{(2)} = \bar{\kappa}/10^{-20}\,J$
DPPC	2,12	-11.0 ± 1	5 ± 2	-0.22 ± 0.14		-1.1 ± 0.5
	47 CG		5 ± 2	-0.02 to -0.05		
DMPC	48 CG	-0.8 ± 0.1	4.6 ± 0.4	-0.018 ± 0.003		-2.3 ± 0.1
POPC	12	-12.0 ± 1	5 ± 2	-0.26 ± 0.17		-0.8 ± 0.5
DOPC	2	-16.1 ± 2	5 ± 2	-0.32 ± 0.22	-0.05 to -0.11[98,101]	-1.1 ± 0.8
	47 CG		5 ± 2	-0.07 to -0.15		
PLPC	12	$-15.3\ 5 \pm 2$	5 ± 2	-0.31 ± 0.20		-0.5 ± 0.5
PAPC	12	-14.4 ± 2	5 ± 2	-0.29 ± 0.19		-0.7 ± 0.5
PDPC	12	-15.0 ± 2	5 ± 2	-0.3 ± 0.2		-0.5 ± 0.5
DPPE	47 CG		5 ± 2	-0.12 to -0.28		
DOPE	47 CG		5 ± 2	-0.15 to -0.35	-0.35 to -0.37[98,102]	
DPPC–CHOL	2	-6.7 ± 2	5 ± 2	-0.13 ± 0.10		-2.3 ± 0.2
DPPC–DESMO	2	-20.4 ± 2	5 ± 2	-0.41 ± 0.29		-2.8 ± 0.8
DPPC–7-DHC	2	-7.6 ± 2	5 ± 2	-0.15 ± 0.08		-1.9 ± 0.3
DPPC–KETO	2	-9.2 ± 2	5 ± 2	-0.18 ± 0.11		-1.9 ± 0.3
DOPC–CHOL	2	-19.2 ± 2	5 ± 2	-0.38 ± 0.27	-0.37[98]	-2.6 ± 0.6
DOPC–DESMO	2	-18.3 ± 2	5 ± 2	-0.37 ± 0.25		-2.8 ± 0.7
DOPC–7-DHC	2	-18.9 ± 2	5 ± 2	-0.38 ± 0.26		-2.2 ± 0.6
DPPCeth	46	-15.0 ± 2	5 ± 2	-0.34 ± 0.10		-1.1 ± 0.7
PDPCeth	46	-13.0 ± 2	5 ± 2	-0.28 ± 0.10		-1.0 ± 0.7
SM	1	-14.3 ± 2	5 ± 2	-0.29 ± 0.18		-3.2 ± 0.6

[a]'CG' stands for coarse-grained, 'expt' stands for experiment and 'eth' refers to ethanol. The data given correspond to atomistic simulations, unless mentioned otherwise. The values for the bending modulus κ^{expt} are suggestive and do not account for the fact that it likely depends on lipid content.[33] However, since the data for the lipid composition dependence of κ are incomplete, we have assumed the bending modulus to be constant in the analysis shown.

and DOPE bilayers, we see that, as expected, the spontaneous curvature is larger with a PE headgroup. However, the effect of cholesterol on spontaneous curvature is more difficult to realize from the simulation results. Spontaneous curvatures for DOPC–cholesterol mixtures are slightly larger than for pure DOPC, while for DPPC–cholesterol mixtures this effect is not as evident. Ethanol seems to have a different effect on saturated and polyunsaturated monolayers: for DPPC it increases the negative curvature whereas for PDPC the effect is the opposite. In general, all the above findings are in qualitative agreement with topological phase studies except for the case where cholesterol was included in a DPPC bilayer.

However, in the above comparisons it should be kept in mind that the error bars are relatively high due to the uncertainty in the bending modulus. Further, the dependence of bending modulus on lipid content is not taken into account here since the data are simply incomplete. Due to these approximations, these results should be treated with some caution. To improve the analysis, one could calculate the bending modulus from simulations and use that value to calculate the spontaneous curvature. This is likely soon to be feasible despite the large system sizes needed for that purpose.[1,99,100]

Experimental measurements of the Gaussian bending modulus are sparse[33,94,97] and only the ratio with bending modulus has been reported.[33] Gaussian bending moduli for different bilayers calculated from molecular dynamics simulations are presented in Table 2.1. The coarse-grained study by Orsi *et al.*[48] suggests that the relation between the Gaussian bending modulus and the bending modulus is $\bar{\kappa}/\kappa \approx -0.5$, which is close to the experimental result $\bar{\kappa}/\kappa \approx -0.8$.[33] The calculation of this ratio is not reasonable from other simulation results presented in Table 2.1, because the value of the bending modulus is uncertain. However, qualitative predictions can be made. Gaussian bending moduli for polyunsaturated bilayers seem to be lower than for saturated bilayers, although the difference is within error bars. Gaussian bending moduli for DPPC–cholesterol and DOPC–cholesterol mixtures are larger compared with pure bilayers and the largest bending modulus is found for the pure sphingomyelin bilayer. This is in agreement with a picture that bending is generally more difficult for bilayers with high order. Inclusion of ethanol does not seem to have an influence on the Gaussian bending modulus.

2.4.6 Free Energy of Protein Activation and Lateral Pressure Profile

Lateral pressure profiles determined from atomistic molecular dynamics simulations have been used to calculate free energy differences associated with conformational changes of membrane proteins.[1,2,44] The studies have typically used simplified models for structures of proteins in two possible states (active *versus* inactive) to describe a transition where a protein essentially changes structure from a cylindrical to a conical shape. As an example, consider Figure 2.10, which describes the transition in a (large) mechanosensitive channel MscL. The radius

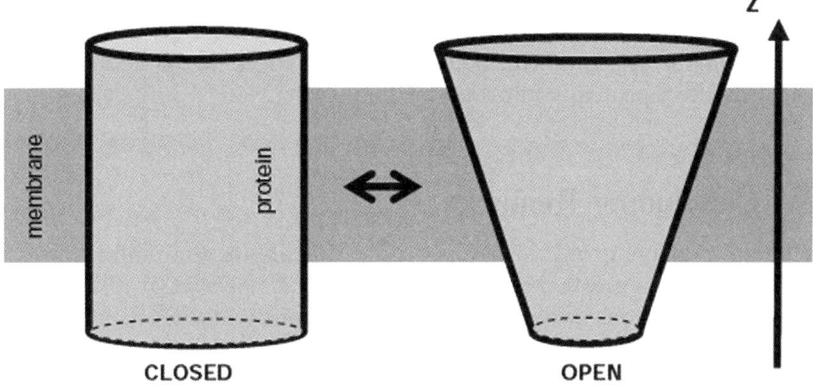

Figure 2.10   Schematic description of the two conformational states of MscL using a cone model. Adapted from reference 103.

of the cone in the center of the bilayer (R) is kept constant and the radius as a function of the normal coordinate z is then written as $r(z) = R + sz$, where s is the slope of the cone. For $s = 0$ one finds a cylinder, whereas a choice such as $s = 0.2$ describes aconical shape. Substituting these into equation (2.10), one obtains

$$\Delta W = \int_{-h}^{h} \mathrm{d}z p(z)[\pi R^2 - \pi(R + sz)^2] = -2\pi R s \tau_b^{(1)} - \pi s^2 \tau_b^{(2)} \qquad (2.14)$$

This description has been discussed as a simple model for MscL.[44] Note that in this case for a symmetric pressure profile $[p(z) = p(-z)]$ the first moment $\tau^{(1)}$ vanishes, while the second moment $\tau^{(2)}$ is non-zero. Hence for this kind of model the free energy difference of a conformational change between two states is small compared with, for example, an hourglass shape where the first moment would also be non-zero. Hence energies calculated for these kinds of models likely underestimate the free energy contribution due to the pressure profile.

Gullingsrud and Schulten[44] used this model together with lateral pressure profiles determined from atomistic molecular dynamics simulations to study the dependence of ΔW on lipid composition using DLPE, DLPC, POPE and POPC lipid matrices. They concluded that changes between these lipid types yield a change of about $(2$–$4)k_{\mathrm{B}}T$ in the free energy barrier of the conformational change. Ollila and co-workers considered free energy differences in DPPC–sterol and DOPC–sterol bilayers with a number of different sterols[2] and in POPC–CHOL–SM ternary membranes.[1] Free energy differences in PC bilayers with different sterols varied between $0.8k_{\mathrm{B}}T$ and $3.3k_{\mathrm{B}}T$, whereas in the ternary mixtures they were considerably larger, about $(4$–$11)k_{\mathrm{B}}T$. All reported values are clearly lower than the total free energy barrier [about $(20$–$50)k_{\mathrm{B}}T$][44,104] associated with gating of MscL. However, the value found in the ternary system is about 20–50% of the total barrier, implying that the contribution due to the pressure profile can be really significant. What is more, as in the above model the first moment $\tau^{(1)}$ disappears, these values can be considered as a lower limit for the free energy change due to the pressure profile. Summarizing, one can conclude that changes in the lateral pressure profile can affect protein conformational equilibria.

2.5 Concluding Remarks

The lateral pressure profile, or stress profile, describes a non-uniform pressure distribution across a soft interface such as a cell membrane. It results from a variety of interactions contributing inhomogeneously inside a membrane. As the changes in pressure inside membranes can be as large as 1012 bar m^{-1}, there is reason to assume that they could have an important role to play in membrane protein activation. Further, as essentially all elastic coefficients of membranes can be derived from the pressure profile, it is obviously the central quantity with regard to membrane elasticity.

In this chapter, we have discussed how the pressure profile depends on membrane composition, including effects of both the variety of lipid components and the partitioning of anesthetics into membranes. The results indicate that even seemingly minor changes in molecular composition may influence pressure profiles substantially, as is the case when cholesterol is replaced by its precursors, that is, desmosterol and 7-dehydrocholesterol. In ordered lipid domains rich in cholesterol and sphingolipids, the nature of the pressure profile is distinctly different from the corresponding profiles in disordered domains rich in polyunsaturated lipids. The effect of anesthetics on the pressure profile is most pronounced close to the water–membrane interface, where anesthetics essentially smooth the profile considerably, reducing pressure profile peaks by ~ 500 bar. As further considerations also show that membrane elasticity and pressure profile are closely coupled, there is reason to stress the importance of a better understanding the role of the pressure profile as a mechanism for activating membrane proteins.

However, there is still a great deal of work to be done. It is important to understand that there is a coupling between stress arising from protein–lipid interactions, and this contribution has been neglected so far. To take these contributions into account, there is a need for extensive studies of pressure profiles in the presence of membrane proteins in various lipid membrane environments. These studies should be carried out in three dimensions, yielding insight into the pressure field surrounding the protein from all directions. Recent work by Ollila *et al.*[9] is a significant step in this direction. They calculated the full 3D pressure distribution for lipid vesicles, a bilayer with coexistence of liquid and gel phases, and in particular a membrane-embedded protein inside a lipid bilayer. Considering the progress in this field, there is therefore reason for optimism. It is likely that future work will soon provide a great deal of insight into the implications of lipid–protein interactions on the pressure profile and on the coupling of the pressure profile on membrane protein functions.

2.6 Abbreviations

CG	coarse-grained
CHOL	cholesterol
CTP	cytidine triphosphate
DESMO	desmosterol
DHA	docosahexaenoic acid
7DHC	7-dehydrocholesterol
DOPC	dioleoylphosphatidylcholine
DOPE	dioleoylphosphatidylethanolamine
DPA	docosapentaenoic acid
DPPC	dipalmitoylphosphatidylcholine
GABA	γ-aminobutyric acid
KETO	ketosterol
MD	molecular dynamics
MscL	large mechanosensitive channel
NMR	nuclear magnetic resonance
PAPC	palmitoylarachidonicphosphatidylcholine
PC	phosphatidylcholine
PDPC	palmitoyldocosahexaenoicphosphatidylcholine
PE	phosphatidylethanolamine
PLPC	palmitoyllinoleoylphosphatidylcholine
PME	Particle-Mesh Ewald
POPC	palmitoyloleoylphosphatidylcholine
PSM	palmitoylsphingomyelin
SDPC	stearoyldocosahexaenoicphosphatidylcholine
SM	sphingomyelin

Acknowledgements

We wish to thank Scott Feller, Mauricio Carrillo Tripp and Perttu Niemela for sharing data that we have used in this chapter for illustration. We also thank Antti Lamberg Mario Orsi, Derek Marsh, Maria-Nefeli Tsaloglou, George Attard and Luca Monticelli for fruitful discussions. The Academy of Finland and the Nanoscience Graduate School are acknowledged for funding and the Finnish IT Centre for Science and the HorseShoe cluster in the University of Southern Denmark are thanked for computing resources.

References

1. P. S. Niemelä, O. H. S. Ollila, M. T. Hyvönen, M. Karttunen and I. Vattulainen, *PLoS Comp. Biol.*, 2007, **3**, 304.
2. O. H. S. Ollila, T. Rog, M. Karttunen and I. Vattulainen, *J. Struct. Biol.*, 2007, **159**, 311.

3. T. S. Wiedmann, R. D. Pates, J. M. Beach, A. Salmon and M. F. Brown, *Biochemistry*, 1988, **27**, 6469.
4. N. J. Gibson and M. F. Brown, *Biochemistry*, 1993, **32**, 2438.
5. M. F. Brown, *Chem. Phys. Lipids*, 1994, **73**, 159.
6. R. S. Cantor, *Biochemistry*, 1997, **36**, 2339.
7. R. S. Cantor, *J. Phys. Chem. B*, 1997, **101**, 1723.
8. D. Marsh, *Biochim. Biophys. Acta*, 1996, **1286**, 183.
9. O. H. S. Ollila, J. Risselada, M. Louhivuori, E. Lindahl, I. Vattulainen and S. J. Marrink, *Phys. Rev. Lett.*, 2009, **102**, 078101.
10. J. N. Israelachvili, *Intermolecular and Surface Force*, Academic Press, London, 1985.
11. J. N. Israelachvili, S. Marcelja and R. G. Horn, *Q. Rev. Biophys.*, 1980, **13**, 121.
12. S. Ollila, I. Vattulainen and M. T. Hyvönen, *J. Phys. Chem. B*, 2007, **111**, 3139.
13. N. J. Gibson and M. F. Brown, *Biochemistry*, 1993, **32**, 2438.
14. M. F. Brown, *Chem. Phys. Lipids*, 1994, **73**, 159.
15. R. S. Cantor, *Biophys. J.*, 1999, **76**, 2625.
16. R. S. Cantor, *Chem. Phys. Lipids*, 1999, **101**, 45.
17. E. Perozo, *Curr. Opin. Struc. Biol.*, 2003, **13**, 432.
18. J. H. A. Folgering, J. M. Kuiper, A. H. de Vries, J. B. Engberts and B. Poolman, *Langmuir*, 2004, **20**, 6985.
19. M. Kuiper, *PhD Thesis*, Rijksuniversiteit Groningen, 2005.
20. E. van den Brink-van der Laan, V. Chupin, J. A. Killian and B. de Kruijff, *Biochemistry*, 2004, **43**, 5937.
21. E. van den Brink-van der Laan, J. A. Killian and B. Kruijff, *Biochim. Biophys. Acta*, 2004, **1666**, 275.
22. A. V. Botelho, T. Huber, T. P. Sakmar and M. F. Brown, *Biophys. J.*, 2006, **91**, 4464.
23. G. S. Attard, R. H. Templer, W. S. Smith, A. N. Hunt and S. Jackowski, *Proc. Natl. Acad. Sci. USA*, 2000, **97**, 9032.
24. G. Shearman, G. Attard, A. Hunt, S. Jackowski, M. Baciu, S. Sebai, X. Mulet, J. Clarke, R. Law and C. Plisson et al., *Biochem. Soc. Trans.*, 2007, **35**, 498.
25. S. A. Safran, *Statistical Thermodynamics of Surfaces, Interfaces and Membranes*, Addison-Wesley, Reading, MA, 1994.
26. J. M. Seddon and R. H. Templer, *Structure and Dynamics of Membranes*, Elsevier, Amsterdam, 1995, pp. 97–160.
27. J. M. Seddon, *Biochim. Biophys. Acta*, 1990, **1031**, 1.
28. T. J. McIntosh and S. A. Simon, *Annu. Rev. Biophys. Biomol. Struct.*, 2006, **35**, 177.
29. R. H. Templer, S. J. Castle, A. R. Curran, G. Rumbles and D. R. Klug, *Faraday Discuss.*, 1999, **111**, 41.
30. T. Kamo, M. Nakano, Y. Kuroda and T. Handa, *J. Phys. Chem. B*, 2006, **110**, 24987.
31. J. Curdova, P. Capková, J. Plasek, J. Repáková and I. Vattulainen, *J. Phys. Chem. B*, 2007, **111**, 3640.

32. S.-S. Feng, *Langmuir*, 1999, **15**, 998.
33. D. Marsh, *Langmuir*, 2006, **22**, 2916.
34. S.-S. Feng, *Langmuir*, 2006, **22**, 2920.
35. D. Harries and A. Ben-Shaul, *J. Chem. Phys.*, 1997, **106**, 1609.
36. T. Xiang and B. D. Anderson, *Biophys. J.*, 1994, **66**, 561.
37. S. I. Mukhin and S. Baoukina, *Phys. Rev. E*, 2005, **71**, 061918.
38. R. Goetz and R. Lipowsky, *J. Chem. Phys.*, 1998, **108**, 7397.
39. M. Venturoli and B. Smit, *Phys. Chem. Commun.*, 1999, **2**, 45.
40. M. Carrillo-Tripp and S. E. Feller, *Biochemistry*, 2005, **44**, 10164.
41. J. Sonne, F. Y. Hansen and G. H. Peters, *J. Chem. Phys.*, 2005, **122**, 124903.
42. E. Lindahl and O. Edholm, *J. Chem. Phys.*, 2000, **113**, 3882.
43. J. Gullingsrud and K. Schulten, *Biophys. J.*, 2003, **85**, 2087.
44. J. Gullingsrud and K. Schulten, *Biophys. J.*, 2004, **86**, 3496.
45. M. Patra, *Eur. Biophys. J.*, 2005, **35**, 79.
46. E. Terämä, O. H. S. Ollila, E. Salonen, A. C. Rowat, C. Trandum, P. Westh, M. Patra, M. Karttunen and I. Vattulainen, *J. Phys. Chem. B*, 2008, **112**, 4131.
47. S. Marrink, H. Risselada, S. Yefimov, D. Tieleman and A. de Vries, *J. Phys. Chem. B*, 2007, **111**, 7812.
48. M. Orsi, D. Y. Haubertin, W. E. Sanderson and J. W. Essex, *J. Phys. Chem. B*, 2008, **112**, 802.
49. B. Griepernau and R. A. Böckmann, *Biophys. J.*, 2008, **95**, 5766.
50. P. Schofield and J. R. Henderson, *Proc. R. Soc. London, Ser. A*, 1982, **379**, 231.
51. H. J. C. Berendsen, *Simulating the Physical World,* Cambridge University Press, Cambridge, 2007.
52. L. Mistura, *Int. J. Thermophys.*, 1987, **8**, 397.
53. B. Hafskjold and T. Ikeshoji, *Phys. Rev. E*, 2002, **66**, 011203.
54. E. Wajnryb, A. R. Altenberger and J. S. Dahler, *J. Chem. Phys.*, 1995, **103**, 9782.
55. S. Morante, G. C. Rossi and M. Testa, *J. Chem. Phys.*, 2006, **125**, 034101.
56. R. Lovett and M. Baus, *J. Chem. Phys.*, 1997, **106**, 635.
57. B. D. Todd, D. J. Evans and P. J. Daivis, *Phys. Rev. E*, 1995, **52**, 1627.
58. J. H. Irving and J. G. Kirkwood, *J. Chem. Phys.*, 1950, **18**, 817.
59. A. Harasima, *Adv. Chem. Phys.*, 1958, **1**, 203.
60. J. S. Rowlinson and B. Widom, *Molecular Theory of Capillarity,* Clarendon Press, Oxford, 1982.
61. M. Patra, M. Karttunen, M. T. Hyvönen, E. Falck, P. Lindqvist and I. Vattulainen, *Biophys. J.*, 2003, **84**, 3636.
62. M. Patra, M. Karttunen, M. Hyvönen, E. Falck and I. Vattulainen, *J. Phys. Chem. B*, 2004, **108**, 4485.
63. M. Patra, M. Hyvonen, E. Falck, M. Sabouri-Ghomi, I. Vattulainen and M. Karttunen, *Comput. Phys. Commun.*, 2007, **176**, 14.
64. I. Szleifer, D. Kramer, A. Ben-Shaul, W. M. Gelbart and S. A. Safran, *J. Chem. Phys.*, 1990, **92**, 6800.

65. G. Brannigan and F. L. H. Brown, *Biophys. J.*, 2006, **90**, 1501.
66. D. Marsh, *Biophys. J.*, 2007, **93**, 3884.
67. R. S. Cantor, *Biophys. J.*, 2002, **82**, 2520.
68. D. van der Spoel, E. Lindahl, B. Hess, B. Groenhof, A. E. Mark and H. J. C. Berendsen, *J. Comput. Chem.*, 2005, **26**, 1701.
69. E. Lindahl, B. Hess and D. van der Spoel, *J. Mol. Model.*, 2001, **7**, 306.
70. S. Marrink, A. de Vries and A. E. Mark, *J. Phys. Chem. B*, 2004, **108**, 750.
71. J. Repáková, P. Capková, J. M. Holopainen and I. Vattulainen, *J. Phys. Chem. B*, 2004, **108**, 13438.
72. J. Repáková, J. M. Holopainen, M. R. Morrow, M. C. McDonald, P. Capková and I. Vattulainen, *Biophys. J.*, 2005, **88**, 3398.
73. K. Gawrisch and L. L. Holte, *Chem. Phys. Lipids*, 1996, **81**, 105.
74. D. Marsh, *Biochim. Biophys. Acta*, 1996, **1279**, 119.
75. B. J. Litman and D. C. Mitchell, *Lipids*, 1996, **31**, S193.
76. S.-L. Niu, D. C. Mitchell and B. J. Litman, *J. Biol. Chem.*, 2002, **277**, 20139.
77. D. C. Mitchell, M. Straume, J. L. Miller and B. J. Litman, *Biochemistry*, 1990, **29**, 9143.
78. M. C. Pitman, F. Suits, A. D. MacKerell Jr and S. E. Feller, *Biochemistry*, 2004, **43**, 15318.
79. S. R. Wassall and W. Stillwell, *Chem. Phys. Lipids*, 2008, **153**, 57.
80. S. Vainio, M. Jansen, M. Koivusalo, T. Rog, M. Karttunen, I. Vattulainen and E. Ikonen, *J. Biol. Chem.*, 2005, **281**, 348.
81. O. G. Mouritsen and M. J. Zuckermann, *Lipids*, 2004, **39**, 1101.
82. E. Hovenkamp, I. Demonty, J. Plat, D. Lutjohann, R. P. Mensink and E. A. Trautwein, *Prog. Lipid Res.*, 2008, **47**, 37.
83. V. Piironen, D. G. Lindsay, T. A. Miettinen, J. Toivo and A. M. Lampi, *J. Sci. Food Agric.*, 2000, **80**, 939.
84. K. Simons and E. Ikonen, *Nature*, 1997, **387**, 569.
85. J. F. Hancock, *Nat. Rev. Mol. Cell Biol.*, 2006, **7**, 456.
86. L. Miao, M. Nielsen, J. Thewalt, J. H. Ipsen, M. Bloom, M. J. Zuckermann and O. G. Mouritsen, *Biophys. J.*, 2002, **82**, 1429.
87. Y.-W. Hsueh, K. Gilbert, C. Trandum, M. Zuckermann and J. Thewalt, *Biophys. J.*, 2004, **88**, 1799.
88. D. Marsh, *Biochim. Biophys. Acta*, 2008, **1778**, 1545.
89. A. G. Lee, *Biochim. Biophys. Acta*, 2004, **1666**, 62.
90. R. S. Cantor, *Toxicol. Lett.*, 1998, **100–101**, 451.
91. A. L. Frischknecht and L. J. D. Frink, *Biophys. J.*, 2006, **91**, 4081.
92. H. V. Ly, D. E. Block and M. L. Longo, *Langmuir*, 2002, **18**, 8988.
93. H. V. Ly and M. L. Longo, *Biophys. J.*, 2004, **87**, 1013.
94. D. P. Siegel and M. M. Kozlov, *Biophys. J.*, 2004, **87**, 366.
95. D. P. Siegel, *Biophys. J.*, 2008, **95**, 5200.
96. J. Wohlert, W. K. den Otter, O. Edholm and W. J. Briels, *J. Chem. Phys.*, 2006, **124**, 154905.
97. R. H. Templer, B. J. Khoo and J. M. Seddon, *Langmuir*, 1998, **14**, 7427.
98. Z. Chen and R. P. Rand, *Biophys. J.*, 1997, **73**, 267.

99. S. Baoukina, L. Monticelli, H. J. Risselada, S. J. Marrink and D. P. Tieleman, *Proc. Natl. Acad. Sci. USA*, 2008, **105**, 10803.
100. J. Wong-Ekkabut, S. Baoukina, W. Triampo, I.-M. Tang, D. P. Tieleman and L. Monticelli, *Nat. Nanotechnol.*, 2008, **3**, 363.
101. J. A. Szule, N. L. Fuller and R. P. Rand, *Biophys. J.*, 2002, **83**, 977.
102. E. E. Kooijman, V. Chupin, N. L. Fuller, M. M. Kozlov, B. de Kruijff, K. N. J. Burger and P. R. Rand, *Biochemistry*, 2005, **44**, 2097.
103. P. S. Niemelä, M. T. Hyvönen and I. Vattulainen, *Biochim. Biophys. Acta*, 2008, **1788**, 122–135.
104. S. I. Sukhraev, W. J. Sigurdson, C. Kung and F. Sachs, *J. Gen. Physiol.*, 1999, **113**, 525.

Coarse-grained Molecular Dynamics Simulations of Membrane Proteins

SARAH ROUSE, TIMOTHY CARPENTER AND MARK S. P. SANSOM

Department of Biochemistry, University of Oxford, South Parks Road, Oxford, OX1 3QU, UK

3.1 Introduction

Membrane proteins perform many roles in the biology of cells, including transport, signalling and cell–cell interactions. Consequently, *ca.* 25% of genes encode membrane proteins.[1] The majority of membrane protein structures are based upon bundles of transmembrane (TM) α-helices. Membrane protein folding may be described in terms of two stages: translocon-mediated insertion, followed by lateral association of TM α-helices.[2,3] Helix dimerization provides a simple model of lateral association in folding. It is also important in mechanisms of signalling across membranes by membrane-bound receptors, in which ligand-binding ectodomains are linked to intracellular signalling domains via single TM helices.

Several computational methods permit modelling and simulation of membrane proteins.[4–6] Molecular dynamics (MD) simulations are a valuable approach for modelling membrane proteins and may be divided into those which employ a full atomistic (AT) representation and those which employ a simplified, coarse-grained (CG) model of the protein and of the bilayer lipids.

RSC Biomolecular Sciences No. 20
Molecular Simulations and Biomembranes: From Biophysics to Function
Edited by Mark S.P. Sansom and Philip C. Biggin
© Royal Society of Chemistry 2010
Published by the Royal Society of Chemistry, www.rsc.org

AT-MD simulations are accurate but computationally demanding, whereas the more approximate nature of CG-MD simulations allow longer simulation times ($>1\,\mu s$) and larger systems ($>10^5$ atoms) to be explored with modest computational resources.

In this chapter, we review some of our recent studies of CG-MD and related methods to model TM helix oligomerization, focusing on two test systems for which a range of experimental data are available: (i) glycophorin A (GpA), a red blood cell membrane protein,[7] and (ii) M2 channels from influenza viruses A and B.[8]

As mentioned above, a limitation of AT-MD simulations is that extended ($>1\,\mu s$) simulations of membrane systems require *substantial* computer resources. CG simulations address the time scale and/or system size issues *via* simplification of the representation of the component biomolecules.[9–13] For example, in one CG approach (using the MARTINI coarse-grained force field[14,15]) that has been applied to simple membranes.[10] atoms are grouped together to form particles, each particle corresponding to about four non-H atoms and new 'artificial' bonded and non-bonded interactions are parameterized to reproduce thermodynamic properties such as oil–water partition coefficients of building block molecules. Not only does this lead to an order of magnitude fewer interactions, but also the removal of the fastest degrees of freedom additionally makes it possible to take much longer time steps (typically 40 fs), which together with the reduced interaction density provides a 2–3 orders of magnitude speed-up compared with atomistic simulations.[14] Although it is still under debate how quantitative the resulting predictions are, it is making entirely new spatial and temporal scales accessible to simulations. Thus, as will be seen below, CG simulations yield overall peptide–bilayer system configurations consistent with, *e.g.*, solid-state NMR data[16] and may be used to assemble more complex protein–bilayer systems.[17]

3.2 Coarse-grained Simulations: Methodology

In this section, we describe the development of coarse-grained models for membranes based on the MARTINI model of Marrink and colleagues.[10,14,15] This and related CG models have been used in a number of simulation studies of membrane proteins and peptides.[5] However, the reader should note that a number of other groups have developed comparable CG models for membranes,[9,18–20] and these have been used to explore, *e.g.*, simple models of membrane protein pores.[21,22]

3.2.1 CG-MD and Lipid Bilayers

The CG model that was originally developed by Marrink *et al.*[10] for lipid systems has recently been updated and expanded to include membrane proteins.[14,15] It is based on a four-to-one mapping, *i.e.* four atoms are represented by a single particle. In the original model, only four particle types were defined,

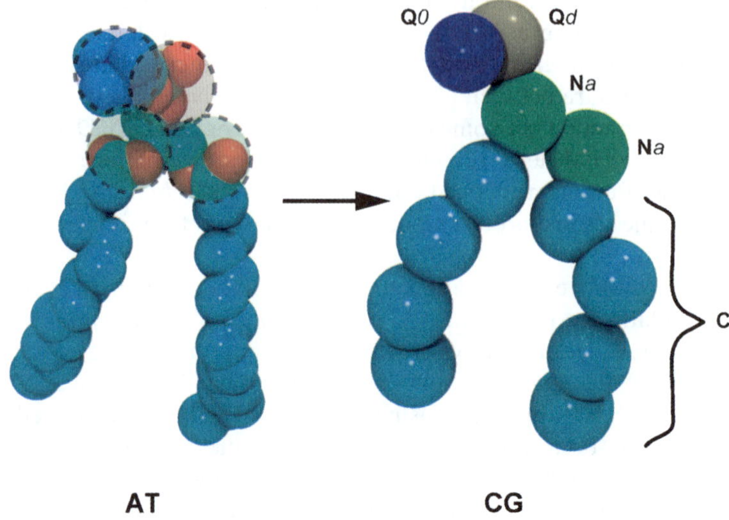

Figure 3.1 Mapping from an atomistic (AT) to a coarse-grained (CG) model for a phospholipid (DPPC). The two charged headgroup particles of choline and phosphate are types Q0 (blue) and Qd (bronze), respectively. The two glycerol moieties are type Na (green), and the four particles in each hydrocarbon tail are type C (cyan).

each having a mass of 72 amu (*i.e.* the mass of four water molecules) and an effective radius of 2.35 Å. The particle types are (i) P – polar (neutral groups), (ii) N – non-polar (mixed polar–apolar groups), (iii) C – apolar (hydrophobic groups) and (iv) Q – charged (ionized groups).

Further subtypes are included to allow the effects of hydrogen bonding to be described. Thus, particles N and Q are further described by the subgroups *0* (no hydrogen bonding capability), *a* (a hydrogen-bond acceptor), *d* (a hydrogen-bond donor) and *da* (could act as either a hydrogen-bond acceptor or a donor). Thus, hydrogen bonding is modelled by moderating the Lennard-Jones (LJ) interactions (see below) between particles. An example of CG mapping of a lipid [dipalmitoylphosphatidylcholine (DPPC)] molecule is shown in Figure 3.1.

Bonding interactions between particles are modelled by a weak harmonic potential. Interactions between non-bonded particles *i* and *j* are described by a standard L) potential:

$$U_{\mathrm{LJ}}(r) = 4\varepsilon_{ij}\left[\left(\frac{\sigma_{ij}}{r}\right)^{12} - \left(\frac{\sigma_{ij}}{r}\right)^{6}\right]$$

where ε_{ij} represents the strength of their interaction and σ_{ij} the effective minimum distance between them. Just five levels of interaction are defined: attractive ($\varepsilon = 5\,\mathrm{kJ\,mol^{-1}}$), semi-attractive ($\varepsilon = 4.2\,\mathrm{kJ\,mol^{-1}}$), intermediate ($\varepsilon = 3.4\,\mathrm{kJ\,mol^{-1}}$), semi-repulsive ($\varepsilon = 2.6\,\mathrm{kJ\,mol^{-1}}$) and repulsive ($\varepsilon = 1.8\,\mathrm{kJ}$

mol^{-1}). The energy levels and scales of the LJ parameters were calibrated using an alkane–water system as a reference point: the attractive level models the strong polar interactions that occur in bulk water; the intermediate level models non-polar interactions between aliphatic chains; and the repulsive level is used to model the interaction between polar and non-polar phases. The LJ potentials are cut off at a distance of 12 Å. Charged particles (Q) interact via an electrostatic Coulombic interaction, shifted to zero from 0 to 12 Å. For ions, a reduced charge of 0.5 is used to model hydration shell effects.

3.2.2 CG-MD and Membrane Peptides and Proteins

The CG lipid model has been extended to membrane peptides and proteins, both in the MARTINI force field[15] and in a parallel modification and extension (CGv2).[16,23,24] MARTINI and CGv2 are very similar in nature and so we will focus on the latter as it has been tested against experimental data for bilayer interactions of a number of membrane peptides and proteins.[16,17]

The types of CG particle assigned to each amino acid were initially based on the partial charges and hydrogen bonding potentials of the constituent atoms of the amino acid. These were refined (in CGv2) to match better thermodynamic data on transfer free energies of amino acid side chains between polar and apolar solvents[23] to yield the CG side chain assignments shown in Table 3.1. The backbone particle subtype depends on the presence of H-bonds within the backbone of the starting atomistic structure. The bond lengths and angles between side chain particles were chosen such that the CG side chain groups matched the size and surface area of their atomistic equivalents. Figure 3.2A shows the coarse-grained representation of side chains of selected amino acids alongside their atomistic counterparts.

Amino acid bond lengths and angles were treated with harmonic potentials. Main chain particle bond potentials had an equilibrium length of 3.6 Å, corresponding to the mean distance between Cα atoms from several high-resolution X-ray structures of membrane proteins. To mimic α-helical

Table 3.1   Side chain particle assignments in CGv2.

Amino acid	*Side chain particle assignment*
Ala, Ile, Leu, Pro, Val	C
Phe	C + C
Cys, Met	Nd
Ser, Thr	Nda
Asn, Gln	P
Tyr	C + P
His	Nd + Nda
Trp	C + Nd
Asp, Glu	Qa (charge − 1)
Lys	C + Qd (charge + 1)
Arg	Nd + Qd (charge + 1)

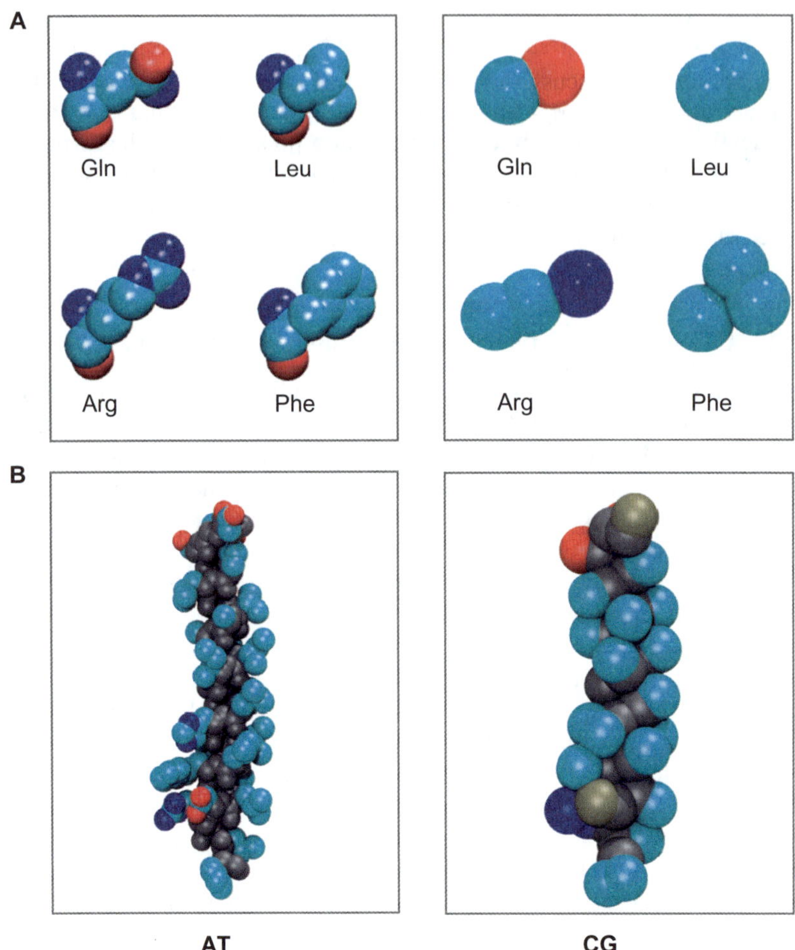

Figure 3.2   AT to CG mapping for amino acid side chains (A) and for an α-helix (B). In (A), examples of hydrophobic (Leu, Phe) and polar (Gln, Arg) side chains are given in AT and CG format. The CG particles are coloured as follows: cyan = non-polar/apolar; red = polar; blue = positively charged; bronze = negatively charged. Note that the non-polar particles will differ in their properties through the different sub-level of H-bonding capabilities, as described in the text. In (B), a model of the transmembrane α-helix of the influenza A M2 channel protein is shown in AT and CG format. Modified from reference 41.

secondary structure, a harmonic restraint, with an equilibrium length of $6\,\text{Å}$ and force constant of $10\,\text{kJ}\,\text{mol}^{-1}\,\text{Å}^{-2}$, is applied to helix backbone particles $i - (i+4)$ [except when $(i+4)$ is proline]. This maintains an α-helical structure (Figure 3.2B). The force constant was chosen such that the root-mean-squared fluctuations (RMSFs) of the protein backbone particles in CG simulations were similar to those of corresponding atomistic simulations.

3.3 Evaluation of CG-MD: Model Membrane Peptides

In this section, we discuss the application of CG-MD to studying α-helical membrane peptide folding. Models of α-helical peptide folding suggest that it may occur in two main stages.[2,25] The first of these stages is the insertion of membrane peptides into a bilayer and formation of *independently stable* transmembrane (TM) α-helices. The second stage is the subsequent interaction of these helices in which they self-associate to pack into the tertiary structure of the protein.

The implications of this model for computational studies of membrane protein folding are that individual TM helices can be inserted into a bilayer and remain stable throughout an MD simulation. Additionally, over time, these helices should move laterally through the bilayer and contact each other. The helices should sample conformational space within the bilayer until the native conformation of the bundle is found. Thus, if the transmembrane regions of a protein are known, the final bundle conformation can be modelled. The microsecond time scales required for the oligomerization process are beyond those currently feasible for conventional AT-MD. However, this process occurs within a time scale that may be achieved using coarse-grained methodology.

For CG-MD to be used to model these approaches, it must first be demonstrated to be capable of modelling the bilayer to a suitable degree of accuracy and to be able to describe the protein–lipid interactions successfully. Here, we discuss the development of the CG-MD approach that allowed this level of simulation study to be achieved.

Initial assessments of the suitability of CG models to describe lipid–water systems were performed during parameterization of such models via comparison with atomistic simulations of lipid bilayers.[9,10] Analysis of simulations of various lipid systems has shown that CG models are able to replicate key experimental and/or atomistic simulation data.[26] Thus CG models appear to be accurate to an at least semiquantitative level for structural and dynamic behaviour of lipid bilayer and related systems.

The next level of system complexity to be evaluated for the CG-MD methodology was the behaviour of model membrane proteins in a lipid bilayer environment. The first studies of this were of the transmembrane (TM) domains of the α-helical membrane protein (a TM helix homodimer) glycophorin A (GpA) and of the β-barrel outer membrane protein OmpA. These proteins were chosen since they represent two classes of membrane proteins and both atomistic simulations and a substantial body of biophysical data were available for comparison with the results of the CG-MD simulations.[24]

For both GpA and OmpA, CG-MD simulations of the self-assembly of these TM domains in a lipid bilayer and in a detergent micelle were performed.[24] In each case, the initial setup consisted of a CG model of the protein (obtained from the NMR structure of GpA, PDB i.d. 1AFO, or from the X-ray structure of OmpA, PDB i.d. 1BXW) placed in a simulation box with randomly positioned and oriented detergent or lipid molecules, plus water particles and counterions. The resultant CG simulations of DPC micelles self-assembling around GpA and

OmpA TM domains compared well with the corresponding AT simulations, reproducing the key steps of protein–detergent micelle self-assembly. Self-assembly simulations of OmpA and of GpA in lipid bilayers in each case yielded a system with the TM domain inserted into the bilayer in a location and orientation consistent with experimental and atomistic simulation data.

The success of these early CG studies of protein–bilayer self-assembly prompted further validation and subsequent extension to a wider range of membrane peptide and protein systems.[16] In particular, a number of membrane-interacting α-helical peptides were simulated, for which biophysical data on location and orientation relative to the bilayer are available, These included two relatively simple, 'designed' peptides, namely WALP23[27,28] and LS3[29,30] (see Figure 3.3). In each case, the interaction of an α-helix with a bilayer was modelled using a CG simulation system consisting of 256 randomly placed DPPC lipids plus an α-helical peptide, along with ~ 3000 water particles.

The WALP peptides are a series of designed peptides that consist of a poly(Leu-Ala) core flanked by Trp residues, with the general formula GWW-(LA)$_n$-WWA. These peptides have been extensively studied and provide a 'benchmark' for simulations of TM α-helices and their insertion into lipid bilayers. In the CG simulations of DPPC bilayer self-assembly in the presence of WALP23, the peptide adopted a stable TM orientation in all simulations. This was due to the interaction of the hydrophobic core of WALP23 with the lipid tails whilst the terminal Trp (W) side chains interacted with the lipid headgroups. This was subsequently confirmed in CGv2 free energy profile (potential of mean force) studies of the thermodynamics of WALP23 into a lipid bilayer.[23]

To assess the ability of CG-MD in modelling peptides that adopt an interfacial orientation, a similar simulation was set up with the designed peptide LS3. LS3 is a synthetic peptide with sequence (LSSLLSL)$_3$ and forms an α-helix that adopts an interfacial orientation, parallel to the bilayer surface, when the peptide is in a monomeric form. In the CG bilayer self-assembly simulations the major orientation adopted by the LS3 helix was indeed interfacial, parallel to the bilayer. In some simulations the peptide was initially inserted in the bilayer in a TM orientation for a short period (~ 25 ns), although by the end of the simulation the interfacial orientation was adopted. The interfacial orientation allows the hydrophobic Leu side chain particles to interact with the acyl lipid tails while the polar Ser side chain particles interact favourably with the glycerol and headgroup particles. One should note that in experiments with a transbilayer voltage difference applied, *multiple* LS3 peptide helices may oligomerize to generate ion channels formed by bundles of parallel TM α-helices. Such bundles have been modelled in CG simulations,[31] although a transbilayer voltage difference was not applied during the simulations.

Overall, simulations of these and related peptides demonstrate that the CG-MD method not only is able to reproduce characteristics of lipid bilayers, but also can successfully model the interactions between peptides and lipid bilayers. This in turn suggests that CG-MD self assembly simulations allow one to model the first stage, helix insertion, of membrane protein folding.

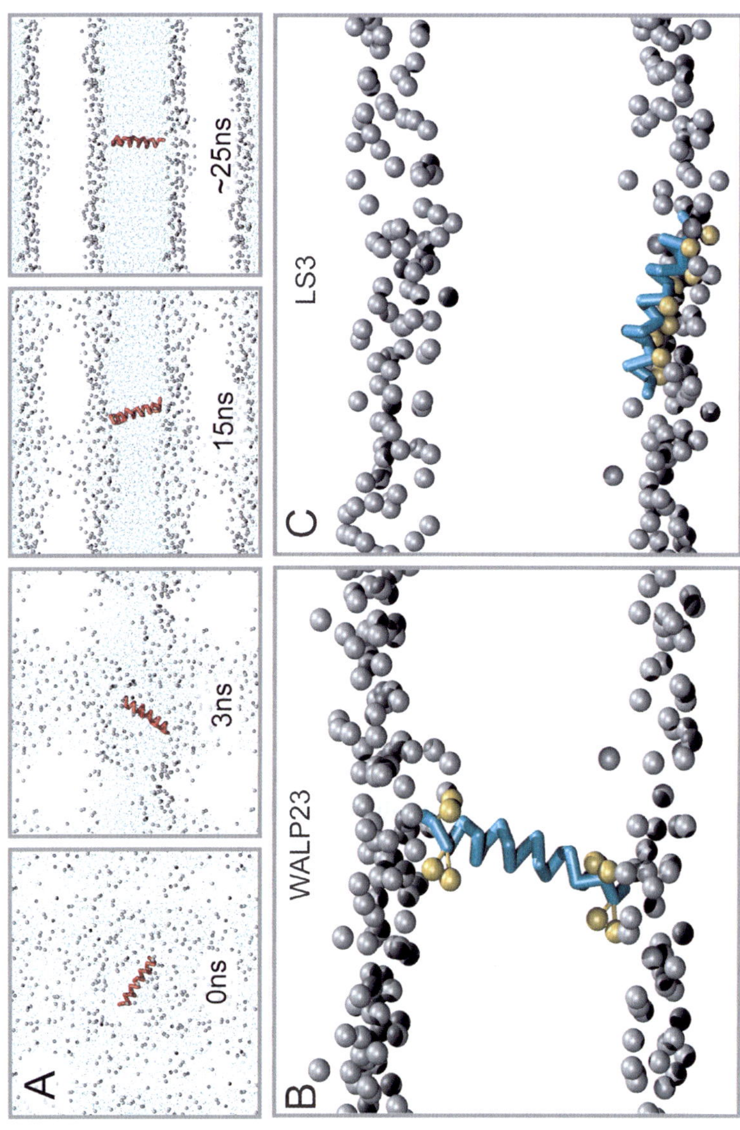

Figure 3.3 CG-MD self-assembly simulations of α-helical peptide/lipid bilayer systems. In (A), successive snapshots over the course of a self-assembly simulation of WALP23 are shown, with the helix as a red Cα trace, the lipid phosphate particles as grey spheres and the lipid tail particles as small blue spheres. Water particles are omitted for clarity. In (B) and (C), structures from the end of a CG-MD self-assembly simulation are shown for (B) WALP23, a model transmembrane α-helix, and (C) LS3, an amphipathic α-helix which adopts an interfacial location when present in a monomeric form. In each case lipid phosphate particles are shown as grey spheres and the peptide is shown in pale blue Cα trace. Key side chain particles are shown in yellow: in (B) the tryptophans of WALP23 and in (C) the serines of LS3. Data for the figure were kindly provided by Alan Chetwynd.

3.4 Simulation Studies of Membrane Peptide Oligomerization

In this section, we focus on the proposed second stage of membrane protein folding: the self-association of inserted helices to yield a tertiary structure of a membrane protein. We describe results for two well-characterized test systems: glycophorin A, which forms a TM helix dimer, and the influenza M2 proton channel, which forms a TM helix tetramer.

3.4.1 Glycophorin A

Glycophorin A (GpA) is a small TM protein found in red blood cell membranes. GpA forms a homodimer; each 131-residue monomer spans the membrane with a 25-residue TM α-helical region which is responsible for dimerization. The interaction interface between the two TM helices in the dimer has been extensively studied, both experimentally and computationally, as a model for helix–helix interactions in membrane protein folding.[2] The TM region of GpA contains a GxxxG sequence motif, found in numerous membrane proteins,[32] that is known to be responsible for right-handed packing of helices (corresponding to a negative crossing angle, $\Omega < 0$; see Figure 3.4). Mutational studies of GpA[33] have established key residues in the GxxxG-containing motif which are important for dimerization, and an NMR structure of the TM helix dimer (in detergent micelles) has been determined.[7] A solid-state NMR-based model of the TM helix in lipid bilayers is also available.[34] Overall, these studies demonstrate that a seven-residue motif $L^{75}IxxGVxxGVxxT^{87}$ in the core of the TM helix is the key to stabilization of the right-handed packing of the α-helices.

Multiple extended ($3\,\mu s$) simulations of the dimerization of wild-type (WT) and mutant GpA TM helices within lipid bilayers have been performed.[35] In these simulations, the starting configuration was two GpA α-helices inserted in a parallel fashion into a pre-formed DPPC bilayer at an interhelix separation of $\sim 55\,\text{Å}$. In the GpA-WT simulations, a long-lasting right-handed α-helix dimer was formed. Analysis of helix–helix contacts in the simulations of GpA-WT confirmed that the main inter-helix interactions were formed by the key residues of the $L^{75}IxxG^{79}VxxG^{83}VxxT^{87}$ motif. This study also explored whether CG-MD could be used to compare the relative stabilities of different mutants of GpA. Five mutants were chosen: three (G79L, G83L, T87F) which were known experimentally to disrupt dimerization and two (A82W and I85F) which have negligible effects on dimer formation. In the simulations, the non-disruptive mutants had little effect on either the structure (right-handed crossing) or stability of the TM helix dimer. In contrast, the disruptive mutants perturbed both the structure and stability of TM helix dimers. Thus, the dimers were no longer predominantly right-handed. Instead, the crossing angles distribution of the disruptive G83L and G79LG83L mutants were bimodal (*i.e.* both right- and left-handed crossing were present), suggesting that these mutants 'soften' the interaction interface. The decreased dimer stability for the disruptive mutants was shown by a decrease in the fraction of time spent in the dimeric

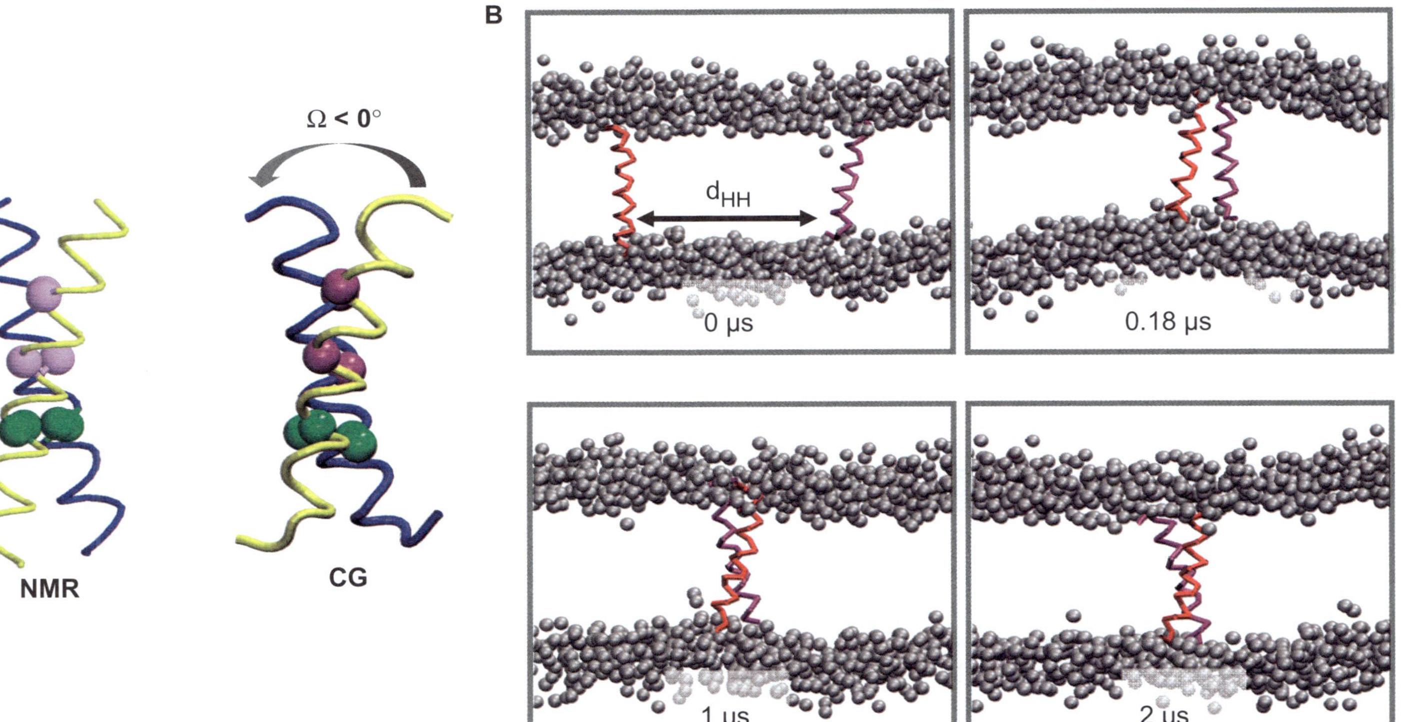

Figure 3.4   CG-MD simulations of the dimerization of the TM α-helix of glycophorin A (GpA). (A) Comparison of structures of the GpA TM helix dimer from NMR (in a detergent micelle; PDB i.d. 1AFO) and from a CG-MD simulation. In each case, the helix crossing angle is negative ($\Omega < 0°$). (B) Progress of a CG simulation starting with distant ($d_{HH} \approx 55\,\text{Å}$) helix monomers in a pre-formed lipid (DPPC) bilayer (headgroups only are shown for the lipids; choline, phosphate and glycerol particles are shown in silver). The two α-helices associate to form a long-lasting TM α-helix dimer. Modified from reference 35.

state, *i.e.* there was increased dissociation of the dimers to yield monomeric helices. From this latter information, the stability of the mutant dimers was calculated relative to the WT. The disruptive mutants had $\Delta\Delta G$ values ranging from 9 to 11 kJ mol^{-1}, corresponding to a *ca.* 40-fold decrease in the equilibrium constant for association.

From these studies, it can be seen that CG-MD is able to distinguish between mutants of GpA in a semiquantitative manner. This work provides a foundation for the use of CG-MD in modelling helix dimerization for a range of membrane proteins.[36]

3.4.2   Influenza M2 Channels

The influenza A and B viruses contain small membrane proteins (A/M2 and BM2, respectively), which form proton-selective ion channels.[8] A/M2 is a 97-residue protein that comprises three domains: a 24-residue extracellular N-terminal region, a 19-residue TM region and a 54-residue cytoplasmic C-terminal domain. The protein forms a homotetramer which functions as a low pH-activated proton-selective channel. A/M2 plays a vital role in the life cycle of the influenza virus: without it, the virus is unable to replicate. A/M2 is therefore of interest as a potential drug target. The proton channel behaviour of A/M2 may be inhibited by the drug adamantine and its derivatives. BM2 is also a proton channel found in influenza virus B. It has low sequence identity to A/M2, but is also homotetrameric and has the same domain structure and topology. Within the TM domain, both A/M2 and BM2 contain an HxxxW sequence motif, thought to play a key role in the gating and permeation mechanisms of the channels. The two channels have different sensitivities to adamantine.

The A/M2 proton channel has been extensively studied both experimentally[8] and computationally[37,38] in recent years, particularly in the form of constructs corresponding to just the TM region. Two high-resolution structures of the A/M2 TM domain (by X-ray diffraction[39] and one by NMR[40]) have recently been determined. The diversity and range of data available for this protein make it a useful test case for the CG-MD method.

Following the dimerization study of GpA, CG-MD simulations have been evaluated for the more complicated case of tetramer formation using A/M2 and BM2 as examples. Similarly to the starting configuration of the GpA dimerization simulations, four CG A/M2 TM helix monomers were placed in a preformed CG DPPC bilayer, the system was solvated and the appropriate number of ions added to restore electrical neutrality. This starting point was the basis for five CG-MD simulations, each of 5 μs duration. In each of the simulations, the A/M2 TM helices diffused laterally through the bilayer, eventually forming a tetrameric bundle via a dimeric or trimeric intermediate. Cluster analysis of the simulations revealed a single, converged structure of the A/M2 TM region tetramer.[41]

From visual inspection, the helix bundle is left-handed (*i.e.* with a positive crossing angle between the helices), in agreement with previous models and with the two experimental structures. During the simulation, the average tilt angle of the helices was 17° ($\pm$4°). The bundle is conical in structure, with the helices packed more tightly at their N-termini. The C-termini are less tightly packed as they must accommodate the larger His and Trp particles. This is also observed in the X-ray structure.

Having shown that CG-MD may be used to self-assemble an A/M2 TM tetramer, for which experimental structures are known, we have also used it to predict the structure of the BM2 tetramer.[42] As for A/M2, an α-helical model of the predicted TM region was built and converted to a CG representation. Helix insertion CG simulations of a single BM2 helix confirmed that it adopts a TM orientation in a bilayer. Following this, four BM2 TM helices were inserted in a parallel fashion, with inter-helix separations of $\sim$45 Å, into a pre-formed DPPC bilayer, and this configuration was used as the basis for five independent 5 µs CG-MD simulations. As in the A/M2 simulations, in each simulation the helices self-associated to form a left-handed tetramer, which cluster analysis demonstrated to correspond to a single converged structure. The orientation of the helices within the tetramer was such that the proposed pore-lining residues[43] faced the centre of the bundle. The average tilt angles were slightly lower for BM2 (13°) than A/M2 (17°). Thus, the CG-MD method may be used to obtain a converged tetrameric structure model the BM2 TM domain (see Figure 3.5).

In summary, CG-MD simulations have been used to achieve a converged structure of the TM domains of A/M2 and of BM2, by modelling the second stage of the process of membrane protein folding. This structure compares favourably with experimental structural data (for A/M2) and with predictions of functionally important residues (for BM2).

3.5 Coarse-grained MD: Larger Systems

3.5.1 Vesicle Simulations

CG methods also enable simulation methods to be applied to much larger membrane systems, *e.g.* lipid bilayer vesicles containing membrane proteins. These have the potential to bridge more directly between simulations and experimental biophysical studies of membrane proteins. There have been a number of CG simulations of membrane vesicle systems, *e.g.* to examine the interactions of antimicrobial peptides with lipid vesicles[44] and to explore lipid raft formation in ternary lipid systems.[45]

Here we exemplify CG simulations of lipid vesicles with simulations of phospholipid vesicles containing the influenza A/M2 channel. A number of system sizes have been explored, ranging from 72 to 265 Å radius, with $\sim$1000–21 000 lipid molecules. These should be compared with the size of the influenza virus particle, the lipid bilayer element of which has a radius of $\sim$250 Å.[46]

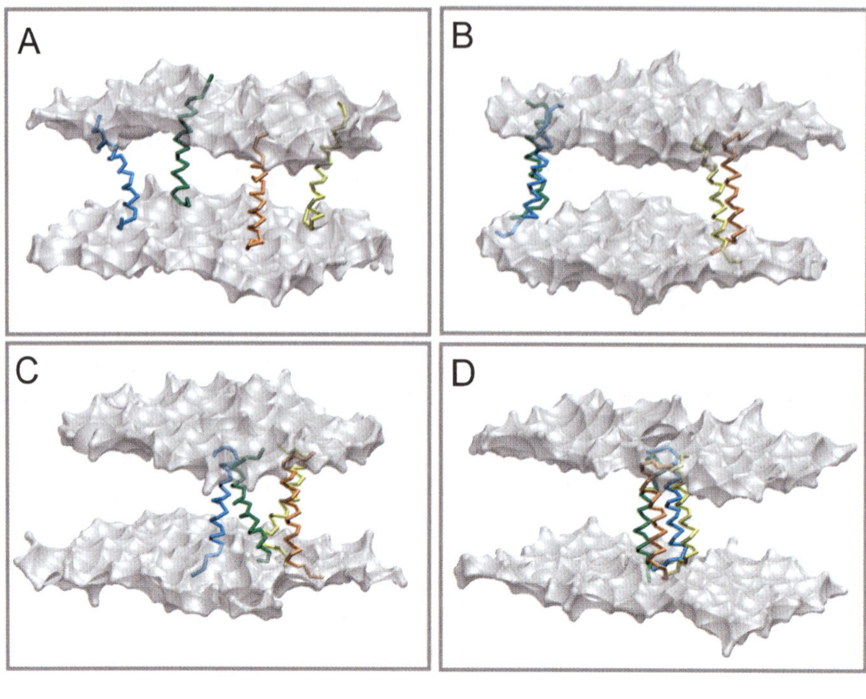

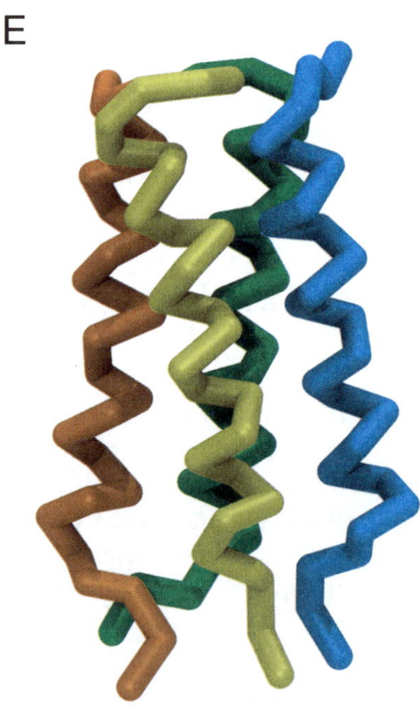

Figure 3.5   Snapshots illustrating the formation of an influenza BM2 transmembrane α-helix tetramer, at times of (A) 0, (B) 18, (C) 62 and (D) ∼ 500 ns. The Cα traces of the four helices are shown in blue, green, yellow and pale brown. The bilayer surface (defined by the choline, phosphate and glycerol particles) is shown in grey. The final converged structure is shown in (E).

Thus, the largest vesicles explored overlap in size with the influenza virus particle. It should be noted that even with CG, the largest system (265 Å radius) would require $\sim 2.5 \times 10^6$ water particles if simulated in a conventional simulation box. However, this can be reduced to $\sim 1 \times 10^6$ particles by use of a truncated octahedral box. Even so, this remains a challenging simulation system.

These vesicle systems can be used to simulate AM2 tetramer formation in a larger bilayer, starting with four distantly placed AM2 TM helices, and also to explore the presence of multiple AM2 channels (*i.e.* TM helix tetramers) within a single vesicle, as is the case in the virus particle.

Tetramerization was explored in a 72 Å vesicle by positioning four copies of the AM2 TM helix at random locations within the vesicle, but with all helices oriented with their C-terminus facing the interior of the of the vesicle (as is the case *in vivo*). The system was simulated for 7 μs (Figure 3.6), during which the individual helices moved around the vesicle via lateral diffusion. Within ~ 0.3 μs, two helices formed a dimer. Within ~ 0.5 μs the other two helices also dimerized. These two dimers then diffused within the vesicle membrane for a further 2 μs before they contact one another, first to form a linear array of four helices which subsequently (after > 2 μs) collapsed into a left-handed tetrameric bundle conformation with helix packing very similar to that described above for the (much smaller) tetramerization simulations. This allows us to demonstrate that the M2 helix bundle seen in the earlier simulations was not biased by the small simulation system size or the initial orientations of the constituent helices.

The largest (radius 265 Å) vesicle system was used for a short (0.5 μs) simulation of five pre-formed tetrameric A/M2 TM helix bundles (thus mimicking the A/M2 channel within the viral membrane particles). The time scale is too short to test for whether the M2 A/channels (*i.e.* tetramers) remain apart or aggregate together. However, this may be considered as the first step towards simulating the viral membrane of a single viral particle.

3.5.2 More Complex Membrane Proteins

CG simulations may also be extended to more complex membrane proteins, such as ion channels[47,48] and G-protein coupled receptors.[49] We have used CG simulations as a high-throughput method for assessing and comparing the lipid–protein interactions of a wide range of membrane proteins.[50] Thus, CG-MD was used for insertion into bilayers of membrane proteins of known structure *via* bilayer self-assembly in the presence of a protein of known structure. In these simulations, the basic CG model described above was supplemented by an elastic network between Cα particles in order to model and maintain the protein tertiary structure.

This method was used to predict the membrane insertion of ~ 100 membrane proteins, representing the majority of known folds of membrane proteins. A good correlation with experiment data was seen for those proteins for which

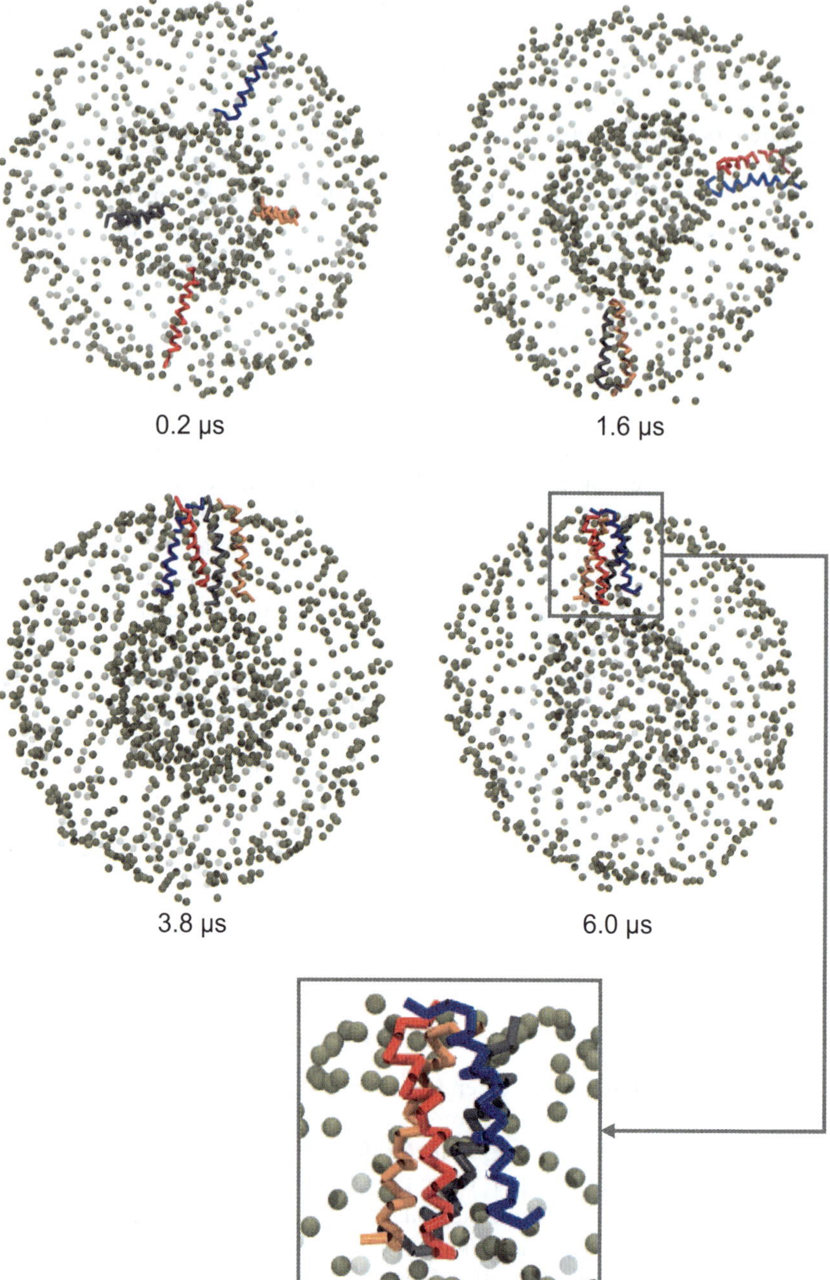

Figure 3.6 Formation of an influenza A/M2 TM helix tetramer within a lipid vesicle (radius 72 Å) showing snapshots of the vesicle (lipid phosphate particles only are shown) and the four TM helices which first form two dimers (1.6 μs) which then aggregate (3.8 μs) before forming a stable tetramer (6.0 μs and inset).

data on bilayer location and lipid contacts were available.[17] This method continues to be applied to membrane proteins as further structures are determined and the results are made available *via* a website (http://sbcb.bioch.ox.-ac.uk/cgdb).[51]

As an example of the use of CG simulations to predict interactions with a lipid of a complex membrane, in Figure 3.7 we show the results of a bilayer self-assembly simulation with the bacterial channel protein ELIC,[52] a homologue of mammalian neurotransmitter receptors of the nicotinic receptor superfamily. It can be seen that the CG procedure orients the protein such that the bundle of α-helices adopts a transmembrane orientation (as expected) while the non-helical receptor domain is exposed to water.

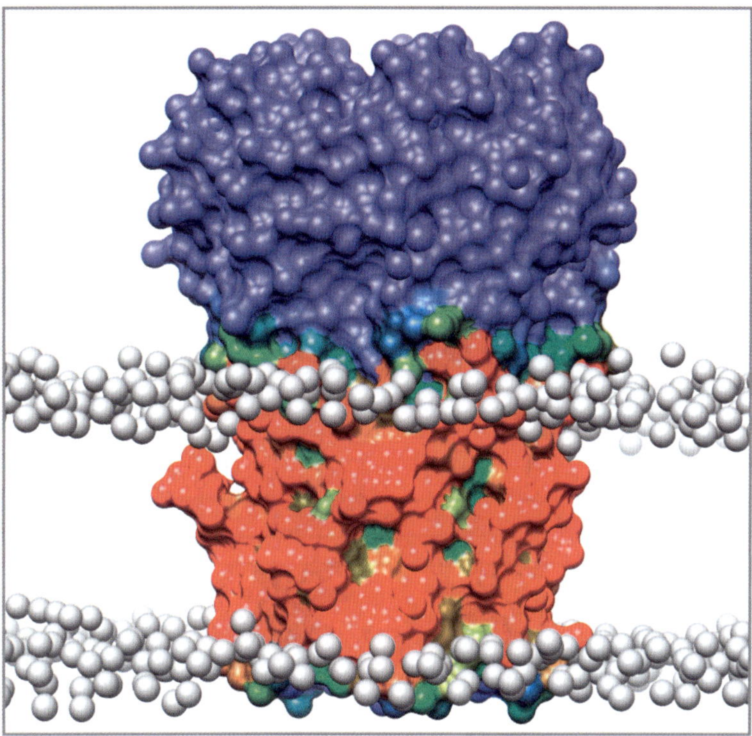

Figure 3.7 A complex membrane protein (the bacterial ion channel ELIC, pdb i.d. 2VL0) in a phospholipid bilayer, generated by self-assembly of the bilayer around the membrane protein. Only the phosphate particles (grey spheres) of the DPPC molecules are shown. The protein is coloured according to the amount of time each residue is in contact with a lipid (blue indicates a residue which has been in contact with a lipid for 0% of the analysed simulation, green corresponds to 50% and red to 100%). See www:// sbcb.bioch.ox.ac.uk/cgdb and the literature[17,51] for further details.

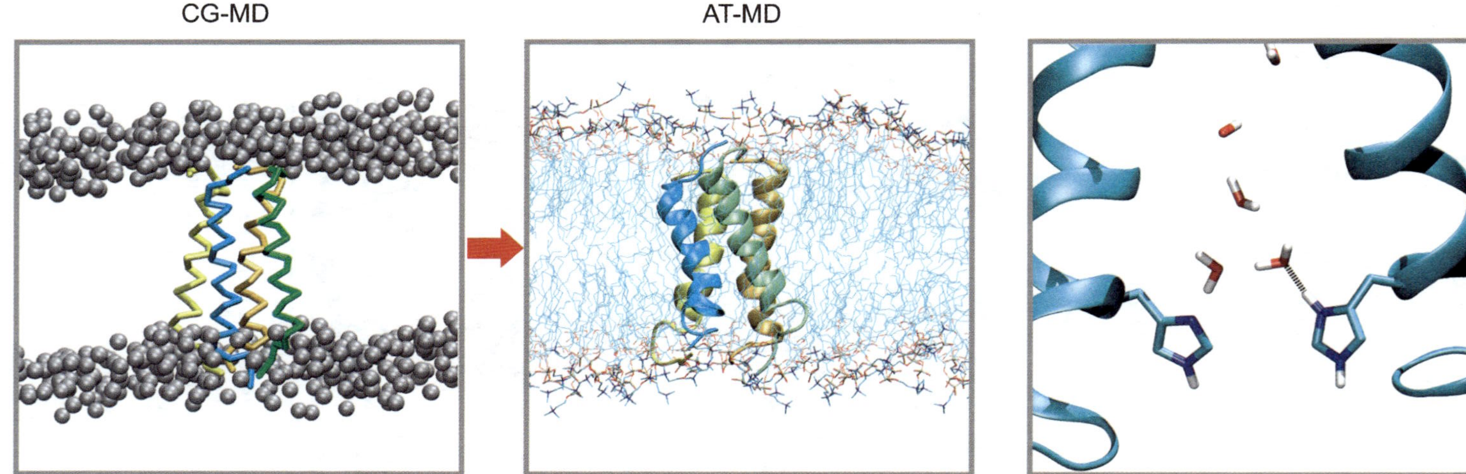

Figure 3.8   A multi-scale simulation of the influenza BM2 TM helix tetramer. A CG-MD tetramerization simulation yields a structure which is then refined by AT-MD, revealing *e.g.* details on H-bonding of water molecules to the functionally important histidine side chains.

3.6 Concluding Remarks and Future Directions

We have shown how CG-MD simulations may be applied to both simple TM helices and to more complex membrane proteins and used to study processes of membrane insertion and TM helix oligomerization. The resultant CG models seem to agree well with available experimental data. However, inevitably they lack the resolution required to answer certain questions concerning, *e.g.*, structure–function relationships. One approach to the latter is to convert models from CG to atomistic (AT) resolution. The resultant AT models may then be refined by AT simulations in what may overall be described as a serial multi-scale approach.[53] For example, this has been used to refine models of BM2 TM helix tetramer to explore the structure and dynamics of water within the resultant transbilayer channel (Figure 3.8).

These and related studies indicate that in the future *multi-scale* simulations based on CG simulations will be used to model simple membrane proteins. Future refinements to the multi-scale simulation approach will expand the range of applications and the accuracy of the resultant models.

Acknowledgements

Work in M.S.P.S.'s laboratory is funded by the BBSRC and the Wellcome Trust.

References

1. J. Nilsson, B. Persson and G. von Heijne, *Proteins Struct. Funct. Bioinf.*, 2005, **60**, 606.
2. J. L. Popot and D. M. Engelman, *Annu. Rev. Biochem.*, 2000, **69**, 881.
3. J. U. Bowie, *Nature*, 2005, **438**, 581.
4. W. L. Ash, M. R. Zlomislic, E. O. Oloo and D. P. Tieleman, *Biochim. Biophys. Acta*, 2004, **1666**, 158.
5. E. Lindahl and M. S. P. Sansom, *Curr. Opin. Struct. Biol.*, 2008, **18**, 425.
6. F. Khalili-Araghi, J. Gumbart, P. C. Wen, M. Sotomayor, E. Tajkhorshid and K. Schulten, *Curr. Opin. Struct. Biol.*, 2009, **19**, 128.
7. K. R. MacKenzie, J. H. Prestegard and D. M. Engelman, *Science*, 1997, **276**, 131.
8. L. H. Pinto and R. A. Lamb, *J. Biol. Chem.*, 2006, **281**, 8997.
9. J. C. Shelley, M. Y. Shelley, R. C. Reeder, S. Bandyopadhyay and M. L. Klein, *J. Phys. Chem. B*, 2001, **105**, 4464.
10. S. J. Marrink, A. H. de Vries and A. E. Mark, *J. Phys. Chem. B*, 2004, **108**, 750.
11. T. Murtola, E. Falck, M. Patra, M. Karttunen and I. Vattulainen, *J. Chem. Phys.*, 2004, **121**, 9156.
12. S. O. Nielsen, C. F. Lopez, G. Srinivas and M. L. Klein, *J. Phys. Condens. Matter*, 2004, **16**, R481.

13. M. J. Stevens, *J. Chem. Phys.*, 2004, **121**, 11942.
14. S. J. Marrink, J. Risselada, S. Yefimov, D. P. Tieleman and A. H. de Vries, *J. Phys. Chem. B*, 2007, **111**, 7812.
15. L. Monticelli, S. K. Kandasamy, X. Periole, R. G. Larson, D. P. Tieleman and S. J. Marrink, *J. Chem. Theor. Comput.*, 2008, **4**, 819.
16. P. J. Bond, J. Holyoake, A. Ivetac, S. Khalid and M. S. P. Sansom, *J. Struct. Biol.*, 2007, **157**, 593.
17. K. A. Scott, P. J. Bond, A. Ivetac, A. P. Chetwynd, S. Khalid and M. S. P. Sansom, *Structure*, 2008, **16**, 621.
18. L. Whitehead, C. M. Edge and J. W. Essex, *J. Comput. Chem.*, 2001, **22**, 1622.
19. M. J. Stevens, J. H. Hoh and T. B. Woolf, *Phys. Rev. Lett.*, 2003, **91**, 188102.1.
20. M. Orsi, D. Y. Haubertin, W. E. Sanderson and J. W. Essex, *J. Phys. Chem. B*, 2008, **112**, 802.
21. C. F. Lopez, S. O. Nielsen, B. Ensing, P. B. Moore and M. L. Klein, *Biophys. J.*, 2005, **88**, 3083.
22. S. O. Nielsen, B. Ensing, V. Ortiz, P. B. Moore and M. L. Klein, *Biophys. J.*, 2005, **88**, 3822.
23. P. J. Bond, C. L. Wee and M. S. P. Sansom, *Biochemistry*, 2008, **47**, 11321.
24. P. J. Bond and M. S. P. Sansom, *J. Am. Chem. Soc.*, 2006, **128**, 2697.
25. J. L. Popot and D. M. Engelman, *Biochemistry*, 1990, **29**, 4031.
26. S. J. Marrink, A. H. de Vries and D. P. Tieleman, *Biochim. Biophys. Acta*, 2009, **1788**, 149.
27. M. R. R. de Planque, J. A. W. Kruijtzer, R. M. J. Liskamp, D. Marsh, D. V. Greathouse, R. E. Koeppe, B. de Kruijff and J. A. Killian, *J. Biol. Chem.*, 1999, **274**, 20839.
28. J. A. Killian and G. von Heijne, *Trends Biochem. Sci.*, 2000, **25**, 429.
29. J. D. Lear, Z. R. Wasserman and W. F. DeGrado, *Science*, 1988, **240**, 1177.
30. K. S. Åkerfeldt, J. D. Lear, Z. R. Wasserman, L. A. Chung and W. F. DeGrado, *Acc. Chem. Res.*, 1993, **26**, 191.
31. P. Gkeka and L. Sarkisov, *J. Phys. Chem. B*, 2009, **113**, 608.
32. A. Senes, M. Gerstein and D. M. Engelman, *J. Mol. Biol.*, 2000, **296**, 921.
33. M. A. Lemmon, H. R. Treutlein, P. D. Adams, A. T. Brunger and D. M. Engelman, *Nat. Struct. Biol.*, 1994, **1**, 157.
34. S. O. Smith, R. Jonas, M. Braiman and B. J. Bormann, *Biochemistry*, 1994, **33**, 6334.
35. E. Psachoulia, P. J. Bond, P. W. Fowler and M. S. P. Sansom, *Biochemistry*, 2008, **47**, 10503.
36. E. Psachoulia, D. Marshall and M. S. P. Sansom, *Acc. Chem. Res.*, 2009, **43**, 388.
37. H. N. Chen, Y. J. Wu and G. A. Voth, *Biophys. J.*, 2007, **93**, 3470.
38. E. Khurana, M. Dal Peraro, R. DeVane, S. Vemparala, W. F. DeGrado and M. L. Klein, *Proc. Natl. Acad. Sci. USA*, 2009, **106**, 1069.

39. A. L. Stouffer, R. Acharya, D. Salom, A. S. Levine, L. Di Costanzo, C. S. Soto, V. Tereshko, V. Nanda, S. Stayrook and W. F. DeGrado, *Nature*, 2008, **451**, 596.
40. J. R. Schnell and J. J. Chou, *Nature*, 2008, **451**, 591.
41. T. Carpenter, P. J. Bond, S. Khalid and M. S. P. Sansom, *Biophys. J.*, 2008, **95**, 3790.
42. S. L. Rouse, T. Carpenter, P. J. Stansfeld and M. S. P. Sansom, *Biochemistry*, 2009, **48**, 9949.
43. C. L. Ma, C. S. Soto, Y. Ohigashi, A. Taylor, V. Bournas, B. Glawe, M. K. Udo, W. F. DeGrado, R. A. Lamb and L. H. Pinto, *J. Biol. Chem.*, 2008, **283**, 15921.
44. P. J. Bond, D. L. Parton, J. F. Clark and M. S. P. Sansom, *Biophys. J.*, 2008, **95**, 3802.
45. H. J. Risselada and S. J. Marrink, *Proc. Natl. Acad. Sci. USA*, 2008, **105**, 17367.
46. A. Harris, G. Cardone, D. C. Winkler, J. B. Heymann, M. Brecher, J. M. White and A. C. Steven, *Proc. Natl. Acad. Sci. USA*, 2006, **103**, 19123.
47. S. Yefimov, E. van der Giessen, P. R. Onck and S. J. Marrink, *Biophys. J.*, 2008, **94**, 2994.
48. W. Treptow, S. -J. Marrink and M. Tarek, *J. Phys. Chem. B*, 2008, **112**, 3277.
49. X. Periole, T. Huber, S. J. Marrink and T. P. Sakmar, *J. Am. Chem. Soc.*, 2007, **129**, 10126.
50. M. S. P. Sansom, K. A. Scott and P. J. Bond, *Biochem. Soc. Trans.*, 2008, **36**, 27.
51. A. P. Chetwynd, K. A. Scott, Y. Mokrab and M. S. P. Sansom, *Mol. Membr. Biol.*, 2008, **25**, 662.
52. R. J. C. Hilf and R. Dutzler, *Nature*, 2008, **452**, 375.
53. G. A. Ayton, W. G. Noid and G. A. Voth, *Curr. Opin. Struct. Biol.*, 2007, **17**, 192.

Passive Permeation Across Lipid Bilayers: a Literature Review

MARIO ORSI AND JONATHAN W. ESSEX

School of Chemistry, University of Southampton, Highfield, Southampton SO17 1BJ, UK

4.1 Introduction

Transport phenomena across biomembranes are crucial processes in cellular biology, and they are also becoming increasingly important in many medical, pharmaceutical and environmental technologies.[1] For example, drug permeation is crucial for the effective delivery to intracellular targets, and is at the basis of the technology of liposomal transport systems.[2] Although important permeation mechanisms, such as those responsible for the translocation of sugars and amino acids, are actively controlled by proteins, passive permeation is the most common way by which solutes cross cell membranes. Most small molecules (such as water and oxygen) and drugs are passively transported.

The fundamental principle of passive permeation is contained in Fick's first law of diffusion: a substance diffuses in the direction that eliminates its concentration gradient, at a rate proportional to the magnitude of this gradient. The permeability coefficient P, representing this proportionality constant, can be calculated as

$$P = \frac{J}{A\Delta C} \tag{4.1}$$

RSC Biomolecular Sciences No. 20
Molecular Simulations and Biomembranes: From Biophysics to Function
Edited by Mark S.P. Sansom and Philip C. Biggin
© Royal Society of Chemistry 2010
Published by the Royal Society of Chemistry, www.rsc.org

where J is the solute's flux, ΔC its concentration gradient across the interface and A the interface area. While experiments can measure overall permeability coefficients, the exact mechanism of unassisted transmembrane transport is still not fully understood, as local membrane–solute interactions are very difficult to probe. In fact, the current understanding of membrane permeability is still influenced by the simple theory developed over a century ago by Overton, who proposed that the membrane permeability coefficient of a solute is correlated to its oil–water partition coefficient.[3] This observation led to the crude representation of the membrane as a homogeneous oil slab.[4] In more recent years, experiments have clearly established that lipid membranes are highly heterogeneous systems, very different from uniform oil phases: for instance, density distributions, order parameters and diffusion in lipid membranes show characteristic properties that are not present in bulk oil systems. It is therefore not surprising to observe experimental deviations from Overton's rule. An understanding of membrane permeability should ideally require the knowledge of how structural and dynamic properties of lipids vary across the bilayer, hence across a very thin ($\sim 5\,\text{nm}$) region. Bilayers are also highly 'disordered' systems, characterized by local motions of lipid segments and also long-range diffusion of individual lipid molecules. These features render experimental investigation extremely challenging. A brief account of experimental studies of transbilayer permeation is given in Section 4.2. Subsequently, the simple solubility–diffusion model, based on Overton's findings, and the more general inhomogeneous solubility–diffusion model, which accounts also for intrabilayer heterogeneities, are described in Section 4.3.

Particle-based simulations can provide insights into the understanding of transport phenomena across lipid bilayers with atomic-level resolution, thereby characterizing local membrane heterogeneities currently inaccessible by experimental means. In principle, simulations can directly reproduce the spontaneous passive permeability phenomenon. For example, using a simplified, 'coarse-grained' model, we were able to calculate the transmembrane permeability coefficient of water from the direct observation of translocation events over a simulation lasting for almost $1\,\mu\text{s}$.[5] However, standard atomic-level membrane simulations cannot currently reach such a time scale, and indeed passive water transport has never been quantified with these traditional models. In a recent 'state-of-the-art' molecular dynamics study,[6] four phosphatidylcholine bilayers, each comprising 128 lipids, were simulated for $50\,\text{ns}$. In the four simulations, two, four, six and seven crossing events of water molecules were observed, respectively. It is evidently not possible to attempt an estimation of the permeability coefficient on the basis of such limited statistics. Moreover, for many other important solute molecules, such as large hydrophilic drugs, the time scales required to observe directly a statistically significant number of translocation events are expected to be in the range of (at least) milliseconds, hence far beyond the capabilities of any particle-based simulation model. Fortunately, there are indirect techniques that can be used to overcome these difficulties;[2] the most popular of these methods is described in Section 4.3.1. Thanks to these methodological advances, in recent years standard molecular dynamics simulations have indeed been successfully employed to predict permeability coefficients and to investigate the general mechanism of passive transport across membranes. We will review the most important simulation studies in the literature, categorized according to the permeant type; in particular, Section 4.4 is devoted to small molecules, Section 4.5 to drugs and Section 4.6 to fullerene molecules. Results are further discussed in Section 4.7, along with limitations and issues of the simulation methodology, and the main conclusions are summarized in Section 4.8.

The research area of membrane permeability simulations has also been reviewed in an excellent article published in 2006 by Xiang and Anderson.[2] In the following, we will place special focus on material not already covered in Xiang and Anderson's review;[2] particular attention will therefore be devoted to the most recent work appearing in the literature.

4.2 Experimental Methods

This section summarizes the most popular experimental methods employed to measure transmembrane spontaneous permeabilities. It should be noted that there is a large scatter of experimental values for the permeability coefficients through membranes. Whereas the relative permeabilities are typically well reproduced, absolute data measured in different laboratories by different techniques can vary across orders of magnitude. This can be due to difficulties in calibrating the measurements and to perturbations caused by the specific method.[7]

4.2.1 Water and Small Organic Molecules

There are two main experimental configurations that can be employed to measure the permeability coefficient of water and small organic solutes, based on either planar lipid bilayers[8–11] or lipid vesicles.[7,12,13] The former method has received the most attention. In particular, permeability coefficients for small organic molecules can be measured across planar lipid bilayers formed on a $\sim 1\,\mathrm{mm}^2$ hole in a polyethylene or Teflon partition separating two magnetically stirred water-jacketed chambers.[8–11] Permeability coefficients P are calculated from the rate of change of the receiver concentration with the following equation:

$$P = s \times \frac{V_{\mathrm{chamber}}}{A C_{\mathrm{donor}}} \tag{4.2}$$

where s is the slope of the receiver concentration *versus* time interval plot, V_{chamber} the volume of the aqueous solution in each chamber, A the bilayer area and C_{donor} the concentration of the solute in the donor chamber.[8]

4.2.2 Drugs

Permeability coefficients of drug molecules are not usually measured from isolated lipid bilayers, but instead using more complex model systems that more closely correlate with the observed physiological drug absorption data. The two most common *in vitro* permeability assays are Caco-2 cell monolayers and the parallel artificial membrane permeability assay (PAMPA); brief descriptions are given below.

4.2.2.1 Caco-2 Cell Monolayers

Caco-2 cells are human colorectal carcinoma cells, characterized by morphological and functional similarities to the small intestinal epithelium cells. Caco-2 experiments allow the study of all major absorption routes: passive transcellular and paracellular transport and active carrier-mediated mechanisms. In typical experiments, a monolayer of cells is grown on a filter separating two stacked microwell plates. The compound under investigation is then introduced on one side of the filter. Concentrations are monitored by ultraviolet spectroscopy or a combination of liquid chromatography and mass spectrometry; alternatively, radiolabelled compounds can be used in the first place. Permeability coefficients are eventually calculated by applying equation (4.2) given above for the case of small solutes. During the past few years, Caco-2 monolayers have been widely accepted by pharmaceutical companies and by regulatory authorities as a standard permeability-screening assay for the prediction of drug intestinal permeability.[14] However, since Caco-2 cells inevitably contain endogenous transporter and efflux systems, reproducibility and data interpretation can be difficult.[15] An additional drawback is that Caco-2 experiments are time consuming, requiring up to 30 days for the preparation of stable monolayers.[15,16]

4.2.2.2 PAMPA

The parallel artificial membrane permeability assay (PAMPA), introduced by Kansy *et al.*[17] in 1998, has since been gaining acceptance in pharmaceutical research as a less expensive alternative to Caco-2.[15,18–21] A PAMPA 'sandwich' is prepared from two plates that are similar to those used for traditional Caco-2 experiments. One plate contains a porous filter disk at the bottom of each well, whereas the other is a reservoir plate that is moulded to sit precisely under the filter plate. The filter is coated with a solution of lipid material in an inert organic solvent to prepare the artificial membrane. Filters are typically $\sim 100\,\mu$m thick.[16] The wells of one plate are then filled with donor solution (drug) and the other with acceptor solution (buffer); the plates are then stacked to create the sandwich and incubated. Incubation times can vary between 15 min for highly permeable molecules and 15 h for poorly permeable molecules. The sandwich is eventually separated and both the donor and acceptor compartments are assayed for the amount of material present. As for the Caco-2 method, measurements are performed by ultraviolet spectroscopy or liquid chromatography combined with mass spectrometry. Clearly, PAMPA assays only measure passive permeation, thus eliminating the possible active contribution which can affect Caco-2 results.

Another advantage of PAMPA is that experiments are much quicker to carry out than with Caco-2. However, the phase formed by lipids in PAMPA membranes is unknown.

4.3   The Solubility–Diffusion Model

Overton's observation that the membrane permeability coefficient of a solute is correlated with its oil–water partition coefficient[3] led to the simplified model of the membrane as an homogeneous oil slab; on this basis, the simple bulk solubility–diffusion model of membrane permeability was proposed.[4] According to this model, the permeability coefficient P can be obtained simply as

$$P = \frac{KD}{h} \tag{4.3}$$

where K is the bulk solvent (oil)–water partition coefficient, D the solute's diffusion coefficient in the solvent and h the membrane thickness.

In more recent years, experiments have clearly established that lipid membranes are highly heterogeneous systems, very different from uniform oil phases: for instance, density distributions, order parameters and diffusion in lipid membranes show characteristic properties that are not present in bulk oil systems. Moreover, the polar components of lipid bilayers, such as the headgroup and the glycerol/ester regions, do not have any counterpart in the simple oil solvents assumed in the solubility–diffusion model. To tackle these issues, the inhomogeneous solubility–diffusion model was proposed;[22–24] this model relates the permeability coefficient of a solute to an integral of depth-dependent parameters across the membrane. In particular, the overall membrane permeability coefficient P is expressed as

$$P = \frac{1}{\int_{z_1}^{z_2} R(z)\mathrm{d}z} = \frac{1}{\int_{z_1}^{z_2} \frac{\exp[\Delta G(z)/k_B T]}{D_z(z)} \mathrm{d}z} \tag{4.4}$$

where $R(z)$, $\Delta G(z)$ and $D_z(z)$ are the solute resistance, free energy of transfer and diffusion coefficients along the z direction, respectively, at position z along the direction normal to the membrane interfacial plane. The free energy difference $\Delta G(z)$ can be related to the partition coefficient $K(z)$, via $K(z) = \exp[-\Delta G(z)/RT]$. The integration extremes z_1 and z_2 are taken in the water phases at the two sides of the membrane, so that the integration is performed over the entire bilayer.

Experimentally, the free energy $\Delta G(z)$ and diffusion coefficient $D_z(z)$ are very difficult to resolve as a function of the bilayer depth z. The free energy of transfer from water to the most favourable region in the bilayer can be obtained from membrane binding experiments, but it is not possible to study the variation of this property across the membrane. As for the diffusion coefficient along z, there are no reliable experimental methods that can be employed.[2] Therefore, it has proved difficult so far to test and validate experimentally the inhomogeneous solubility–diffusion model.

On the other hand, a number of simulation techniques are capable of yielding quantitative data on the depth-dependent solubility and diffusion across model membranes; the most popular of these methods, the so-called z-constraint technique, widely used in conjunction with the inhomogeneous solubility–diffusion model, is described in the following section. While other methodologies, such as Widom's particle insertion,[25] the mean residence time method,[26] umbrella sampling[27] and thermodynamic integration,[28] are able to access the free energy $\Delta G(z)$, only the z-constraint method can estimate the diffusion coefficient $D_z(z)$ and hence eventually the permeability coefficient. Since in this chapter we are focusing on simulation studies of permeability, we will only consider the z-constraint technique. The other methods mentioned above are summarized in Xiang and Anderson's review[2] and are described in detail in the original references.

4.3.1 The z-Constraint Method

The quantities featuring in the inhomogeneous solubility–diffusion model [equation (4.4)] can be obtained from simulations by applying the z-constraint method.[24] The z-constraint technique involves constraining the mass centre of a chosen permeant molecule to fixed positions along the bilayer normal, which typically coincides with the z-axis of the system frame of reference. The permeant solute remains free to move in the xy plane. Both $\Delta G(z)$ and $D_z(z)$ can then be simply calculated from the constraining force $f_z^c(z)$ required to keep the solute mass centre fixed at the selected z locations. In particular, the free energy of transfer $\Delta G(z)$ from water into the membrane is computed as

$$\Delta G(z) = \int\limits_{\text{water}}^{z} \langle f_z^c(z') \rangle \mathrm{d}z' \qquad (4.5)$$

where $<f_z^c(z')>$ is the average constraint force at position z' over the total simulation time. The local diffusion coefficient $D_z(z)$ along the z-dimension is calculated as[29]

$$D_z(z) = \frac{(k_\mathrm{B}T)^2}{\int_0^\infty \langle \Delta f_z^c(z,t)\Delta f_z^c(z,0) \rangle \mathrm{d}t} \qquad (4.6)$$

where k_B is the Boltzmann constant, T the temperature and $\Delta f_z^c(z,t)$ the 'random' force, defined as the deviation of the instantaneous force from the average force acting on the solute: $\Delta f_z^c(z,t) = f_z^c(z,t) - <f_z^c(z)>$.

4.4 Small Molecules

In the early 1990s, as soon as the first particle-based models for 'pure' lipid bilayers were successfully simulated by molecular dynamics, membrane force fields started to be extended to incorporate simple small solute molecules, with the ultimate aim of characterizing the transmembrane permeability process.

Seminal permeability simulations were performed by Bassolino-Klimas *et al.* to study the diffusion of benzene in a lipid bilayer.[30] Despite the small system size (comprising only 36 lipid molecules) and short simulation time (1 ns), the authors were able to observe that benzene molecules diffuse by a 'rattling and jumping' behaviour of the kind predicted for lipid molecules.[31]

Marrink and Berendsen[24] reported the first calculation of the transbilayer permeability coefficient by molecular dynamics. In particular, they applied the *z*-constraint algorithm to quantify the permeability of water through a DPPC lipid bilayer; the results were comparable to the corresponding experimental data.[24] Marrink and Berendsen subsequently studied the permeability of oxygen and ammonia through a 64-DPPC bilayer hydrated by 736 water molecules.[32] Interestingly, in this study it was decided to divide the lipid charges by a factor of 2, to account for the insufficient shielding properties of the SPC water model.[32]

These initial investigations have been followed in more recent years by simulation studies that, thanks to hardware progress, can afford more realistic conditions, in terms of system size and simulation time.

Shinoda *et al.*[28] simulated the transbilayer permeation of seven solutes: H_2O, NH_3, O_2, CO, NO, CO_2 and $CHCl_3$. All molecules were treated as rigid bodies. Two different lipid bilayers were employed, one composed of 'standard' DPPC lipids and the other of branched-chain DPhPC. They observed that water diffusion was reduced in the branched DPhPC bilayer compared with the DPPC system; as a result, the permeability coefficient in DPhPC was 30% lower than in DPPC.

Sugii *et al.*[33] studied the effect of the lipid hydrocarbon chain length on the permeability of H_2O, O_2, CO and NO. They showed that the lipid membranes with longer chains display larger and wider free energy barriers. They also estimated the water permeability coefficient and found that it decreases slightly with increasing chain length.[33]

The refinement of force fields for different molecules has also allowed the calculation of permeability coefficients for several small organic molecules. In particular, Bemporad *et al.*[34,35] employed the *z*-constraint method to calculate the permeability of the following solutes, representing the most common chemical functional groups: acetamide, acetic acid, benzene, ethane, methanol, methyl acetate, methylamine and water. The permeability coefficients obtained are generally one order of magnitude larger than corresponding experimental data, but the relative permeabilities are well reproduced.[34,35]

For obvious reasons, the most studied 'small molecule' has been water. In particular, a number of research groups have calculated the free energy barrier for permeation of water through lipid bilayers. The values obtained are collected in Table 4.1.

Table 4.1 Comparison between different simulation studies regarding the free energy of transfer ΔG_z from the bulk water phase to the bilayer centre for permeating water molecules.

Reference	*System*	$\Delta G_z / kJ\,mol^{-1}$
Marrink and Berendsen[24]	64 DPPC + 736 H_2O	26
Jedlovszky and Mezei[36]	50 DMPC + 2033 H_2O	54
Shinoda *et al.*[28]	72 DPPC + 2088 H_2O	26
Bemporad *et al.*[35]	72 DPPC + 2094 H_2O	23
Sugii *et al.*[33]	64 DPPC + 1600 H_2O	23

It can be noted that all published studies agree on a free energy barrier of $\sim 25\,kJ\,mol^{-1}$, with the exception of Jedlovszky and Mezei,[36] who obtained the substantially higher value of $54\,kJ\,mol^{-1}$.

4.5 Drugs

The ability of drugs to permeate through biological membranes is a crucial factor in drug delivery. In particular, passive transmembrane permeability is known to be one of the major mechanisms for drug absorption.[14] Lipophilic compounds can rapidly partition into the cell membrane thanks to their affinity for the hydrocarbon core of lipid bilayers, and hence they typically exploit the transcellular pathway. In between cells, small water-filled pores (also called 'tight junctions') are normally present; despite the much smaller surface area offered by these pores compared with the overall epithelium cell surface, it is reasonable to expect that (small) hydrophilic drugs can also diffuse via such a paracellular route. Alternative drug transport processes, including protein-assisted and vesicle-mediated transport, are less frequently observed than the unassisted mechanisms. From a technological perspective, passive drug permeation is at the basis of liposome drug delivery systems.[2]

It is therefore evident that understanding transmembrane permeation is crucial for rational drug design. Computational methods to predict trans-membrane permeability coefficients of drugs before synthesis are increasingly desirable to minimize the investment in pharmaceutical design and development.[2] However, the accurate prediction of drug permeability represents a great challenge for *in silico* models, due to the complexity of the underlying physiological mechanism.[37] The simplest theoretical tools to predict drug permeability are the quantitative structure–activity relation (QSAR) models. QSAR models relate numerical properties of the drug molecular structure to its activity *via* a mathematical expression.[38] An early example was the discovery by Meyer and Overton[56–58] of a correlation between anaesthetic potency of a compound and its oil–water partition coefficient. Partition coefficients are also traditionally correlated to permeation: within chemical series, this is often verified. However, computational models based on molecular properties typically fail when large sets of diverse compounds are analysed.[15]

The recent increase in computer power has also allowed the calculation of permeability coefficients for drug molecules through molecular dynamics simulations.[39–42] Studying the permeation process by particle-based computer models is particularly attractive: simulations have the potential to investigate this mechanism with atomic detail, while also providing accurate estimates of the permeability coefficient. However, since most drugs are large (MW > 100) and flexible molecules, their simulation proves computationally challenging. An early attempt to simulate the drug permeation process was performed by Alper and Stouch,[43] who studied a nifedipine analogue in a lipid bilayer. Owing to limited computer resources, the system was simulated for only 4 ns. Such a short simulation time nonetheless proved sufficient to study the drug orientation and diffusion and to analyse the role of hydrogen bonding.[43] Transbilayer permeation has been recently simulated, using the z-constraint method, for the amphiphilic drug valproic acid,[40] β-blockers[41,44] and psoralen derivatives.[42] Free energy profiles, diffusion coefficients and eventually permeability coefficients were calculated.[40–42,44] In general, the results from these studies are qualitatively consistent with experiments, in that the relative permeabilities, and hence the ranking orders, are reproduced. However, the actual values for the calculated permeabilities coefficients cannot easily be compared to the corresponding experimental figures, because of the different systems employed. As an example, the data obtained from simulations of β-blocker drugs and a number of corresponding experimental measurements are collected in Table 4.2.

It can be seen that the permeability coefficients from simulations are typically several orders of magnitude larger than those from experiments. In fact, this is to be expected considering the differences between the simulation and the experimental materials and conditions. Simulations are conducted on simple, 'minimal' pure lipid bilayers. Experiments are instead carried out on layers of entire cells (Caco-2) or on thick solutions of lipids of unknown phase (PAMPA). In both experimental systems, solutes must cross a much thicker barrier compared with that represented by the single lipid bilayer in the simulations. It is therefore reasonable to observe much larger permeability coefficients in simulations compared with experiments. It is also worth noting how the experimental measurements show great variability even amongst each other. Again, this is not surprising considering how

Table 4.2   β-Blocker permeability coefficients/cm s^{-1}.

Method (reference)	Alprenolol	Atenolol	Pindolol
Simulation, DPPC bilayer[44]	10 ± 10	$(3.3 \pm 4.2) \times 10^{-1}$	3.0 ± 3.5
Experiment, Caco-2[53]	4.8×10^{-3}	3.7×10^{-5}	3.0×10^{-4}
Experiment, Caco-2[54]	2.4×10^{-4}	1.0×10^{-6}	5.1×10^{-5}
Experiment, Caco-2[16]	1.8×10^{-2}	8.8×10^{-5}	2.6×10^{-3}
Experiment, PAMPA[16]	1.0	8.6×10^{-6}	1.8×10^{-2}
Experiment, PAMPA[55]	1.1×10^{-5}	–	4.9×10^{-6}
Experiment, PAMPA[18]	2.5×10^{-3}	–	1.4×10^{-3}

Table 4.3 Relative permeability coefficients.

Method (reference)	Alprenolol	Atenolol	Pindolol
Simulation, DPPC bilayer[44]	30	1	9
Experiment, Caco-2[53]	130	1	8
Experiment, Caco-2[54]	240	1	51
Experiment, Caco-2[16]	204	1	30
Experiment, PAMPA[16]	120 790	1	2091

sensitive permeability coefficients are to slightly different conditions and setup details. However, it is most important to look at relative permeabilities; this also has a high practical value for drug design, as it is the ranking order among a set of compounds which is crucial, rather than the knowledge of the absolute individual magnitudes. Relative permeability coefficients, calculated for each complete set of data with respect to the permeability coefficient of atenolol (the slowest permeant), are collected in Table 4.3.

The ranking orders obtained by experimental and simulation methods ($P_{\text{alprenolol}} > P_{\text{pindolol}} > P_{\text{atenolol}}$) are fairly consistent with each other. In particular, there is an overall good consistency between the simulation data and the Caco-2 experimental measurements, whereas the PAMPA experiment considered reports larger differences between the relative values.

4.6 Fullerene

The interaction between nanomaterials and biological materials is becoming an increasingly important research subject, mainly due to the implications in biomedical technologies. Also, the general use of nanomaterials in industry is rapidly growing, raising health and environmental concerns which demand quantitative assessment.

Special attention has been paid to fullerene (C_{60}) and its derivatives, which constitute an important subset of nanomaterials. Fullerenes play a role in a wide range of potential biomedical applications, such as anti-HIV drugs, skin cancer treatments, DNA cleavage agents, antioxidant drugs and contrast agents for X-ray and magnetic resonance imaging;[45,46] promising future applications involve the use of fullerenes as drug carriers for selective tissue targeting.[45] However, carbon nanoparticles often display some degree of toxicity. For example, Sayes *et al.*[47] observed that fullerene C_{60} causes membrane leakage. The concerns raised by such findings are accentuated by the known ease with which fullerenes diffuse throughout the body. Oberdörster *et al.*[48] showed that fullerene aggregates, despite their large size, can even cross the blood–brain barrier. The exact mechanisms by which fullerenes cross and disrupt the membrane are not yet understood. Recent molecular dynamics studies have started to shed some light on the permeation processes of fullerene and derivatives across lipid bilayers.[49–51]

Qiao *et al.*[49] employed the z-constraint method to study the permeability characteristics of fullerene C_{60} and its derivative $C_{60}(OH)_{20}$ across a DPPC bilayer modelled with a united-atom force field. Their simulation results indicate that fullerene C_{60} possesses a typical hydrophobic character; it preferentially partitions inside the bilayer hydrophobic core and its overall free energy difference from the bilayer centre to the outside water phase is negative.[49] The fullerene derivative $C_{60}(OH)_{20}$ displays instead a hydrophilic behaviour, as it preferentially partitions at the headgroup–water interface and is characterized by an overall positive free energy barrier between the bilayer core and the water phase.[49] The permeability coefficient of fullerene C_{60} is therefore predicted to be several orders of magnitude higher than that of its derivative $C_{60}(OH)_{20}$; these findings might therefore explain the reduced toxicity of functionalized fullerene on the basis of its reduced tendency to penetrate cell membranes.[49]

Wong-Ekkabut *et al.*[50] simulated the effect of high fullerene concentrations in lipid membranes (up to one fullerene per lipid) using a simplified, coarse-grained (CG) model, where groups of 4–6 atoms are grouped into single interaction centres;[52] such simplifications allowed large aggregates to be simulated for the necessary amount of time. Since no evident damage to the bilayer structure was observed, the authors concluded that the mechanism of fullerene toxicity is unlikely to involve mechanical damage.

Bedrov *et al.*[51] studied the permeability of fullerene across a DMPC bilayer using an all-atom model. They obtained results qualitatively similar to those reported in the investigations summarized above; also in this study fullerene was found to favour strongly the hydrophobic bilayer core with respect to bulk water.

However, in general, these three studies[49–51] report values for the preferred location, maximum free energy difference and permeability coefficient which are rather different from each other; these data are collected in Table 4.4.

A number of possible reasons behind these disagreements might be noted. Two of these studies[49,51] employed traditional atomic-level models; owing to the high simulation cost of these models, simulations could be performed for only $\sim 10\,\mathrm{ns}$. It is not clear whether such short simulation times are sufficient to yield converged results. The other study[50] was conducted using a CG model for both lipids and fullerenes. While in this case simulation times of the order of

Table 4.4   Fullerene C_{60} simulation data.[a]

Reference	Force field	Z/nm	$\Delta G_z/kJ\,mol^{-1}$	$P/cm\,s^{-1}$
Qiao *et al.*[49]	United-atom	1.1	~ -35	–
Wong-Ekkabut *et al.*[50]	Coarse-grained	0.9	~ -100	0.06
Bedrov *et al.*[51]	All-atom	0.6	~ -90	~ 100

[a]Z represents the distance from the bilayer centre, along the z-axis, at which the solute preferentially partitions; this corresponds to the location of the minimum in the free energy profile. ΔG_z is the free energy difference between the reference value of zero in the outer water phase and the minimum at distance Z from the bilayer centre. P is the permeability coefficient.

microseconds could be achieved, the model employed[52] relies on a number of assumptions which may undermine the reliability of the results obtained. In particular, this CG model[52] contains a very simplified description of electrostatic interactions; moreover, the CG water is represented by generic neutral Lennard-Jones macrospheres, thus lacking any charge accounting for the highly dipolar nature of real water.

4.7 Discussion

Particle-based computational methods, in particular molecular dynamics, have been increasingly employed over the past decade to the modelling of passive permeation processes across lipid bilayers. It is reassuring to observe that many simulations tend to yield consistent results for a number of systems investigated. For example, several simulations from different research groups have been carried out to study the permeability of water. As can be seen from Table 4.1, all published studies (with one exception[36]) agree on a free energy barrier of $\sim 25\,\mathrm{kJ\,mol^{-1}}$. However, a somewhat less clear picture emerges from the calculation of diffusion coefficients. Marrink and Berendsen[24] reported that the diffusion coefficient for water in the bilayer centre is about two times higher than in the outer bulk phase.[24] However, Shinoda *et al.*[28] and Bemporad *et al.*[35] observed an opposite trend, the diffusion coefficient of water being about two times lower in the bilayer centre with respect to bulk phase. As already noted elsewhere,[2] this disagreement might be due to the use of the (less accurate) united-atom force field by Marrink and Berendsen[24] as opposed to the all-atom models employed by Shinoda *et al.*[28] and Bemporad *et al.*[35]

In recent years, it has been possible to run permeability simulations for large molecules such as drugs[39–42] and carbon nanoparticles.[49,51] In general, these investigations have been extremely useful in understanding many aspects of bilayer permeation with atomic resolution. However, the huge computational cost associated with the simulation of atomic-level models results in a series of limitations and issues. For example, obtaining well-converged data is often problematic, as series of long simulations are required for every solute. Also, bilayer sizes must be rather small to be computationally amenable; this can induce artefacts, especially when large solutes are inserted into the membrane. In fact, the results are sometimes controversial; for example, as highlighted in Table 4.4 for the case of fullerene, different research groups using somewhat different methods can produce very different results. Furthermore, in general, the number of solutes that can be investigated in a reasonable amount of time is extremely limited; this seriously hinders applications in the context of drug design, where screenings of large sets of candidate compounds are normally required.

4.8 Conclusions

A large number of simulation studies appearing in the literature over the past 15 years have demonstrated the potential of molecular dynamics to predict transbilayer permeability coefficients and provide atomic-level insights into the translocation mechanism. Although simulating spontaneous permeation is still computationally unfeasible, special techniques (particularly the z-constraint method) can be successfully applied to calculate not only the overall permeability coefficient, but also free energies of transfer and diffusion coefficients at different depths across the bilayer. It is therefore also possible to predict the preferred partitioning location across the membrane, in addition to the most frequently occurring orientations of the solutes. The importance of simulation in the context of membrane permeability is particularly significant, because of the well-known experimental difficulties associated in general with the investigation of membrane systems. Given the continuous increase in available computational power, it is foreseeable that molecular dynamics simulation will play an ever more important role in the investigation of passive permeation of solutes across biological membranes, especially in industrial contexts such as drug design.

References

1. A. Pohorille and M. A. Wilson, *Cell. Mol. Biol. Lett.*, 2001, **6**, 369.
2. T.-X. Xiang and B. D. Anderson, *Adv. Drug Deliv. Rev.*, 2006, **58**, 1357.
3. E. Overton, *Vierteljahrsschr. Naturforsch. Ges. Zurich*, 1896, **41**, 383.
4. A. Finkelstein, *J. Gen. Physiol.*, 1976, **68**, 127.
5. M. Orsi, D. Y. Haubertin, W. E. Sanderson and J. W. Essex, *J. Phys. Chem. B*, 2008, **112**, 802.
6. S. Ollila, M. T. Hyvönen and I. Vattulainen, *J. Phys. Chem. B*, 2007, **111**, 3139.
7. D. Huster, A. J. Jin, K. Arnold and K. Gawrisch, *Biophys. J.*, 1997, **73**, 855.
8. T.-X. Xiang, X.-L. Chen and B. D. Anderson, *Biophys. J.*, 1992, **63**, 78.
9. A. Finkelstein, *J. Gen. Physiol.*, 1976, **68**, 127.
10. E. Orbach and A. Finkelstein, *J. Gen. Physiol.*, 1980, **75**, 427.
11. A. Walter and J. Gutknecht, *J. Membr. Biol.*, 1984, **77**, 255.
12. S. Paula, A. G. Volkov, A. N. Vanhoek, T. H. Haines and D. W. Deamer, *Biophys. J.*, 1996, **70**, 339.
13. T.-X. Xiang and B. D. Anderson, *J. Membr. Biol.*, 1995, **148**, 157.
14. P. Shah, V. Jogani, T. Bagchi and A. Misra, *Biotechnol. Prog.*, 2006, **22**, 186.
15. J. Ruell, *J. Mod. Drug Discov.*, 2003, **January**, 28.
16. A. Avdeef, P. Artursson, S. Neuhoff, L. Lazorova, J. Grasjo and S. Tavelin, *Eur. J. Pharm. Sci.*, 2005, **24**, 333.
17. M. Kansy, F. Senner and K. J. Gubernator, *J. Med. Chem.*, 1998, **41**, 1007.
18. A. Avdeef, P. E. Nielsen and O. Tsinman, *Eur. J. Pharm. Sci.*, 2004, **22**, 365.
19. S. Carrara, V. Reali, P. Misiano, G. Dondio and C. Bigogno, *Int. J. Pharm.*, 2007, **345**, 125.

20. R. P. Verma, C. Hansch and C. D. Selassie, *J. Comput.-Aid. Mol. Des.*, 2007, **21**, 3.
21. D. Galinis-Luciani, L. Nguyen and M. J. Yazdanian, *J. Pharm. Sci.*, 2007, **96**, 2886.
22. J. M. Diamond and Y. Katz, *J. Membr. Biol.*, 1974, **17**, 121.
23. J. M. Diamond, G. Szabo and Y. Katz, *J. Membr. Biol.*, 1974, **17**, 148.
24. S.-J. Marrink and H. J. C. Berendsen, *J. Phys. Chem.*, 1994, **98**, 4155.
25. B. Widom, *J. Chem. Phys.*, 1963, **39**, 2808.
26. L. Koubi, M. Tarek, M. L. Klein and D. Scharf, *Biophys. J.*, 2000, **78**, 800.
27. G. M. Torrie and J. P. Valleau, *J. Comput. Phys.*, 1977, **23**, 187.
28. W. Shinoda, M. Mikami, T. Baba and M. J. Hato, *J. Phys. Chem. B*, 2004, **108**, 9346.
29. B. Roux and M. Karplus, *J. Phys. Chem.*, 1991, **95**, 4856.
30. D. Bassolino-Klimas, H. E. Alper and T. R. Stouch, *Biochemistry*, 1993, **32**, 12624.
31. L. C. Vaz and P. F. Almeida, *Biophys. J.*, 1991, **60**, 1553.
32. S.-J. Marrink and H. J. C. Berendsen, *J. Phys. Chem.*, 1996, **100**, 16729.
33. T. Sugii, S. Takagi and Y. Matsumoto, *J. Chem. Phys.*, 2005, **123**, 184714.
34. D. Bemporad, C. Luttmann and J. W. Essex, *Biophys. J.*, 2004, **87**, 1.
35. D. Bemporad, J. W. Essex and C. Luttmann, *J. Phys. Chem. B*, 2004, **108**, 4875.
36. P. Jedlovszky and M. Mezei, *J. Am. Chem. Soc.*, 2000, **122**, 5125.
37. J. E. Penzotti, G. A. Landrum and S. Putta, *Curr. Opin. Drug. Discov. Dev.*, 2004, **7**, 49.
38. A. R. Leach, *Molecular Modelling – Principles and Applications,* Prentice Hall, Englewood Cliffs, NJ, 2001.
39. A. Grossfield and T. B. Woolf, *Langmuir*, 2002, **18**, 198.
40. J. Ulander and A. D. J. Haymet, *Biophys. J.*, 2003, **85**, 3475.
41. D. Bemporad, C. Luttmann and J. W. Essex, *Biochim. Biophys. Acta*, 2005, **1718**, 1.
42. D. J. V. A. dos Santos and L. A. Eriksson, *Biophys. J.*, 2006, **91**, 2464.
43. H. E. Alper and T. R. Stouch, *J. Phys. Chem.*, 1995, **99**, 5724.
44. D. Bemporad, *PhD Thesis*, University of Southampton, 2003.
45. S. Bosi, T. Da Ros, G. Spalluto and M. Prato, *Eur. J. Med. Chem.*, 2003, **38**, 913.
46. E. Nakamura and H. Isobe, *Acc. Chem. Res.*, 2003, **36**, 807.
47. C. Sayes, J. Fortner, W. Guo, D. Lyon, A. Boyd, K. Ausman, Y. Tao, B. Sitharaman, L. Wilson, J. Hughes, J. West and V. Colvin, *Nano Lett.*, 2004, **4**, 1881.
48. G. Oberdörster, Z. Sharp, V. Atudorei, A. Elder, R. Gelein, W. Kreyling and C. Cox, *Inhal. Toxicol.*, 2004, **16**, 437.
49. R. Qiao, A. Roberts, A. Mount, S. Klaine and P. Ke, *Nano Lett.*, 2007, **7**, 614.
50. J. Wong-Ekkabut, S. Baoukina, W. Triampo, I. M. Tang, D. P. Tieleman and L. Monticelli, *Nat. Nanotechnol.*, 2008, **3**, 363.

51. D. Bedrov, G. D. Smith, H. Davande and L. W. Li, *J. Phys. Chem. B*, 2008, **112**, 2078.
52. S.-J. Marrink, H. J. Risselada, S. Yefimov, D. P. Tieleman and A. H. de Vries, *J. Phys. Chem. B*, 2007, **111**, 7812.
53. A. Adson, P. S. Burton, T. J. Raub, C. L. Barsuhn, K. L. Audus and N. F. H. Ho, *J. Pharm. Sci.*, 1995, **84**, 1197.
54. K. Palm, K. Luthman, A. L. Ungell, G. Strandlund and P. Artursson, *J. Pharm. Sci.*, 1996, **85**, 32.
55. M. Fujikawa, R. Ano, K. Nakao, R. Shimizu and M. Akamatsu, *Bioorg. Med. Chem.*, 2005, **13**, 4721.
56. H. H. Meyer, *Arch. Exp. Pathol. Pharmakol.*, 1899, **42**, 109–118.
57. H. H. Meyer, *Arch. Exp. Pathol. Pharmakol.*, 1901, **46**, 338–346.
58. C. E. Overton, *Studien über die Narkose Zugleich ein Beitrag zur allgemeinen Pharmakologie*, 1901, Gustav Fischer, Jena, Switzerland.

Implicit Membrane Models For Peptide Folding and Insertion Studies

MARTIN B. ULMSCHNEIDER[a] AND JAKOB P. ULMSCHNEIDER[b]

[a] Department of Physiology and Biophysics, University of California at Irvine, Irvine, CA 92697, USA; [b] Interdisciplinary Center for Scientific Computing (IWR), University of Heidelberg, Im Neuenheimer Feld 368, D-69120, Heidelberg, Germany

5.1 Introduction

Structure prediction of membrane proteins is of immense biomedical importance, given their role as principal drug targets.[1] Due to major experimental difficulties, only a handful of structures are currently known.[2] Computational approaches offer a promising complementary technique. However, computer simulations of membrane proteins in their native lipid bilayer environment are usually limited to the $<100\,\mathrm{ns}$ time scale, if explicit models for solvent and membrane are used. This is due to the large number of non-bonded interactions that need to be evaluated for such large and complex systems and the short time step needed to follow covalent bond vibrations in detail.[3-6] These time scales can be sufficient to study various short time scale processes such as overall protein stability in a lipid bilayer[4,5] or self-assembly of protein–detergent micelles.[7,8] However, long time scale processes,[9] such as membrane protein adsorption, folding and partitioning, require much more thorough sampling,

RSC Biomolecular Sciences No. 20
Molecular Simulations and Biomembranes: From Biophysics to Function
Edited by Mark S.P. Sansom and Philip C. Biggin
© Royal Society of Chemistry 2010
Published by the Royal Society of Chemistry, www.rsc.org

corresponding to micro- to millisecond simulations using conventional all-atom molecular dynamics techniques.

Recent algorithmic advances and vast improvements in computer hardware (multi-core and single-instruction–multiple-data CPU architectures) now permit fully converged simulation of the folding processes of small membrane proteins using both molecular dynamics (MD) and Monte Carlo (MC) simulations.[10,11] Nevertheless, the computational effort required to study systems in a fully explicit lipid bilayer environment is orders of magnitude more costly than for equivalent implicit membrane methods. Implicit membrane models generally represent the lipid bilayer and surrounding solvent as a continuum of variable polarizability. Several such methods have recently been developed, employing a variety of different algorithms as simplifications.[12–19] Methods based on the generalized Born (GB) solvation model[20] have been particularly successful in predicting the native conformation of membrane associated peptides through *ab initio* partitioning and folding simulations.[21–27] Recent applications include the prediction of oligomeric helical bundle structures in the membrane.[28,29]

Here we concentrate on the application of these methods to simulate the adsorption, interfacial folding and insertion of membrane proteins and associated peptides through implicit membrane models. In general, any protein folding simulation must meet three fundamental requirements:

1. *Atomic resolution.* This is essential for accurate protein structure prediction and sampling of conformational equilibria. Coarse-grain methods, although popular and fast, usually require the secondary structure to be fixed and by their construction cannot capture the intricate hydrogen bond networks (see Figure 5.1).[30] Low-resolution models also suffer from having identical parameters for many side chains (*e.g.* valine, leucine and isoleucine), leading to a poor description of packing properties and conformational specificity.[31]

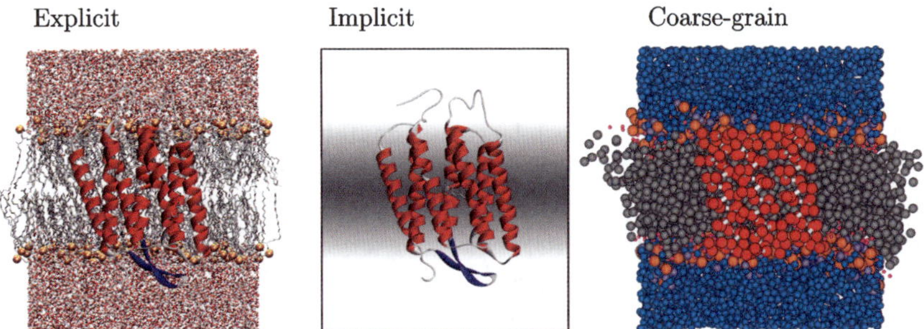

Figure 5.1   Explicit, implicit and coarse-grain membrane models. Implicit models reduce the number of interactions by restricting atomic resolution to the protein and treating the environment as a continuum. Coarse-grain approaches group several atoms into larger pseudoatoms, thus reducing the resolution and number of interactions.

2. *Adequate sampling times.* In general, micro- to millisecond time scales are required for non-equilibrium sampling to find the native state. Subsequent convergence of thermodynamic properties in the equilibrium part of the simulation require similar time scales.

3. *Accurate force fields.* Correct determination of the native state of a protein requires an accurate description of the conformational equilibria. Getting this right is essential. In general, the vast effort that has gone into parameterization of organic compounds in order to develop accurate unbiased force fields can be readily exploited if the simulation technique is compatible with the functional form of the force field.

The key property any implicit membrane model has to reproduce is the strongly hydrophobic core of the lipid bilayer (see Figure 5.2). However, continuum models introduce some general approximations and artefacts into

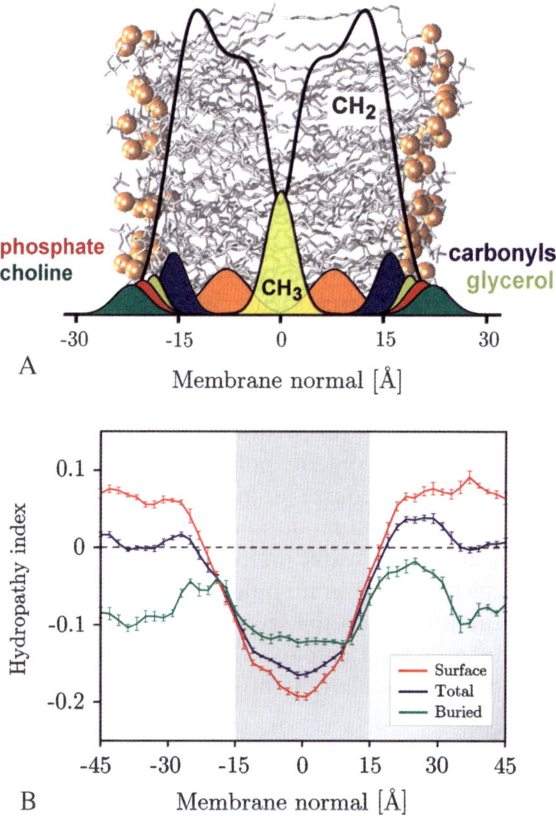

Figure 5.2　(A) Neutron/X-ray scattering profile of a lipid bilayer membrane. (B) The average single-residue hydrophobicity of membrane proteins of known structures along the membrane normal. The hydrophobic core is obvious in both figures.

the simulation system. These limitations and deficiencies have to be taken into account when analyzing and interpreting the results and might need to be compensated through specific adjustment of the model.

One of the chief deficiencies of implicit models is the absence of entropic terms due to lipid order. At present, the effect that this has on simulations is not entirely clear. Related to this is the fact that implicit membranes cannot deform. This means that one cannot use these methods to study peptide-induced bilayer deformations *via* hydrophobic mismatch, pore formation events and membrane fusion or lysis. Neither can the lipid composition of the bilayer be varied. This precludes studies of lipid mixtures or rafts. Generally, the approximations must be justified by a significant increase in computational performance, allowing studies not possible with explicit bilayers. For theoretical studies, the simplified approach can be useful, allowing the effect of specific fundamental properties of membranes to be varied individually.

5.2 Implicit Membrane Models

5.2.1 Overview

Implicit membrane models promise to reduce greatly the large computational cost of studying bilayer phenomena and membrane embedded proteins. Over the years, numerous diverse models have been proposed that span a wide range of techniques and algorithms: knowledge-based methods include energy terms derived from a library of known protein sequences and structures[32–34] (see also reference 35 for an up-to-date review). Coarse-grain or off-lattice models have been successfully employed to study the insertion process of a variety of membrane peptides.[36–38] The peptides are modelled as a linked chain of hard spheres, each representing a residue, whereas the membrane and surrounding aqueous phase are modelled by properties depending only on the membrane normal, *e.g.* fractional water content, polarity and hydrophobicity. If parameterized well, these types of coarse-grain models can yield excellent results. Other, more recent, coarse-grain models do not use special membrane energy terms, but represent lipids and water molecules using coarse-grain particles.[39] One of the problems with coarse-grain models is the reduced representation of the protein, which cannot sustain the secondary structure on its own. In such models, extensive restraints are employed to fix both secondary and tertiary structure, which allows the simulation only of systems of which the structure is known *a priori*.

More rigorous approaches combine standard all-atom force fields with implicit solvent energy terms. Modelling a membrane protein implicitly in such a fashion has clear advantages over fully explicit models for two reasons: First, the solvent and lipid degrees of freedom are removed, thus reducing the number of atoms that have to be explicitly treated by $\sim 80\%$. This results in a significantly lower cost per time step compared with explicit solvent. In addition, the mean-field nature of the solvent/membrane allows for substantially accelerated dynamics per time step so that relevant thermodynamic states of biomolecules are sampled much faster than with explicit solvent. This is especially relevant if solvent friction is omitted. With implicit models, solvent viscosity is optional and can either be adjusted to reproduce kinetic properties accurately or switched off if rapid convergence of thermodynamic properties is desired.

For example, both the aqueous solvent and the lipid phase can be modelled as a lattice of Langevin dipoles mimicking the spatial polarization of the protein environment.[40] Other studies have applied distance-dependent dielectrics and also Gaussian screening functions to model both membrane and aqueous solvation.[13] Another approach builds on atomic solvation parameter methods, which model the solvation terms as an effective surface tension multiplied by the accessible peptide surface area.[15,41] More accurate is the use of the Poisson–Boltzmann (PB) equation in determining the electrostatic free energy of solvation, but its use in molecular dynamics simulations is prohibitively slow.[42,43] A further overview of the large number of present and future applications of implicit membrane models is given in recent reviews.[44–46]

Many older implicit membrane models were designed to predict and model rigid protein segments, such as helices and bundles, where they perform acceptably. However, such models are not designed to study *ab initio* adsorption, folding and insertion of entirely unstructured peptide segments or the oligomerization of folded domains into functional membrane proteins. Typical problems are that models are either too slow, too empirical (*i.e.* have no predictive power), lack atomic resolution of the protein or allow no accurate equilibrium conformational sampling.

5.2.2   Implicit Membrane Models for Studying Membrane Protein Folding

Recently, implicit membrane models have been used for the first time to study the *ab initio* folding and assembly of membrane proteins. In addition to predicting native structures, realistic modelling of biomolecular conformational equilibria in the membrane allows for the study of both the thermodynamics and kinetics of folding. For this, models that provide all-atom accuracy for the protein in addition to very long simulation time scales are required.

Most of the recent successful folding studies[17,23] of membrane proteins have been performed with implicit membrane models derived from GB theory. The immense success of the GB method[20,47] in globular protein and peptide folding simulations (see, *e.g.*, references 48–52) has led to numerous new developments.[46] It has therefore been a promising idea to apply the GB formalism also to represent a membrane environment implicitly.[17,18,21] In its simplest form, the membrane is incorporated into the GB model as a uniform hydrophobic slab. To date, several GB implicit membrane (GBIM) models have been proposed. Although the underlying GB equation is similar, these methods differ in the way Born radii are obtained and the chosen tradeoff between simulation performance and computational accuracy.

5.2.3 The Generalized Born Model

The free energy of solvation ΔG_{sol} of a macromolecule (*e.g.* protein) is the free energy to bring the molecule from vacuum into the solvent. It consists of two parts:

$$\Delta G_{sol} = \Delta G_{pol} + \Delta G_{np} \tag{5.1}$$

where ΔG_{pol} is the electrostatic free energy change from the difference in the polarizable environment and all non-polar effects are summed into ΔG_{np}. One of the computationally fastest ways to obtain ΔG_{pol} is the GB equation. This gives the polarization free energy of a set of charges q_i as

$$\Delta G_{pol} = -c\left(1 - \frac{1}{\varepsilon}\right) \sum_i^n \sum_j^n \frac{q_i q_j}{\left[r_{ij}^2 + \alpha_i \alpha_j \exp\left(\frac{-r_{ij}^2}{4\alpha_i\alpha_j}\right)\right]^{\frac{1}{2}}} \tag{5.2}$$

where charges are in electrons, distances in Ångstroms, energy in kcal mol^{-1} and $2c = 332.06$ kcal Å mol^{-1} e^{-2} is the usual unit conversion factor. The GB algorithm has two stages, with the Born radii α_i calculated first. Using the Coulomb field approximation, which assumes that the electric displacement is Coulombic in form (*i.e.* neglecting boundary effects), the Born radius can be written as an integration over the solute interior region, excluding a radius around the origin:

$$\alpha_i^{-1} = R_i^{-1} - 4\pi \int\limits_{\substack{in \\ r > R_i}} \frac{1}{r^4} dV \tag{5.3}$$

were R_i is the atomic Lennard-Jones (LJ) radius, usually taken as $\sigma/2$ from the molecular mechanics force field. The integral in equation (5.3) can be calculated numerically by constructing a set of concentric spherical shells around atom i and calculating the fractional area of these shells lying inside or outside the van der Waals volume of the other atoms,[20,53] or by using a cubic integration lattice.[54] One of the fastest method is the asymptotic pairwise summation of Qiu *et al.*:[53]

$$\alpha_i^{-1} = \frac{1}{R_i + P_1} - \frac{1}{c}\sum_j^{bonds} \frac{P_2 V_j}{r_{ij}^4} - \frac{1}{c}\sum_j^{angles} \frac{P_3 V_j}{r_{ij}^4} - \frac{1}{c}\sum_j^{nb} \frac{P_4 V_j ccf}{r_{ij}^4} \tag{5.4}$$

where P_1–P_4 are parameters and ccf is a close contact function. Neighbouring 1–2 or 1–3 atoms are treated differently (P_2, P_3). The advantage of this method is the almost negligible cost of calculating Born radii as compared with other alternatives. Summation of the $1/r^4$ terms can be conveniently included in the normal force field Coulombic and LJ evaluations, where $1/r^2$ is already determined. Thus, the only overhead over a vacuum simulation stems from equation (5.2). Although

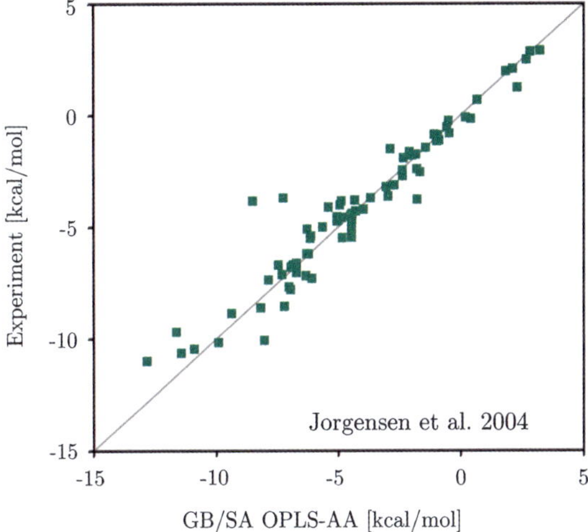

Figure 5.3 Experimental versus computed free energies of solvation for small organic compounds. The figure shows that the GB/SA method produces highly accurate solvation free energies requiring no further parameter improvements.

much more elaborate techniques exist to calculate Born radii, the asymptotic model has been demonstrated to yield excellent results in predicting experimental free energies of solvation as well as hydration effects on conformational equilibria (Figure 5.3).[55]

The success of GB methods has spurred attempts to apply the GB formalism to represent the membrane environment implicitly.[17,18,21] The first application to membranes was the method developed by Spassov *et al.*[18] The membrane is modelled as a planar dielectric slab with thickness $2L$, surrounded by a uniform polar solvent with a dielectric constant $\varepsilon_w = 80$. Both the protein interior and the slab are assumed to have the same interior dielectric constant $\varepsilon_m = 2$. Spassov *et al.*[18] modelled the membrane by restricting the pairwise summation to solute atoms outside the membrane. This was done by switching the volumes $V_i(z_i)$ of atoms i inside the membrane to zero at the interfaces $z = \pm L$. The contribution due to the membrane was modelled using an exponential switching function that changes the self-solvation term of each atom between a buried and a solvated state at the interfaces (see Figure 5.4). Thus the solvation energy for each atom is given by

$$\alpha_i^{-1} = \Gamma(z_i, R_i, L) - \frac{1}{c}\sum_j^{1-2}\frac{P_2 V_j(z_j)}{r_{ij}^4} - \frac{1}{c}\sum_j^{1-3}\frac{P_3 V_j(z_j)}{r_{ij}^4} - \frac{1}{c}\sum_j^{1>4}\frac{P_4 V_j(z_j)ccf}{r_{ij}^4} \quad (5.5)$$

Compared with equation (5.4), the self term has been replaced by the membrane $\Gamma(r, R, L)$ obtained by fitting a smooth function to the polarization

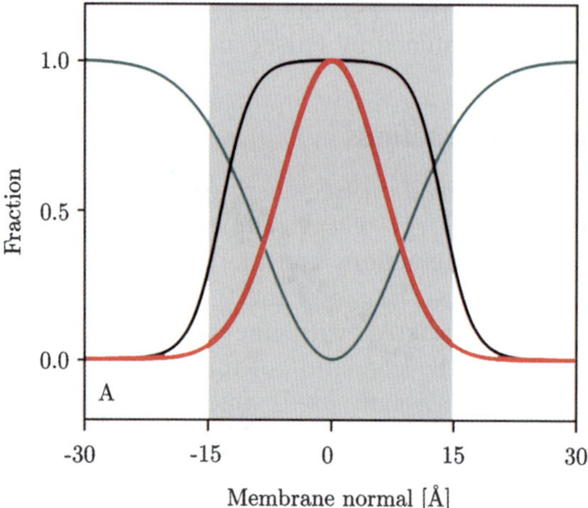

Figure 5.4   Exponential (black) and Gaussian functions with $\gamma = -1.5$ (green), indicating the attractive non-polar term and $\gamma = -3$ (red), indicting the repulsive generalized Born term.

energy of a singly charged ion, determined by a Poisson–Boltzmann solver, as it is moved through the dielectric slab.

Subsequently, Im *et al.* reported a GBIM model where Born radii are calculated using radial integration and a smoothing function.[17] This model has been used successfully to fold small membrane-bound peptides such as WALP and influenza A M2 in the membrane.[23,24] Feig and co-workers reported a model that uses more than two dielectric zones.[12,21] A disadvantage of these models is the much higher computational cost than the model by Spassov *et al.* For long-scale membrane folding simulations, a quick model based on the asymptotic approach [equation (5.5)] would be most desirable. To overcome some of the limitations of the earlier model of Spassov *et al.*, Ulmschneider *et al.* reported an improved asymptotic model.[19] A real membrane is not a hydrophobic slab with a uniform dielectric constant, but rather a heterogeneous medium with a highly non-uniform distribution of charge, density and polarizable solvent. A different approach is to treat the membrane as a region that becomes increasingly apolar (*i.e.* increasingly inaccessible to the solvent) towards the centre of the membrane. The self-solvation terms and the atomic volumes $V(z_i)$ can be modified to vary smoothly between full solvation and a limiting value for burial at the centre of the membrane. The proposed function is of a Gaussian shape:

$$\Gamma(z_i) = g_0 + (g_{\text{centre}} - g_0)e^{\gamma\left(z_i^2/L^2\right)} \tag{5.6}$$

where g_0 is the limiting value of Γ at a large distance from the membrane (*i.e.* $z \gg L$) corresponding to the self-solvation term of the unmodified generalized

Born method, $g_0 = -0.5c/(R_i + P_1)$, and $g_{\text{centre}} = \ln2/L$ is the value of Γ at the membrane centre, as determined by Parsegian.[56]

5.2.4 Non-polar Interactions

The second part of the solvation free energy ΔG_{np} stems from all non-polar effects, such as the free energy of cavity formation, the solvent and solute–solvent van der Waals' interactions and the hydrophobic effect. It is usually proportional to the solute surface area and therefore modelled using an effective surface tension associated with the solvent-accessible surface area (SASA).[53] To avoid a costly calculation of the accurate surface area, a mimic based on the Born radii is usually used, which is much faster.[57] The surface tension γ must be supplied and is obtained by fitting to experimental solvation energies of essentially non-polar compounds, such as saturated hydrocarbons, where $\Delta G_{pol} \approx 0$.

Modelling of the non-polar effects in GB membrane models is an entirely different problem. In the membrane core, the non-polar terms are not due to surface effects, but to partitioning of the peptide into a hydrophobic lipid phase. This makes ΔG_{np} the dominant energetic effect determining the stability of peptides in the membrane. A good membrane model must faithfully capture the free energy of partitioning:

$$\Delta\Delta G_{\text{sol}} = \Delta G_{\text{sol}}^{\text{membrane}} - \Delta G_{\text{sol}}^{\text{solvent}} = \Delta\Delta G_{\text{pol}} + \Delta\Delta G_{\text{np}} \tag{5.7}$$

which is the transfer free energy to move a peptide from the solvent into the membrane. For any set of charges, $\Delta\Delta G_{pol} > 0$ upon burial of a solute in the membrane core, *i.e.* energy must be expended to move charges from a polar to a non-polar medium. Hence substantial attractive hydrophobic effects must compensate if membrane proteins are to reside in the lipid bilayer. Figure 5.5 illustrates schematically the basic energetics involved: burial of a typical ~ 20 amino acid hydrophobic peptide will result in a large desolvation penalty of the polar backbone of $\sim 26\,\text{kcal mol}^{-1}$. This is captured accurately by the ΔG_{pol} term of the GB membrane model. The penalty is overcompensated by the large attractive burial of the hydrophobic side chains, estimated at $\sim -36\,\text{kcal mol}^{-1}$. The result is a net stabilization of $\sim -10\,\text{kcal mol}^{-1}$ for the inserted transmembrane (TM) peptide (see Section 5.3.2.2 below).

In the implicit membrane model, ΔG_{np} is handled by an attractive term acting on all atoms in the peptide. Gaussians have been chosen in good agreement with experimental evidence from lipid distortion,[58,59] X-ray and neutron diffraction experiments on fluid liquid crystalline bilayers[60–62] and partitioning experiments on hexane[63] in lipid bilayers. For distances far from the membrane (*i.e.* $z >> L$), the non-polar contribution is included with the standard surface tension of solvation in water. As it is moved towards the centre of the membrane, the surface energy contribution of each atom vanishes and only the Gaussian attractive term remains.

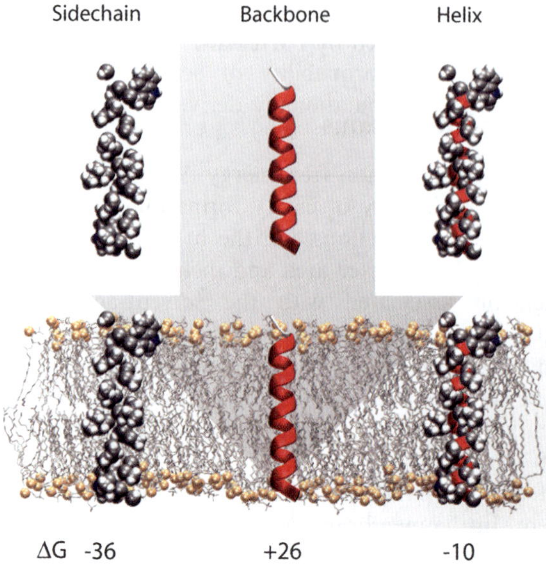

Figure 5.5 Basic energetics governing partitioning of peptides into lipid bilayers. Example of helical peptide of ~ 20 residues consisting entirely of hydrophobic amino acids.

The challenge in developing implicit membrane models is to balance the major opposing contributions to obtain accurate values for $\Delta\Delta G_{\mathrm{sol}}$. It should be noted that current membrane models neglect effects due to differences in lipid composition and charge distribution of the two bilayer leaflets, and also effects due to the transmembrane voltage. However, it is in principle possible to include these properties by replacing the Gaussians with an equivalent nonsymmetric function. There are also GB membrane models that employ more than two dielectric zones, with areas of high polarizability at the interfaces.[12,21]

5.2.5 Accuracy and Partitioning Properties

Implicit membrane models have many adjustable parameters and therefore need to be properly parameterized first. For the model in equations (5.5) and (5.6), these are the width of the membrane ($2L$), the width of the Gaussian curve (γ), the surface tension and g_{centre}. Unfortunately, there is a scarcity of experimental data available to obtain these. Common parameterization strategies have therefore been focused on the partitioning of side-chain analogues in hydrophobic solvents or even matching explicit bilayer simulations.[21]

However, better results are expected if experimental partitioning data for real membrane-bound peptides is used: Hessa *et al.*[64] challenged the endoplasmic reticulum translocon Sec61 with an extensive set of designed polypeptide segments using an *in vitro* assay to measure the efficiency of membrane integration.[65]

The peptides have the general design GGPG–X$_{19}$–GPGG, where the GGPG flanks serve to insulate the central 19-residue stretch from the surrounding sequence by having a low probability of secondary structure formation. Figure 5.6 compares the computationally derived scale with the experimental

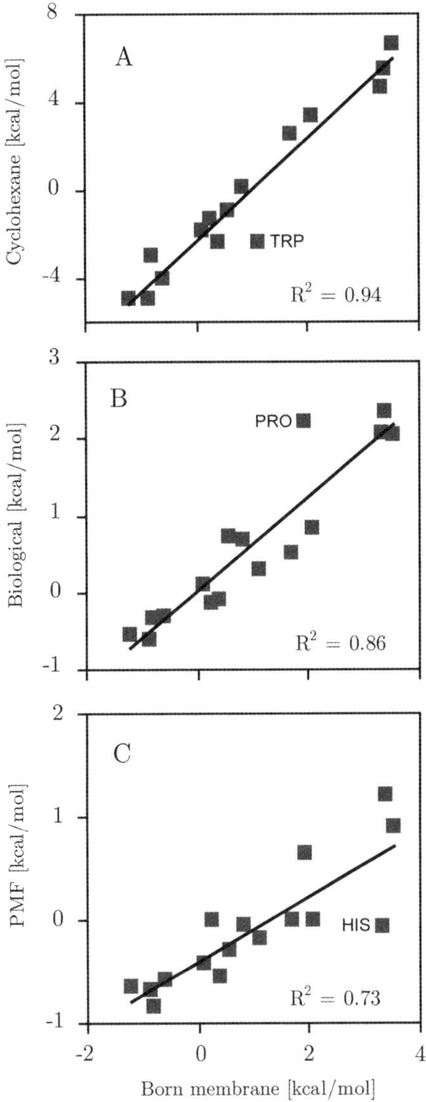

Figure 5.6 (A) Calculated insertion energy of the designed peptides against experimental transfer free energies of side-chain analogues from water into cyclohexane. (B) Insertion energy of the designed peptides: generalized Born membrane *versus* the biological hydrophobicity scale.[64] (C) Calculated insertion energy of the designed peptides against insertion potentials from a statistical membrane potential.[34]

apparent free energies of the biological hydrophobicity scale.[64] All peptides were modelled as perfect α-helices with extended GGPG flanking segments. Pre-equilibrated side-chain conformations were used and the segment termini were acetylated (C-terminus) and methylated (N-terminus). Considering the simplicity of the membrane model, both scales correlate remarkably well ($C = 93\%$). The linear fit has a slope of 0.6 and the scales have identical origins ($r_{square} = 0.86$). Comparison with a statistical scale derived from membrane protein structures[34] gave a correlation of 87%, an offset of $-0.41\,kcal\,mol^{-1}$ and a slope of 0.32 ($r_{square} = 0.73$). The calculated scale also correlates remarkably well ($C = 97\%$) with experimental transfer free energies of side-chain analogues into cyclohexane.[66]

5.2.6 Transmembrane and Surface-bound Helices, Insertion Energy Landscape

As a first test, the implicit membrane model was challenged with the task of correctly identifying the minimum energy conformation of a set of six TM helices and antimicrobial peptides. All TM helices inserted correctly with insertion energies in the range $1–13\,kcal\,mol^{-1}$, while the antimicrobial peptides were found to occupy surface-bound conformations, in agreement with experimental observations.[67–70] The results closely match those from a pre-viously reported statistical membrane potential.[71] Figure 5.7 shows the insertion energy landscapes for a given rigid conformation of the acetylcholine receptor M2 helix (AchR) (PDB: *1a11*, model 1) and magainin (PDB: *2mag*, model 1). The zero point of the potential is at infinity. AchR has four distinct minima, the two deepest corresponding to inserted configurations with the helices approximately parallel to the membrane normal. The other two minima are surface-bound configurations with the helix axis parallel to the plane of the membrane. It should be noted that due to the symmetry of the membrane model, the cytoplasmic and intracellular minima have identical insertion energies, as do the two inserted minima. For magainin, the free-energy land-scape can be seen to differ substantially in topology. There are only two interfacial minima and the membrane region forms a large barrier spanning the entire tilt range of the helix with a $50–100\,kcal\,mol^{-1}$ energy penalty for insertion into the membrane

5.2.7 Thermodynamic Analysis

For partitioning simulations, the peptide can generally be partitioned into four thermodynamic states: S = surface adsorbed unfolded, SF = adsorbed folded, M = membrane inserted unfolded and MF = inserted folded. Each sampled configuration is assigned to one of these states. Inserted configurations can be defined as having a centre of mass position close to the centre of he membrane (*e.g.* $-8\,\text{Å} < z < 8\,\text{Å}$) and conformers are classified as folded if at least 60% of their sequence is helical.

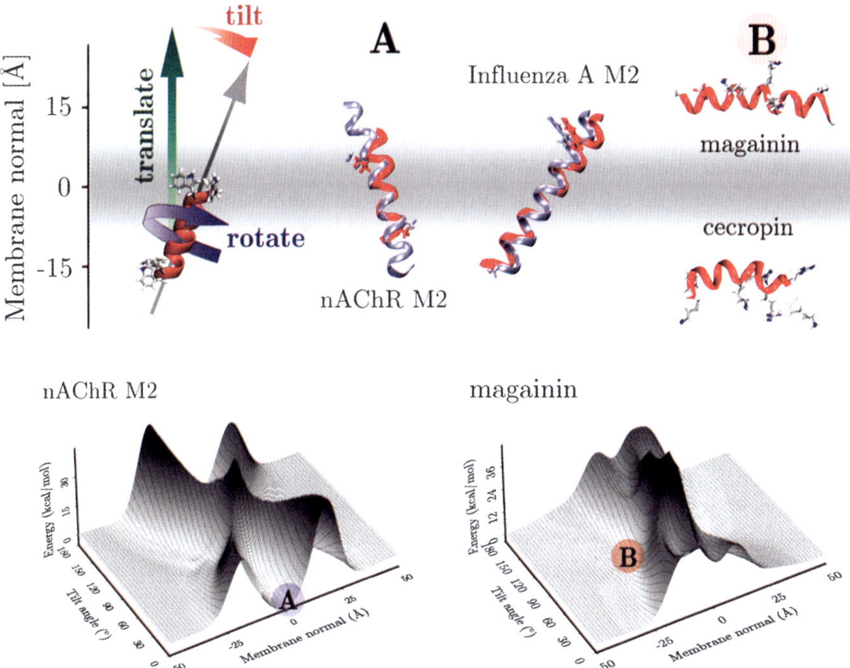

Figure 5.7　Insertion energy surface for a given conformation. If the structure is known from NMR, the location in the membrane can be quickly estimated by performing a rigid body scan in the implicit membrane. This feature allows differentiation between natively inserted peptide (*e.g.* nAChR M2) and peptides that do not insert monomerically (*e.g.* magainin).

The populations p_i of the four states and the free energy differences:

$$\Delta G_{ij} = -kT \ln\left(\frac{p_i}{p_j}\right) \tag{5.8}$$

are calculated for each replica temperature over the equilibrium phase of the simulation. Linear fits of the ΔG_{ij} with respect to T are then used to determine the energetic component ΔE_{ij} and the solute entropic component ΔS_{ij} for each transition ij:

$$\Delta G_{ij} = \Delta E_{ij} - T\Delta S_{ij} \tag{5.9}$$

Note that ΔE_{ij} is not the internal solute energy, but contains the generalized Born solvation free energy, including entropic terms due to the solvent ordering near the solute surface. ΔE_{ij} is also determined by subtracting the bin averages of E for each temperature replica. Finally, the populations of the individual states at each temperature as predicted by the four-state fit are obtained using

$$p_i(T) = \frac{1}{1 + \sum_j \exp[-\beta \Delta G_{ij}(T)]} \tag{5.10}$$

5.3   Simulating Peptide Folding and Partitioning

5.3.1   Summary

The aim of developing implicit membrane algorithms is to study long-term scale events such as protein folding. Since the computational effort and the sampling required both grow dramatically with the size of the protein, folding studies have so far concentrated on small, autonomously folding peptides. These often form part of larger functional assemblies. One of the most widely studied systems is viral channel-forming peptides. We concentrate here on the transmembrane segment of virus protein U from human immunodeficiency virus type 1 (HIV-1). Many other such systems have been studied.

5.3.2   Transbilayer Peptide Folding

If we are mainly interested in predicting the native state and the corresponding equilibrium dynamics, the peptide can be folded from a physically unrealistic starting configuration. The non-equilibrium part of the simulation, where the peptide moves into physically realistic configurations, must then be neglected in the analysis. This method is particularly useful for peptide segments containing charges. These systems tend to be kinetically trapped at the membrane inter-faces, because the large energetic penalty associated with burial of charged residues in the hydrophobic membrane core acts as a barrier to correct insertion or translocation.

5.3.2.1   *Virus Protein U, a Transmembrane Helix*

Virus protein U (Vpu) is an 81-residue membrane protein of HIV-1.[72,73] It consists of one N-terminal hydrophobic membrane helix and two shorter amphipathic helices that remain in the plane of the membrane on the cyto-plasmic side.[74] Two main functions of Vpu are observed. The first, which involves the cytoplasmic domain in the C-terminal half of the protein, is to accelerate the degradation of the CD4 receptor in the endoplasmic reticulum of infected cells.[75,76] Second, Vpu has been shown to amplify the release of virus particles from infected cells, a process that involves the transmembrane domain.[77,78] Vpu and its isolated transmembrane part oligomerize in lipid membranes[79] and show channel activity.[80–84] Here, we focus on the trans-membrane α-helix of Vpu. Its structure has been determined experimentally,[85] and its orientation relative to the plane of the lipid bilayer has been estimated using both NMR spectroscopy[85–88] and Fourier transform infrared (FTIR) spectroscopy.[89]

Several previous simulation studies have been performed on Vpu in explicit bilayers, with either the complete peptide,[90] part of the peptide,[91,92] or only the N-terminal transmembrane helix as monomer or as oligomer.[93–99] However, the short time scale ($\sim$ 1–5 ns) of these simulations was not sufficient to study folding. Longer simulations (200 ns) of Vpu have been performed using a coarse-grain method.[100] However, these models cannot be used for accurate *ab initio* folding (see above). Here, an implicit membrane model is used to predict the structure of the TM helix of Vpu, and also to study its oligomeric structure in the membrane.

5.3.2.2　Insertion Energy Landscape

One of the most useful features of implicit membrane models is that they provide the solvation free energy of any given conformation without sampling. This permits one to obtain a rapid estimate of the insertion energy landscape and the local minimum energy orientations in the implicit membrane potential as a function of position along the membrane normal, tilt and rotation angle of the helix. The rotation angle is generally optimized (*i.e.* the rotation angle for each position and tilt angle is such that the energy is minimal), thus providing a two-dimensional insertion energy landscape. Figure 5.8 shows the resulting insertion energy landscape for a completely helical structure of Vpu. The zero point of the potential was chosen at an infinite distance from the membrane. Vpu has four distinct minima, the two deepest corresponding to inserted configurations with the helices approximately parallel to the membrane normal. The other two minima are surface-bound configurations with the helix axis parallel to the plane of the membrane. It should be noted that due to the symmetry of the membrane model, the cytoplasmic and intracellular minima have identical insertion energies, as do the two inserted minima. The X-shaped pattern is typical of a transmembrane peptide.

Generally, the inserted transmembrane configuration corresponds to the global energy minimum. For Vpu, the insertion energy is –5.5 kcal mol^{-1}, with a tilt angle of 5.6° and a position close to the centre of the membrane, $z_{\mathrm{CM}} = 3.7$ Å. Adsorption of the peptide on the membrane surface is also favourable but to a significantly lesser extent, with an energy minimum of –0.8 kcal mol^{-1} at 20 Å and a parallel orientation with a tilt angle of 80°.

The implicit membrane allows the relative contributions played by the polar and non-polar energy terms to the overall insertion potential of the peptide to be distinguished. Figure 5.8 shows the polar (ΔG_{pol}) and non-polar (ΔG_{np}) contributions to the overall insertion energy landscape. Burial of charged and polar residues in the membrane interior is highly unfavourable and the characteristic X shape of ΔG_{pol} is caused by the position of such residues at the helix termini.[34,101] The hydrophobic effects are the main contributors to helix insertion and, as expected, make the dominant contribution for a completely buried helix parallel

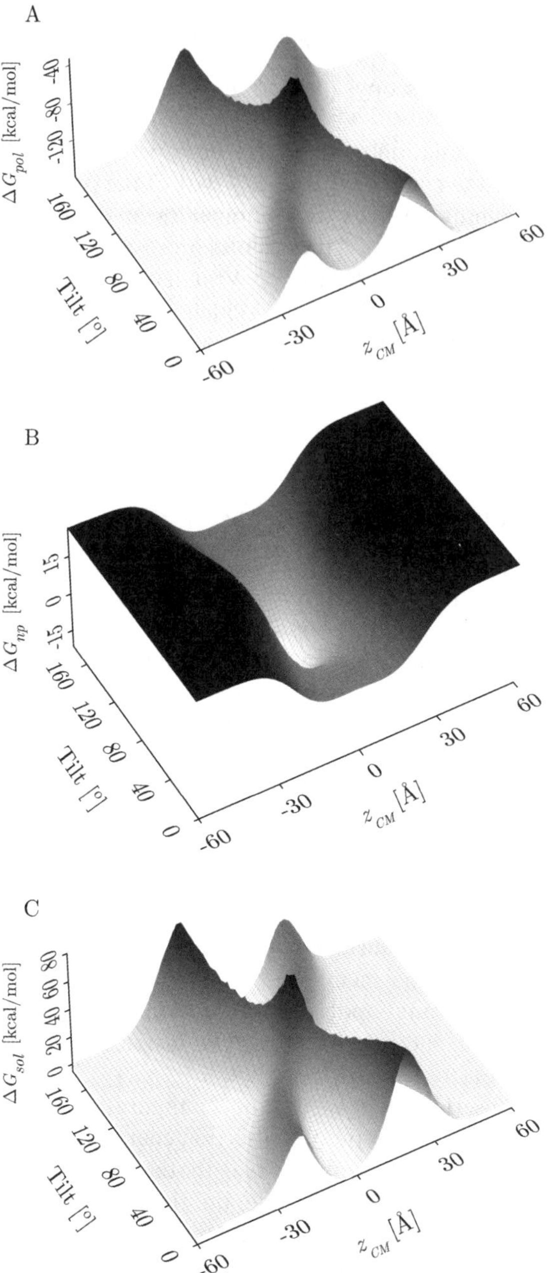

Figure 5.8 Insertion energy landscape of the Vpu helix with charged termini as a function of the helix tilt angle and centre of mass position along the membrane normal for optimized rotation angle (around the long axis of the helix). (A) The contribution of the polarization energy (generalized Born term); (B) the non-polar energetic component (hydrophobic surface area term). (C) The total solvation energy, shifted such that it is zero at an infinite distance from the membrane. For a typical transmembrane peptide, the surface has a characteristic X-shaped pattern with four distinct minima, two for surface-bound configurations with the peptide parallel to the membrane and two membrane-spanning inserted minima, comprising the native state.

and in the centre of the membrane (panel B). It is generally recognized that overall hydrophobicity is the main driving force for the integration of transmembrane helices into the lipid bilayer.[102] Indeed, most residues in transmembrane helices are hydrophobic.[103] Nevertheless, polar, charged and aromatic residues are known to be important for anchoring the helix termini into the lipid headgroup environment at the membrane interfaces.[104–106] The overall potential favours transmembrane orientations since hydrophobic residues strongly prefer an inserted to a surface-bound configuration. For the burial of a typical transmembrane peptide of about 20 residues in the membrane, White *et al.* roughly estimated a hydrophobic contribution of $\sim 40\,\mathrm{kcal\,mol^{-1}}$, offsetting a unfavourable dehydration of the α-helical peptide backbone of about $\sim 30\,\mathrm{kcal\,mol^{-1}}$. This results in a net favourable free energy of about $\sim 10\,\mathrm{kcal\,mol^{-1}}$.[107] The orientational scan of Vpu allows an estimate for the hydrophobic contribution of $-22.5\,\mathrm{kcal\,mol^{-1}}$ and a polarization penalty of $+17\,\mathrm{kcal\,mol^{-1}}$, resulting in the $-5.5\,\mathrm{kcal\,mol^{-1}}$ insertion energy of the helix, roughly the same order of magnitude.

5.3.2.3 Role of Terminal Charges

When calculating the properties of membrane proteins, it is important to take into account the charge state of the amino acid side chains. Implicit membrane models are very sensitive to changes in charge, since burial (*i.e.* desolvation) of charged groups is highly unfavourable. For Vpu, both Glu28 and Arg30 are modelled in their charged state (pH = 7). In addition, the chain ends can be modelled as either charged (NH_+^3/COO^-) or capped with methyl groups, neutralizing the termini. The lack or presence of this additional dipole can have a strong effect on the outcome of simulations. In the capped case, there is no strongly charged group on the N-terminal part of the peptide. Figure 5.9 shows

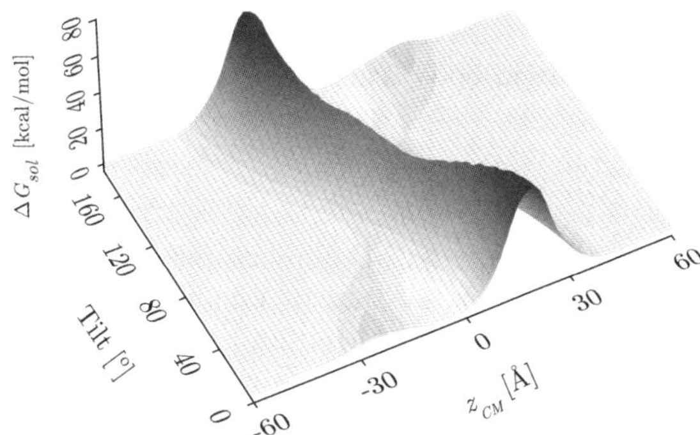

Figure 5.9   Insertion energy profile of the Vpu helix with capped uncharged termini as a function of the helix tilt and centre of mass position along the membrane normal for optimized rotation angle (around the long axis of the helix). The lack of a charged residue on the N-terminal side of the helix results in a weak barrier ($\sim 6\,\mathrm{kcal\,mol^{-1}}$) on one side of the transmembrane-inserted minimum. This barrier is much larger in the uncapped helix (Figure 5.8).

ΔG_{pol} recalculated for the capped system, revealing a markedly different shape than the uncapped system given in Figure 5.8C. The lack of a charged residue on the N-terminal side of the helix results in a considerable lowering of the barrier ($\sim 6\,\text{kcal}\,\text{mol}^{-1}$) on the N-terminal side of the transmembrane inserted minimum. The characteristic X shape is almost lost. While the lowest energy configuration for the peptide remains inserted, the system can ultimately overcome the barrier and exit the membrane.

5.3.2.4 *Insertion Energy Profile from Simulations*

If fully flexible simulations are run sufficiently long (typically $>1\times10^9$ MC steps or $>100\,\text{ns}$ MD), they can be used to calculate converged insertion free energy landscapes from a direct population analysis. For both the capped and uncapped Vpu simulations with a membrane width of $30\,\text{Å}$, a two-dimensional population histogram was calculated as a function of the centre of mass position along the membrane normal and the tilt angle. The negative logarithm of the histogram bins gives the overall solvation free energy profile of the system and is plotted in Figure 5.10. The free energies are relative to the lowest bin, which has been set to zero. A close similarity to the profiles shown in Figures 5.8 and 5.9 is evident and expected. Note that the profiles in Figure 5.10 can only extend over the conformational space that was physically sampled, while the rigid-body scan results above can plot the entire landscape – albeit for a fixed conformation. Figure 5.10A reveals that the uncapped peptide has thoroughly explored the inserted minimum and remains strongly contained by large barriers, as already visible in Figure 5.8. The results for the capped peptide shown in Figure 5.10B are very different. After spending a considerable time sampling the inserted states, the peptide overcomes the weak interfacial barrier (*cf.* Figure 5.9) and exits to the surface of the membrane. The small barrier height – caused by the lack of charged groups on the N-terminal side of the Vpu peptide – is only $\sim 2\,\text{kcal}\,\text{mol}^{-1}$, even smaller than the estimate of $\sim 6\,\text{kcal}\,\text{mol}^{-1}$ from the rigid body scan.

5.3.2.5 *Folding Simulations*

The next step is to demonstrate that the implicit membrane model can predict the experimentally determined native state of Vpu in an *ab initio* protein folding simulation. For this, replica exchange Monte Carlo (REMC) simulations were run with 10 replicas for 1×10^9 MC steps each, starting from completely extended conformations perpendicular to the membrane plane. The simulations were performed both for capped Vpu and for the uncapped system. Figure 5.11 shows the folding progress of the transmembrane system over the course of the simulations. Only the $318\,\text{K}$ replica, the temperature closest to the NMR experiments, is shown. Both capped and uncapped Vpu fold into stable membrane spanning helices within the first $\sim 400\times10^6$ MC steps. Replicas with the higher temperatures contain a large amount of helical secondary structure,

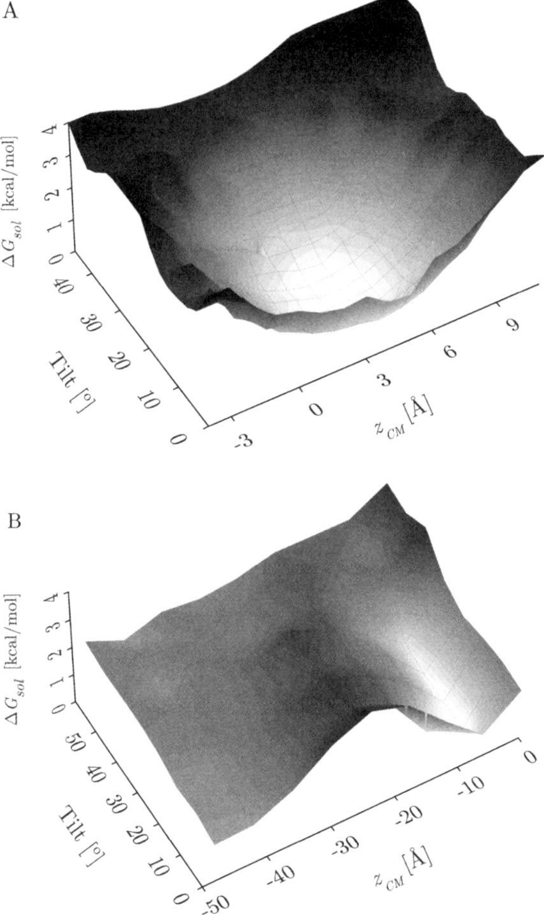

Figure 5.10 Free energy profile of Vpu as calculated from a population analysis for the fully flexible simulations of the inserted conformation. ΔG is plotted as a function of the helix tilt and centre of mass position (z_{CM}) along the membrane normal and is the free energy relative to the lowest bin that has been set to zero. The uncapped system is shown (A), revealing a stable membrane spanning inserted minimum at $z = 4.5$ Å, tilt $= 16.6°$, as found in the rigid-body scan (Figure 5.8). The capped system plotted in (B) shows the same minimum. Due to the strongly lowered barrier in the absence of N-terminal charges, the peptide exits the membrane-inserted state after 500×10^6 MC steps (*cf.* Figure 5.9).

but do not form stable helices. Once formed, the helix shows strong tilting and kinking. To quantify the similarity to the native state, defined as the completely helical structure found in the NMR measurements,[85] one can calculate the overall system helicity as it increases over the course of the simulation and the results are shown in Figure 5.12. After a steady build-up of helical content, a

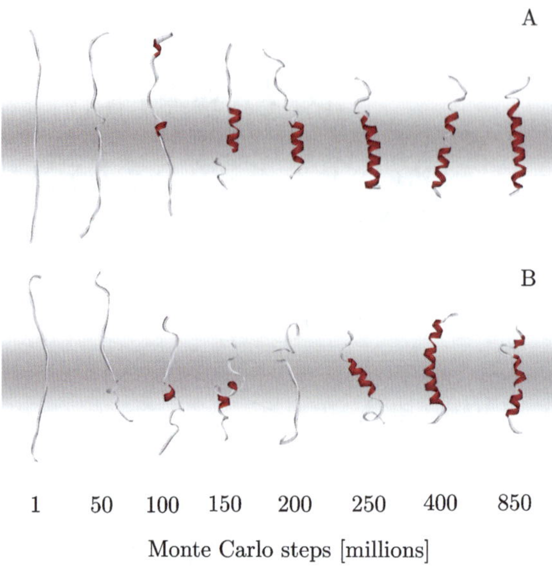

Figure 5.11 Transmembrane folding of Vpu for the 318 K replica of the REMC simulations, showing the capped system (A) and the uncapped system (B). Vpu folds into a stable membrane spanning helices within the first $\sim 400\times 10^6$ MC steps. Higher temperature replicas retain largely extended or coiled conformations.

plateau is reached after $\sim 400\times 10^6$ MC steps. The chain ends are found to be flexible and mostly do not sample helical conformations.

The folding results can be directly compared with the single fully flexible simulations of Vpu starting from the native helical membrane spanning conformation (see Figure 5.12). For both the capped and uncapped peptides, strong tilting and kinking are observed, but the completely helical conformation remains intact, with $79\pm 7\%$ helicity for the uncapped peptide and $67\pm 2\%$ for the capped system. The lower helicity of the capped peptide is due to the more flexible chain termini. While the REMC folding run for the capped peptide reaches the same plateau in helicity as the reference native run, the helicity observed for the REMC folding simulation with the uncapped peptide is lower. This indicates a sampling problem of the more highly polar system, where partly helical structures present in the various replicas persist much longer due to charge–charge interactions than in the case of the capped system, were the complete helix quickly dominates. Such partly helical structures are swapped into the 318 K replica and thus contribute to the overall helicity. In contrast, in the native state simulation of the uncapped peptide, partly helical conformations are not sampled at all. Overall, the results prove that the native state of Vpu can be accurately predicted after a relatively short simulation from a completely random extended conformation. The choice of the chain termini seems to not influence these results. This matches experimental observations that found the effect of terminal caps on the helicity of a designed membrane-inserted peptide to be marginal.[108]

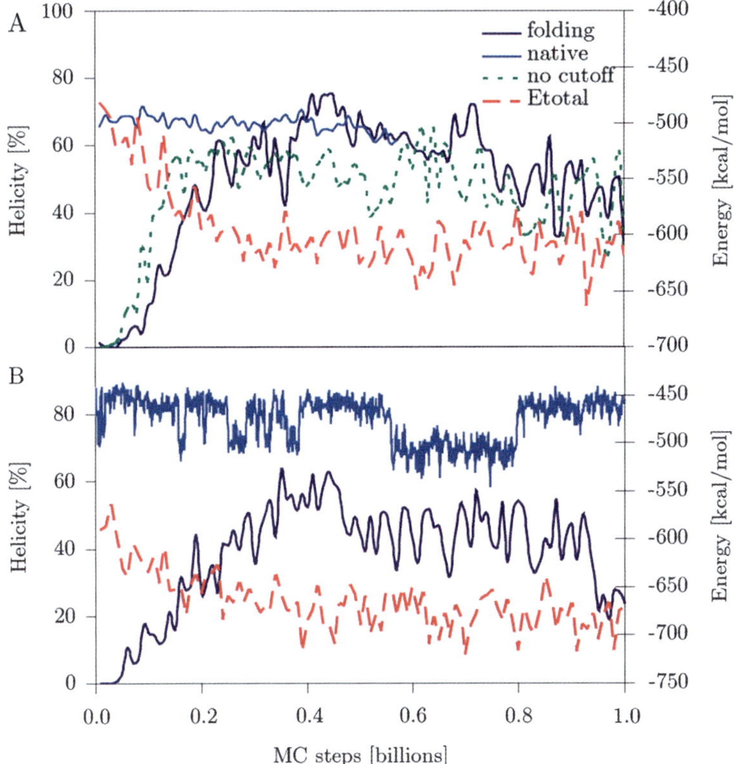

Figure 5.12   This graph shows the build-up of the helicity of Vpu for the 318 K replica during the folding runs of 1×10^9 MC steps (left axis, solid lines) and the change of the total system energy (internal energy plus solvation free energy; right axis, dotted line). The helicity of the single simulations starting from the native helical TM-inserted state is also shown in the same plot for comparison. (A) Capped system. The two folding trajectories are with and without cutoffs, respectively. (B) Uncapped peptide. In both cases, the complete TM helix forms in $\sim 400 \times 10^6$ MC steps. Due to frequent replica swaps to only partly helical structures that contribute to the average, the helicity is lower in the REMC folding runs und fluctuates much more, indicating that longer sampling is still required (*i.e.* not all replicas have folded). In the native MC runs, the helix remains completely stable, but the capped peptide exits the membrane after $\sim 500 \times 10^6$ MC steps.

5.3.3   Peptide Adsorption, Insertion and Folding

Folding and integration of peptides into lipid bilayer membranes remains one of the most intriguing processes in biophysics, as it cannot be directly observed at sufficient temporal and spatial resolutions. Recent experiments applying the translocon machinery to insert designed peptides[64,109] and statistical analyses of membrane protein structures[34] revealed that the distributions of individual amino acid types correlate strongly with their expected solvation energies along the membrane normal.[66] In addition, the Sec translocon structure revealed that nascent peptides are threaded into a narrow water-filled channel, which opens laterally, allowing hydrophobic segments to partition into the bilayer.[110] Many details of this process, including how much folding actually occurs inside the channel, are currently unclear, but direct peptide–bilayer interactions (*i.e.* the solvation free energy) seem to be the key determining the partitioning and folding properties of a particular sequence. From a physical chemistry perspective, transfer of solvated peptides into a hydrocarbon phase should follow a two-stage pathway,[111] where helical segments fold at the phase boundary before inserting. The high cost (estimated at $\sim 4\,\mathrm{kcal\,mol^{-1}}$) of desolvating exposed peptide bonds prevents unfolded insertion.[112,113]

Due to the vast increase in sampling, implicit membrane models can be applied to study directly interfacial absorption, folding and spontaneous partitioning of membrane-associated peptides. Studies to date have concentrated on suitably hydrophobic sequences that have neutral termini. This allows for rapid convergence of partitioning equilibria by avoiding the kinetically trapped states at the membrane interface. Implicit membranes almost certainly overestimate the barrier of insertion of charged residues. In contrast to explicit models, charged residues cannot lower their insertion penalty by taking a water shell with them and no satisfactory methods exists at present that will allow for spontaneous neutralization during the course of a simulation. For neutral peptides, however, these problems do not exist, allowing direct simulation of the folding and partitioning pathway, as the peptide can be started from a realistic configuration in water without becoming kinetically trapped. We concentrate here on the synthetic WALP peptide, which serves as a template for helical membrane-spanning peptides.

5.3.3.1 WALP

WALP is part of a family of artificial membrane-inserting α-helices that were designed by Killian and colleagues to explore how lipid bilayers adapt to accommodate peptides of different hydrophobic lengths.[58,114] The peptides consist of a tryptophan-flanked hydrophobic core, which can be varied in length. Its small size, simplicity and absence of charged groups allow for direct folding and partitioning simulations in implicit and explicit membranes, using replica exchange,[10] or conventional molecular mechanics at elevated temperatures.[11] The explicit systems are extremely costly, requiring multi- to microsecond simulations in order to capture the complete folding process in the membrane.[11] The WALP16 peptide used for the studies below has the sequence Ace–AWW–(LA)$_5$–WWA–Ame, where the C- and N-termini are acetylated (Ace) and amidated (Ame), respectively. Apart from single temperature simulations, two REMC simulations with 20 replicas were performed for 2×10^9 MC steps, giving an accumulated simulation time of 16×10^{10} MC steps, corresponding to $\sim 20\,\mu s$ of molecular dynamics.[115]

5.3.3.2   *The Partitioning Pathway*

Figure 5.13 shows the folding pathways for the transmembrane (TM) and solvent (SB) starting configurations of the WALP16 peptide. In the SB simulation, the peptide is initially located in bulk solvent 35 Å from the membrane centre and oriented parallel to the membrane plane in a fully extended configuration. In the TM system, the peptide initially spans the membrane perpendicularly in a fully extended conformation. Both simulations exhibit rapid folding in the implicit membrane. Three distinct folding stages are observed (coil, semi-helix, full helix), as illustrated in Figure 5.13. Initially the peptide adsorbs rapidly at the membrane interface, either from the bulk solvent (SB) or through expulsion from the membrane (TM), while retaining an extended conformation. Once at the surface, the peptide samples a large variety of conformations. In this 'coil phase', the relative root mean square deviation (RMSD) at time t with respect to a previous time $t - \Delta t$ is large, as can be seen in Figure 5.14, and no helical conformation is yet present. The 'semi-helical' phase is entered after $\sim 4 \times 10^6$ scans in the case of the TM run and $\sim 31 \times 10^6$

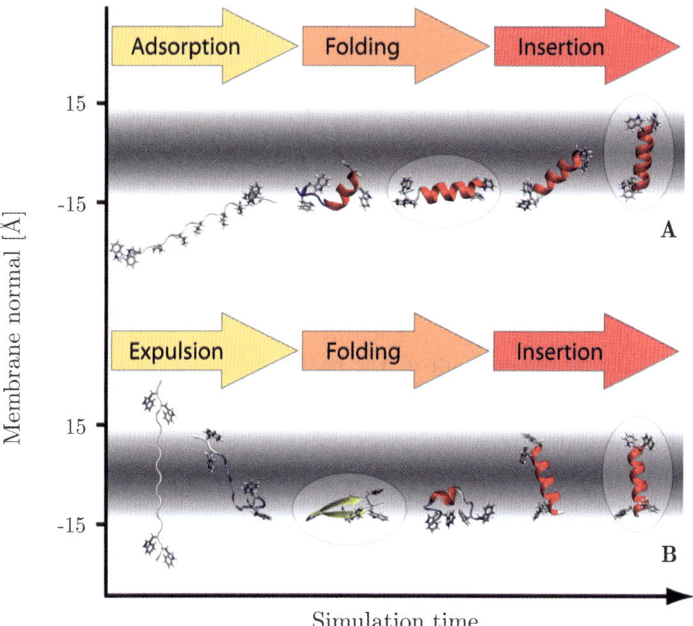

Figure 5.13   Folding of the WALP16 peptide at room temperature. (A) Starting from an extended conformation in solution (35 Å from the centre of the membrane), the peptide rapidly adsorbs on the membrane surface, retaining its extended configuration. Interfacial folding is followed by membrane insertion. (B) Starting from an extended transmembrane configuration, the peptide first exits the membrane to fold at the interface; it subsequently reinserts to form the native state. The three major states of the system are highlighted.

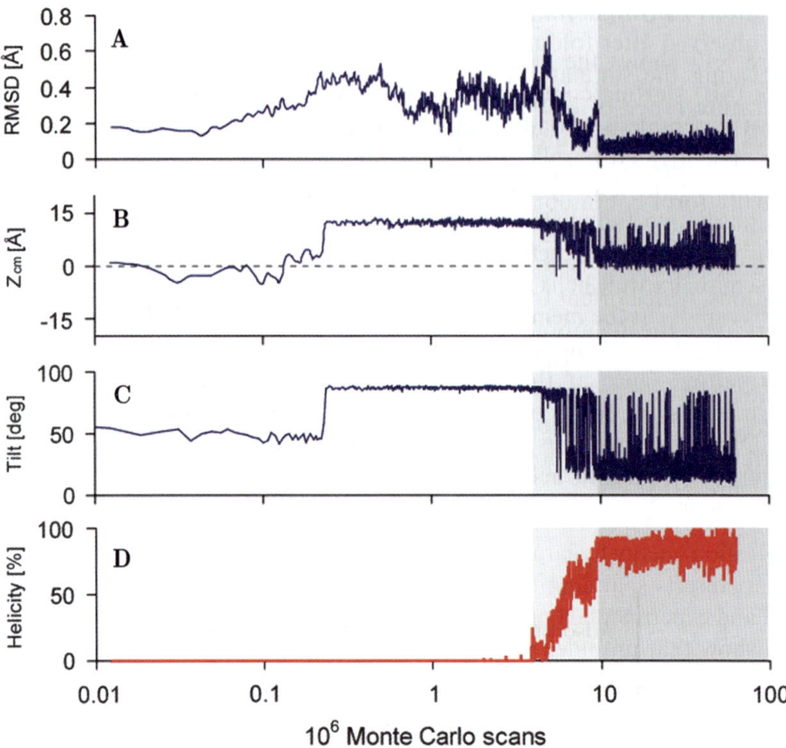

Figure 5.14   Adsorption, folding and insertion of WALP16 in the implicit membrane model starting from an extended membrane-spanning conformation (TM). The corresponding simulation starting from an extended conformation in bulk solvent is identical. (A) Relative backbone RMSD of frame t with frame $t - \Delta t$ ($\Delta t = 10^7$ MC steps). The three major stages of the folding process are indicated. (B) Position of the centre of mass along the membrane normal. The peptide quickly exits the membrane to the surface and only reinserts after folding. (C) Peptide tilt angle. (D) Peptide helicity as a function of simulation time. Note the logarithmic scale of simulation time.

scans for the SB run. This phase is characterized by a substantial build-up of helical structure. This occurs chiefly at the centre of the peptide, while chain ends remain unfolded. Finally, the 'fully folded' helix is reached in the TM run after 9×10^6 scans and in the SB run after 63×10^6 scans. Once fully folded, WALP16 remains stable throughout the rest of the simulation.

In both simulations, folding does not occur in the membrane. Figure 5.14B shows the centre position of WALP16 along the membrane normal. In both the SB and TM runs, the unfolded peptide remains on the surface at $z = 12 \, \text{Å}$. Spontaneous insertion of the helix is observed for fully formed helices only. The 'folded' phase is characterized by a two-stage behaviour: there is a TM-bound helix and a surface-bound helix. Figure 5.14C shows the helix tilt and the correlation with Figure 5.14B is apparent, with only the values of $\sim 20°$ and $\sim 85°$ observed after folding. The TM state is the dominant conformation.

Repeating the simulations using a replica exchange method over a wide temperature range of 300–500 K gives very similar overall results but allows analysis of the temperature dependence of the states. All replicas rapidly adsorb on the membrane surface within 10^6 MC steps. Low-temperature replicas eventually form continuous helices with the long axis parallel the membrane surface, whereas high-temperature replicas remain unfolded or trapped in β-sheet conformations. Spontaneous insertion is observed only for helical conformations, which orient to span the membrane. The helices frequently exit and reinsert into the membrane in a temperature-dependent manner, with the lowest temperature replica having the largest fraction of inserted states. Insertion and exit always occur to the same interface, indicating a translocation barrier. Insertion occurs only from the N-terminal side.

Current peptide partitioning theory strongly suggests a three-stage process (see Figure 5.15),[111,116–118] where the peptide first adsorbs at the membrane interface in an unfolded state. Interfacial helix formation subsequently lowers the solvation energy by largely burying the polar peptide backbone. Once folded, the helix inserts spanning the hydrocarbon core. This model is based on theoretical and experimental considerations derived from bulk hydrophobic solvents, which show that unfolded insertion would entail enormous energy penalties of the order of 6–8 kcal mol^{-1} per residue, associated with desolvating the peptide backbone.[119–121] The helical conformation largely buries the backbone, thus removing the insertion penalty. The implicit membrane partitioning simulations follow this model closely. A similar pathway was also found in a recent study of WALP16. Using a different generalized Born-based implicit membrane algorithm[17] in combination with the param22 all-atom force field,[122,123] Im and Brooks were able to simulate spontaneous insertion and folding of WALP16 by

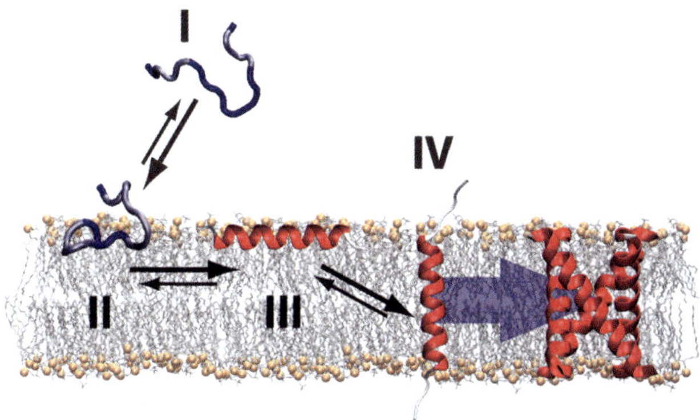

Figure 5.15 Theoretical model of peptide partitioning. The unfolded peptide in solution (I) adsorbs on the membrane surface (II), folds into a helix (III) and subsequently inserts (IV). Final oligomerization allows the formation of quaternary structure.

simulating 32 replicas in temperature range 300–800 K for 14 ns.[24] They found that membrane insertion required the peptide to have at least 80% helicity. Similarly a recent long time scale (600 ns) coarse-grain simulation of helically restrained WALP16 conformers in a 256 DPPC lipid bilayer found the system to oscillate between a transmembrane (population = 41%) and surface-bound state (59%), with average lifetimes of $\sim$ 10 ns each.[100]

5.3.3.3 Membrane Location and Conformation

Figure 5.16 shows the states sampled by the WALP replica exchange simulation (REMC) at 300 K. The figure shows a very strong correlation between tilt angle and insertion depth. Two distinct clusters are observed corresponding to a surface-bound configuration, with the helix parallel to the surface of the membrane and a membrane spanning inserted state. The simulations starting from fully solvated peptides (SB) sample one side of the membrane only (the insertion side), while the simulations starting from a transmembrane conformation sample states on both interfaces since different replicas exit the membrane to different interfaces.

This behaviour corresponds remarkably well to the insertion energy landscape for a helical conformer (see Figure 5.17), since at 300 K the peptide is helical for nearly 80% of the simulation time (*cf.* Figure 5.19). The insertion energy landscape gives barrier estimates, which are difficult or impossible to extract from the folding simulations as they correspond to regions of very low sampling. The surface has four distinctive minima, two corresponding to transmembrane configurations (I and IV) and two representing surface adsorbed helices (II and III). These correspond exactly to the four states sampled in the folding simulations (*cf.* Figure 5.17).

The two inserted configurations are not centred on the membrane but one terminus is always at an interface. hence the helix does not fully span the membrane but is still oriented along the membrane normal. Each inserted configuration is separated from its associated surface-bound state by a $\sim$ 5 kcal mol^{-1} barrier (*e.g.* I $\leftrightarrow$ II), which is much lower than the $\sim$ 17 kcal mol^{-1} barrier to its opposite surface-bound configurations (*i.e.* I $\leftrightarrow$ III). The helix therefore inserts and exits the membrane remaining in one bilayer leaflet only. The profile of the barrier on the tilt/z surface is given in Figure 5.17B. The height of the barrier is also evident from the very low number of translocation events, which restrict sampling to one leaflet (*cf.* Figure 5.16A).

5.3.3.4 Spontaneous Translocation

Two spontaneous translocation events were observed in the solvent starting simulations of the WALP peptide. Translocation occurred at high temperatures (487–500 K), taking less than 10^6 MC steps (see Figure 5.18). In each case, the intermediate membrane-inserted conformers contained significant β-structure and no helical content. Since both translocations occurred very late in the simulation

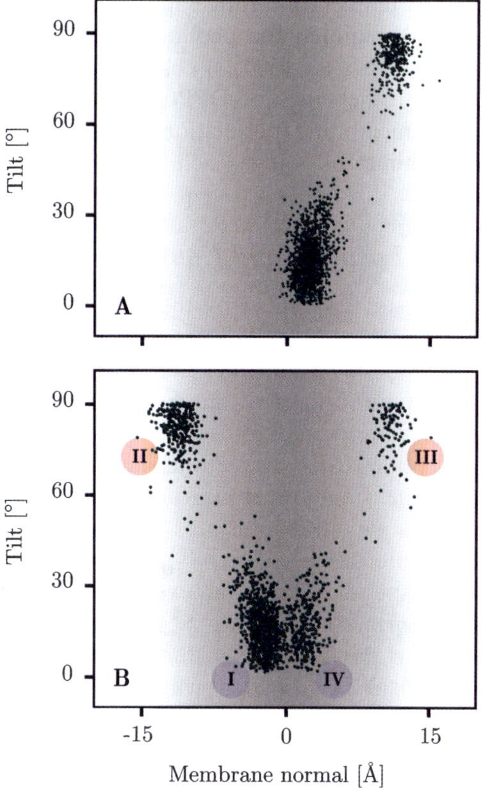

Figure 5.16 Scatter plot showing the distribution of tilt angles and centre of mass position of the WALP 16 peptide along the membrane normal for the 300 K replicas. (A) The surface-bound starting configuration only translocates very late in the simulation and hence only explores one interface. (B) The transmembrane starting configuration has an almost equal probability of exiting the membrane on each leaflet exploring both interfaces. Inserted and surface-bound configurations show clear tilt preferences matching the energy landscape of a rigid body scan (see Figure 5.17A). The roman numerals correspond to locations/orientations with respect to the membrane, depicted in Figure 5.4 (I and IV = M and MF, II and III = S and SF).

(at $\sim 1.9 \times 10^9$ steps), the remaining simulation time was insufficient to allow the cooling required for sampling the low-temperature regions of the second interface (*cf.* Figure 5.16A). However, the events demonstrate that in principle all states of the system are accessible starting from a bulk solvated peptide.

5.3.3.5 Thermodynamic Behaviour

The replica exchange method significantly decreases the simulation time required for the system to reach equilibrium. In addition, it provides thermodynamic

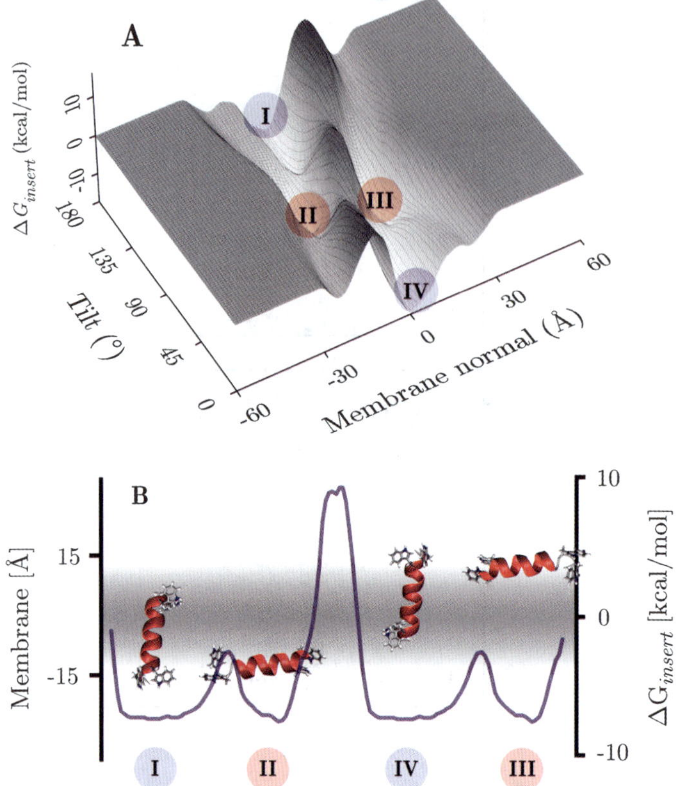

Figure 5.17   (A) Insertion energy landscape of a helical WALP16 conformer. The distinctive four minima are indicated. There are two transmembrane configurations (I and IV), which are separated by a barrier, which is much higher than the barrier between the transmembrane and its associated surface-bound configurations (*i.e.* I↔II and III↔IV). The insertion energy minima match those of the 300 K replicas (*cf.* Figure 5.16). (B) Orientations of the four minima in the membrane; the minimum energy pathway on the tilt/z surface connecting the four states shows the high translocation barrier, separating the states in each membrane leaflet.

information over a wide temperature range. Analysis of the normal and replica exchange trajectories showed that at equilibrium the system could be subdivided into four distinct states regardless of simulation methodology or starting configuration: (1) a surface-adsorbed unfolded conformation (S), (2) a surface-adsorbed α-helix (SF), (3) an unfolded membrane-inserted configuration (M) and (4) a folded transmembrane helix (MF) (see Figure 5.20 for a graphical visualization of each state). Since membrane adsorption is rapid and completed for all replicas after 10^6 MC steps, no state that is significantly distanced from the membrane is observed in the equilibrium phase. Included in the S state are many β-structures. Figure 5.19A shows the average helicity and β-sheet content

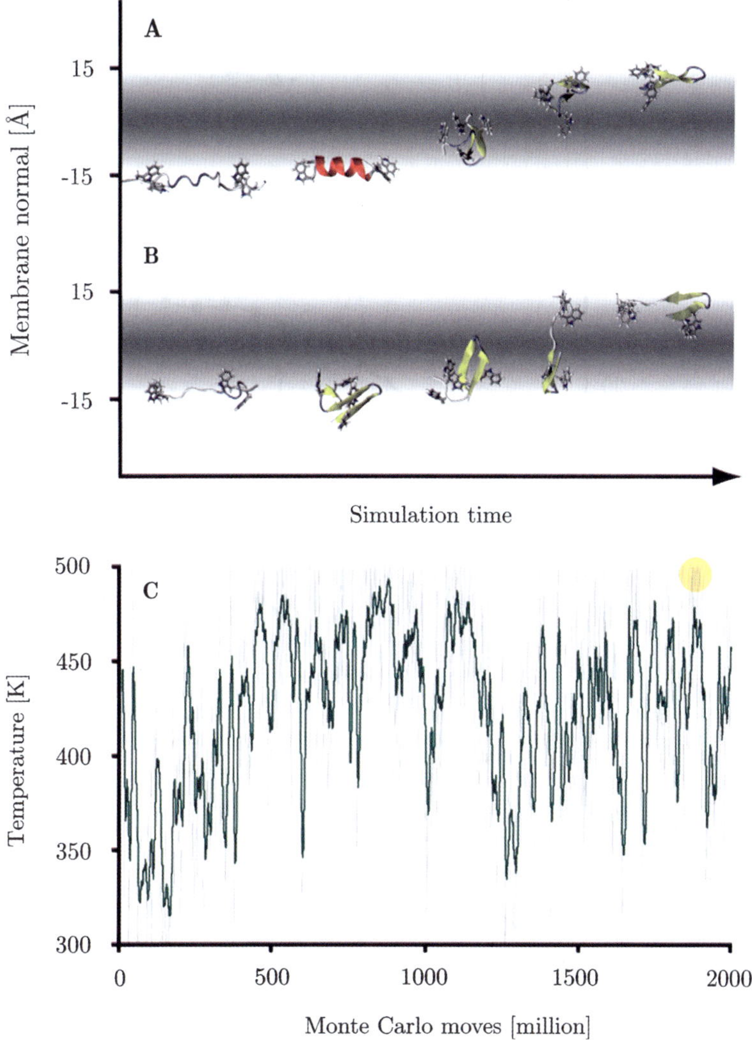

Figure 5.18 Two spontaneous translocation events occurred for the WALP peptide during the surface-bound simulations. In both cases, translocation occurred at high temperatures (487–500 K), taking less than 10^6 MC steps. (A) The 362 K starting simulation; (B) the 461 K starting simulations. Both intermediate conformers contained large amounts of β-structure and no helical content. (C) Simulation temperature (green = running average over 100 frames) of the 362 K starting simulation; the translocation event is indicated by a circle (at MC step 1.89×10^9).

of the WALP peptide as a function of the system temperature. Low temperatures strongly favour helical conformations, whereas β-sheets and unfolded states are the dominant secondary structure at elevated temperatures. Figure 5.19B shows the temperature dependence of the populations of the four

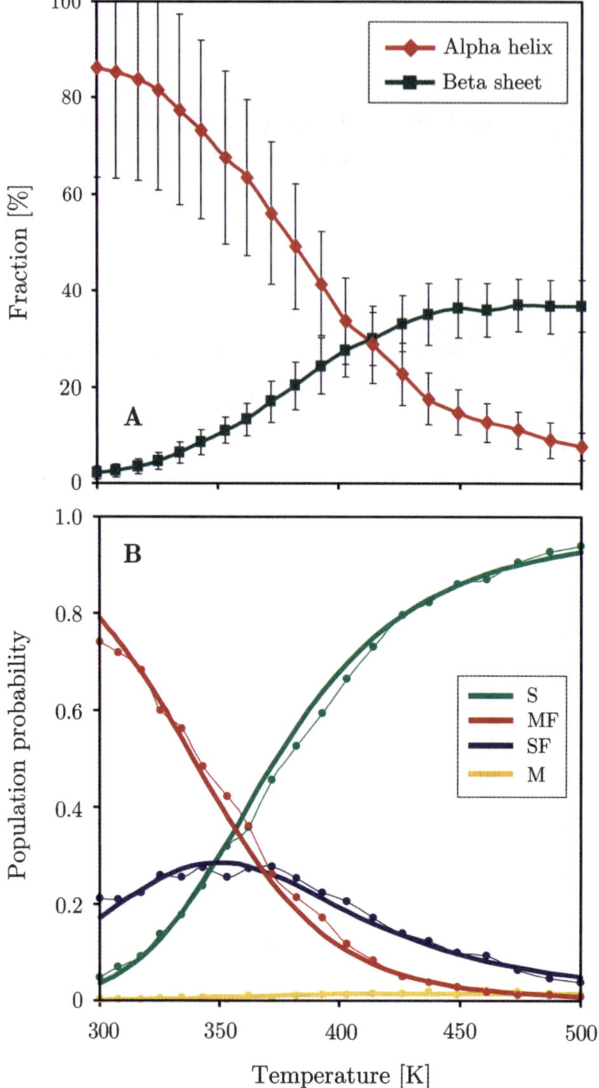

Figure 5.19 Thermodynamic analysis of the transmembrane WALP peptide simulation; the surface-bound simulation shows identical behaviour. (A) Average peptide helicity and β-sheet content as a function of replica temperature (excluding equilibration). (B) Four-state fitting of all sampled states (S = surface unfolded, SF = surface folded, M = membrane unfolded, MF = membrane folded). The M state is hardly populated, reducing the system to three main states: S, SF and MF. The dots represent the direct populations measured for each temperature replica and the thick line is the prediction from the thermodynamic four-state model fit.

states. The transmembrane helix is the dominant conformation for low temperatures. Slight elevation of the temperature increases the fraction of surface bound helices, whereas at high temperatures unfolded surface-bound conformers dominate. The population of unfolded membrane-inserted configurations increases with temperature but is generally very low.

The corresponding probabilities obtained from a four-state thermodynamic fit are shown as thick lines in the Figure 5.19 and the fitted free energies are shown in Figure 5.20A. The quality of the four-state fit is excellent, with $r^2 > 0.98$ and

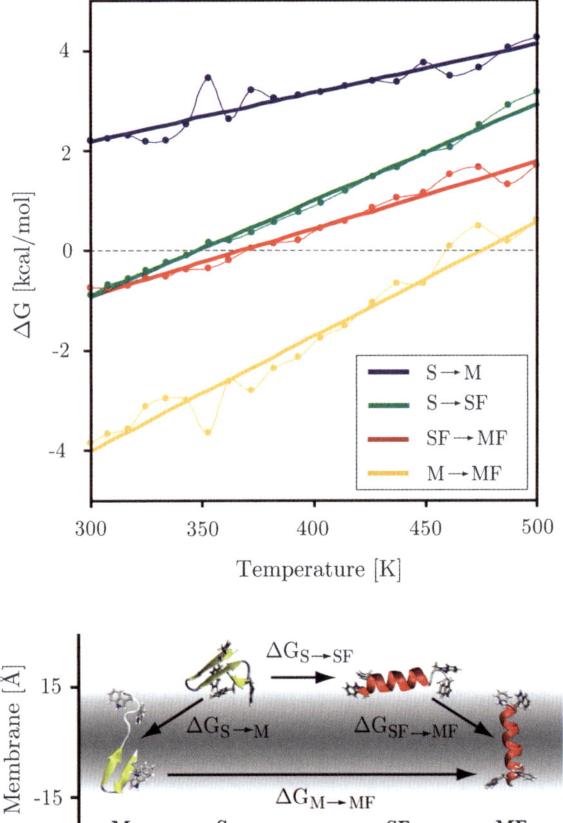

Figure 5.20 Temperature dependence of the transition free energy between the four states (M, S, SF and MF). The four states and main associated transitions are shown at the bottom. Surface folding (S→SF) can be seen to be favourable below ~350 K and the insertion of folded helices (SF→MF) is favourable below ~370 K. Unfolded insertion (S→M) is strongly unfavourable at all temperatures, as is intra-membrane unfolding (MF→M). Dots and thin lines represent the directly measured values and the thick straight lines are the predicted thermodynamic behaviour from the four-state fitting.

Table 5.1 Entropic and energetic contributions to the free energies of the four states (M = membrane-inserted unfolded, S = surface-adsorbed unfolded, SF = surface-adsorbed folded and MF = membrane-inserted folded) of the WALP peptide, as determined from a thermodynamic four-state fit. Shown are the relative energies ΔE (solute internal energy plus generalized Born solvation free energy, including solvent entropy) and solute entropies ΔS. The fitted free energy ΔG is given at 300 K. Also shown is ΔG as determined directly from equation (5.8) and ΔE as determined from direct bin averages. The folded membrane inserted state (MF) has the lowest free energy, energy and entropy and is defined as the reference state zero.

State	$\Delta G_{300\,K}/$ $kcal\,mol^{-1}$	$\Delta G^{fit}_{300\,K}/$ $kcal\,mol^{-1}$	$\Delta E/$ $kcal\,mol^{-1}$	$\Delta E_{fit}/$ $kcal\,mol^{-1}$	$\Delta S_{fit}/$ $cal\,mol^{-1}\,K$
M	3.84	4.01	27 ± 5.3	10.9	22.9
S	1.64	1.83	14 ± 4.0	11.6	32.6
SF	0.75	0.91	4.7 ± 3.5	4.9	13.4
MF	0.00	0.00	0.0 ± 3.7	0.0	0.0

root mean square errors of $\sim 0.15\,kcal\,mol^{-1}$. The results of the fit are shown in Table 5.1. The MF state is stabilized by $\Delta E \approx 5\,kcal\,mol^{-1}$ with respect to the SF state and also has the lowest solute entropy, as expected. Note that ΔE contains both the intra-solute energy and the solute–solvent free energy. Both the S and M states are much higher in energy. The S state has the highest solute entropy, as expected. The values of ΔE were also obtained by averaging E for each of the four states at all 20 temperatures. The match to the estimate of ΔE from the thermodynamic fit is good, except for the poorly sampled M state, which exhibits considerable fluctuation. Table 5.1 also gives the free energy differences at 300 K, both directly calculated by equation (5.8) and as obtained from the fit. At this temperature, the folded states clearly dominate, with a stabilization of $\sim 1\,kcal\,mol^{-1}$ for the MF state over the SF state. This is in good agreement with the values from the insertion landscape (Figure 5.17).

The transition probabilities between the states are highly temperature dependent (see Figure 5.20): unfolded insertion (S → M) is unfavourable for all temperatures, whereas for folded helices insertion is favourable below $\sim 375\,K$ (SF → MF). Helical folding of surface-bound configurations is favourable below $\sim 350\,K$ (S → SF). The lowest curve shows that the penalty for intra-membrane unfolding is extremely high for all but the highest temperatures (M → MF). This thermodynamic description matches the quantitative pathway description presented above. It should be noted that the M state is very sparsely populated, resulting in deviations from the thermodynamic straight-line fit. All data presented here were from the transmembrane simulation. Results for the surface-bound simulation are identical, demonstrating that the system has been thoroughly sampled in both cases.

5.3.3.6 Adaptation to Mismatch

Experimentally, WALP16 is known to form stable membrane-spanning α-helices with low tilt angles[124,125] of <15–$20°$ irrespective of hydrophobic mismatch.[126,127] Several computational approaches have been used to determine the tilt angle and location in the membrane. Poisson–Boltzmann rigid body scans of helical WALP19 and WALP20 peptides, which have slightly longer hydrophobic segments in a 25 Å width low-dielectric slab ($\varepsilon_{membrane} = 2$) gave tilt angles of $\sim 0°$.[42] Short time scale (1.5 ns) molecular dynamics simulations of helical WALP16 conformers in explicit DPPC and DMPC lipid bilayers found mean tilt angles of $\sim 9 \pm 3°$.[128] Im and Brooks reported an average tilt angle of $12 \pm 7°$ for the 300 K WALP16 replica of their implicit membrane simulation, similar to the $16 \pm 9°$ found above.[24] A recent coarse-grain simulation of helically restrained WALP16 conformers in a DPPC lipid bilayer found mean tilt angles of $23 \pm 13°$.[100]

The transmembrane configurations encountered in the study presented above do not fully span the membrane since one pair of tryptophan residues firmly anchors one helix terminus in one interface, restricting the motion of the peptide to one leaflet. This is a result of 'negative' mismatch, where the peptide length is shorter than the hydrophobic membrane core. Similar behaviour was also observed in coarse-grain simulations of WALP16 in a DPPC lipid bilayers,[100] which have a hydrophobic core width of ~ 34 Å, exceeding the ~ 25 Å length of helically constrained WALP16. Interestingly, the study found no translocation for WALP16, whereas simulations of WALP19, which has a longer hydrophobic core, showed several such events, indicating that 'negative' mismatch may cause a barrier for membrane crossing events.

Insertion for WALP always occurs from the N-terminal side, with the C-terminus remaining at the interface. This behaviour was also observed by Im and Brooks using a different implicit membrane method.[24] Since the peptide sequence is completely symmetric, they suggested that the larger dipole moment of the carbonyl groups at the C-terminus might explain this result.

Experimentally, the bilayer width in the vicinity of WALP does not change significantly with hydrophobic mismatch.[126,127] Instead, peptide localization in the membrane seems to be determined chiefly by interfacial tryptophan anchoring.[129] Both the coarse-grain and the implicit membrane simulations seem to reproduce this result. However, the helical restraint used in the coarse-grain simulation may mask other insertion and translocation pathways involving unfolded or partially folded conformers. The present implicit membrane method, on the other hand, samples a large number of structurally diverse protein conformations, but does not model entropy changes of the bilayer itself. The simulations therefore cannot account for local variations in hydrophobic width and changes in lipid ordering in the vicinity of the peptide, which might be important factors in peptide insertion and translocation.[130]

5.3.3.7   Simulation Accuracy and Temperature Dependence

The observed temperature dependence of secondary structure and preferred peptide location is generally in accordance with expected thermodynamic behaviour (see Figure 5.19). Higher replica temperatures exhibit a rapidly decreasing helicity, which drives the peptide out of the membrane due to the high cost of exposing the peptide bond.[118] This interplay between secondary structure and membrane localization causes the overall temperature dependence of the transition free energies between the four chief states of the system. The fact that for WALP both surface and transmembrane starting configurations converge to yield identical and fully converged thermodynamic behaviour (*i.e.* the free energy of each transition varies linearly with temperature) is remarkable and demonstrates that the system has been sufficiently sampled (see Figure 5.20). At equilibrium, the OPLS all-atom force field employed captures a large range of conformations, ranging from unfolded, partial helical folds, misfolded β-sheets to fully folded α-helices, with non-native folds dominating at higher temperatures as expected. This behaviour differs drastically from simulations with a similar implicit membrane model employing the CHARMM force field, which have consistently reported that all replicas rapidly fold continuous helices up to temperatures of 1000 K.[17,23,24] The authors suggested that this is most likely due to a helical bias of the CHARMM force field.[131] Such biases may veil important thermodynamic behaviour and also insertion pathways involving unfolded states.

5.3.4   Comparison with Explicit Methods

Comparing implicit membrane simulations with explicit methods is difficult. However, such comparisons are necessary to evaluate the relative accuracy and performance increases that justify the development and use of continuum approaches. We present here a detailed comparison of the folding behaviour of WALP in explicit membrane environments with the implicit membrane data from the previous section. An elevated temperature of 80 °C was chosen for the simulations since the peptide has significant populations of both surface-bound and inserted states at this temperature, allowing a detailed analysis and quantification of the rate of exploration of phase space.

5.3.4.1   Octane Slab Membranes

Is the implicit generalized Born membrane model just a 'hydrophobic slab'? If yes, results should correspond closely to a membrane mimic such as a layer of octane surrounded by water. Octane is a commonly used membrane mimic, due to its much lower reorientation times compared with explicit lipid bilayers.[132] This leads to faster system equilibration and sampling. Two simulations were performed of the WALP16 peptide in a layer of octane with surrounding water. Similarly to the implicit membrane simulations, the starting conformations were completely extended, one in solvent and the other spanning the

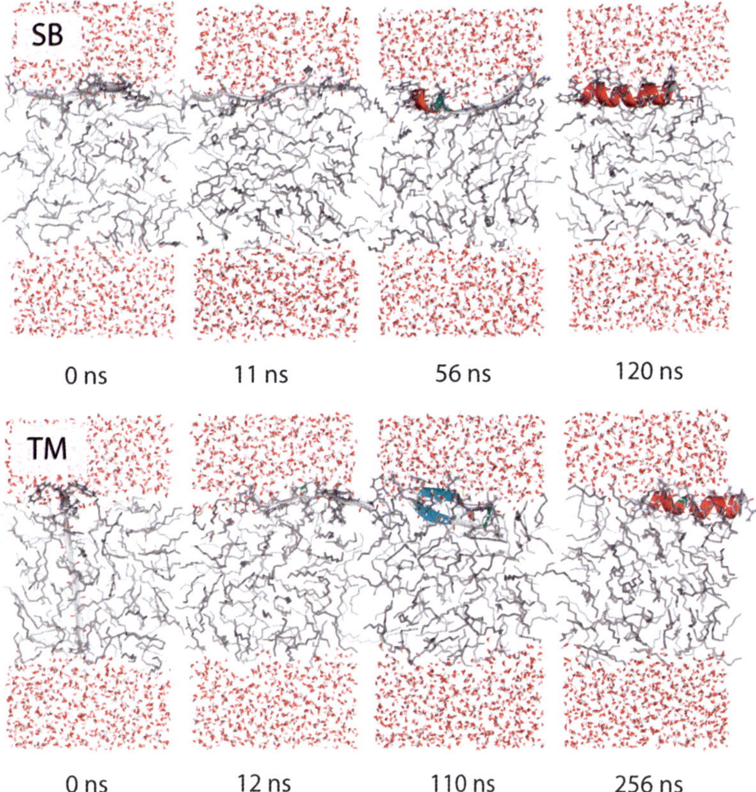

Figure 5.21 Folding simulations of WALP16 in the octane layer that serves as a membrane mimic. The surface-bound (SB) starting run results in a folded α-helix after 120 ns. For the transmembrane-bound (TM) starting run, WALP16 is quickly expelled from the hydrophobic phase at 12 ns. A β-hairpin is formed at the interface, followed by an α-helix. WALP16 remains at the interface during the simulations and no insertion is observed. The helix is stable at 353 K.

membrane. The folding pathway is shown in Figure 5.21 As in the implicit membrane model, the unfolded peptide is quickly expelled (at $\sim 12\,\text{ns}$) from the octane layer, adsorbing at the interface, where it remains for the rest of the simulations. Folding also occurs at the interface. First, a rapid conformational exploration 'coil' phase is seen. Subsequently, the peptide forms a partial helix, followed by a fully folded helix. Once formed, the helix resides stably on the interface with a tilt angle of $\sim 90°$, similar to the implicit membrane simulations. The interfacial folding pathway closely mimics the implicit simulations, but at no point, either during the folding process or when fully helical, is any insertion of WALP16 into the octane layer observed. Clearly, there is a large free energy penalty to bury the peptide in the hydrocarbon phase. In fact, repeating the simulation with a membrane-spanning helix shows that this configuration is not stable, with the helix exiting to the surface after several hundred nanoseconds.[133]

5.3.4.2 *Lipid Bilayer*

Finally, the computationally most demanding comparison entails WALP16 in an explicit lipid bilayer environment. Starting from a fully extended conformation in bulk solvent, the simulation was continued for 3.0 µs in a DPPC lipid bilayer (see Figure 5.22). A second control simulation was run for 500 ns from a TM-inserted helical state. The partitioning behaviour is drastically different from both from the implicit membrane and the octane layer simulations. First, adsorption is slow and the peptide remains fully solvated for ~ 100 ns. After precipitating on the membrane surface, the peptide immediately

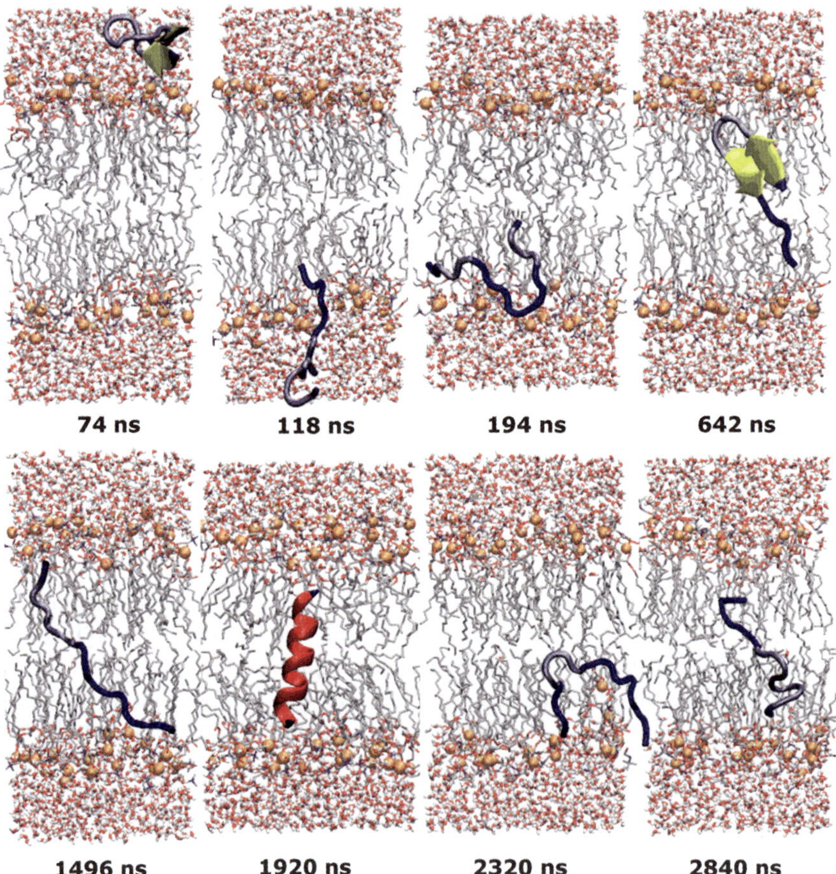

Figure 5.22 Adsorption, unfolded insertion, temporary folding and subsequent unfolding of the WALP peptide in an explicit lipid bilayer.

starts to cross the phosphocholine headgroups. This process is driven by favourable van der Waals interactions and takes ~ 300 ns. During this time, the peptide remains virtually completely unfolded and lost electrostatic interactions with the water molecules are compensated by interactions with the polar headgroups, which are of the same order of magnitude. Subsequently the peptide inserts into the hydrophobic membrane core in an unfolded configuration and for the remaining $\sim 2.5\,\mu$s the peptide oscillates between deeply inserted completely unfolded and misfolded conformations (β-hairpins). There is an absence of both a stable interfacial helix and a stable inserted helix (see Figure 5.23). Instead, the dominant conformations are inserted β-hairpins. Such conformations should be less favourable than helices since only half the backbone hydrogen bonds can be satisfied. Eventually, after $\sim 1.9\,\mu$s rapid formation of a helical conformer is observed from a completely extended membrane-spanning configuration. However, this is not concomitant with

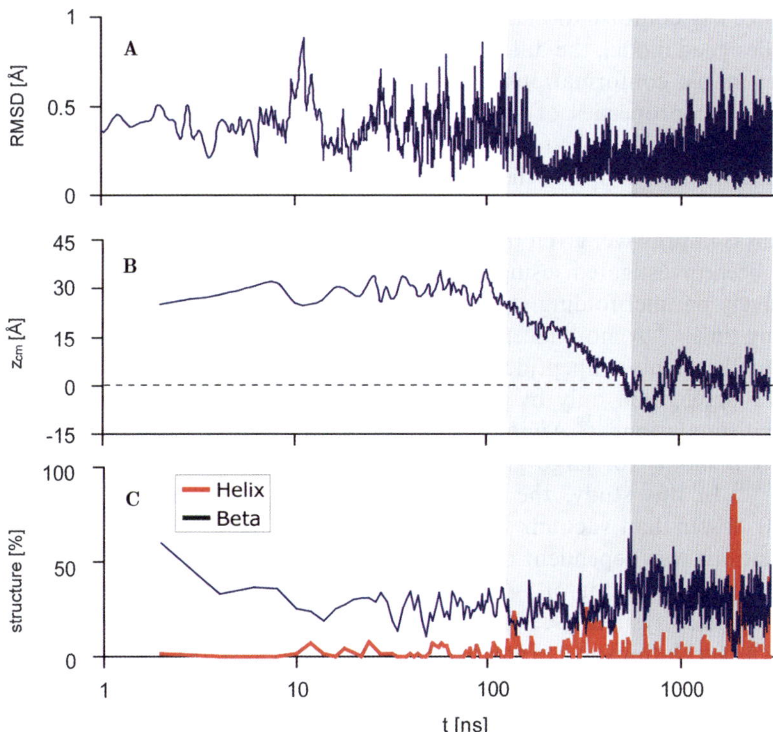

Figure 5.23	Folding and insertion of WALP16 in the DPPC bilayer starting from an extended surface-bound conformation. The peptide first inserts the bilayer before folding to a TM-inserted helix that is not stable at this temperature. (A) Relative backbone RMSD of frame t with frame $t - \Delta t$ ($\Delta t = 2$ ns). (B) Position of the centre of mass along the membrane normal. (C) Helicity and beta content. The main observed conformation is beta-like.

further energetic stabilization and the helix remains stable for only $\sim 200\,\mathrm{ns}$ before unfolding again. The $500\,\mathrm{ns}$ control simulation of a fully folded WALP16 peptide in a membrane-spanning configuration was found to be unstable, unfolding after $\sim 300\,\mathrm{ns}$.

5.3.5 Sampling Performance

Implicit membrane models promise significantly increased sampling time scales compared with explicit bilayers. In the ideal case, the same protein energy surface is sampled, but much faster, *i.e.* the thermodynamics are unaltered but the dynamics are greatly increased. Unfortunately, it is not possible to construct simplified models (*e.g.* implicit solvents, coarse-grain models) that reproduce the same energy landscape as an equivalent full representation. Consequently, none of these models can be 'fairly' compared from a pure sampling perspective. However, it can be compared how quickly the protein explores key conformational states in the different surrounding 'medium'. In a well-designed model, the differences in the energy landscape and the population of key stable conformations are small. For the case of the GB model, thermodynamic properties of small peptides have been compared using replica exchange,[134,135] and also long microsecond-scale simulations.[136] The general conclusion has been that there can be differences in stable conformations, in addition to the predicted native state. However, the performance of the implicit models is impressive, given the simplicity of the GB method. Kinetic properties have been investigated, using solvent viscosity as a tuneable parameter.[137] In massively parallel folding studies, a factor of ~ 20 has been reported for the folding time.[138] A more recent study focusing on conformational transitions in the smaller alanine-dipeptide indicated that barrier crossings with a GB model can be accelerated only by a maximum factor of ~ 2, when compared with explicit simulations.[139] An interesting performance comparison of a commonly used GB model for large globular proteins and DNA was reported by Feig *et al.*[140] In this study, the GB calculations were shown to be a factor of 10–20 slower than vacuum and performance relative to equivalent all-explicit simulations was dependent on the size of the protein/DNA studied. Speed-up ranged from a factor of ~ 10 for small systems (<10 residues) to ~ 2 for medium-sized systems (<100 residues). For large systems (>250 residues), the explicit calculations were found to be significantly faster. Clearly, much better performance is required if it is hoped to use such models more widely.

Simpler GB models, such as the asymptotic approach elaborated above, allow for a more significant speed-up: the GBIM model in equations (5.5) and (5.6) was optimized to be only ~ 2.5 times slower than vacuum calculations. A direct comparison of the sampling efficiency of the different models was performed using various metrics summarized in Table 5.2. This allows for a quantitative estimate of the sampling speed-up of octane and implicit membrane models. It is important to note that a simple comparison of the computation time to perform, *e.g.*, one time step is not indicative of efficiency, but rather the relative

Table 5.2 Sampling performance metrics for the GBIM, octane and DPPC systems. t_{exit}, simulation time of expulsion of the unfolded starting conformer in the TM run from the membrane core. t_{helix}, Time required to reach $>90\%$ helicity; $t_{cluster50}$, time required to reach 50 conformational clusters; τ_{acf}, correlation time of the backbone torsional angles; dt_{RMSD}, average time Δt required to reach a backbone RMSD of 0.15 Å with frame $t - \Delta t$ over the coil phase of the simulations; r_{av}, sampling efficiency is indicated by the average ratio of the metrics of the TM and SB runs.

| Metric | GBIM | | Octane | | | | |
	TM	SB	TM	SB	r_{av} GBIM/octane	DPPC	r_{av} GBIM/DPPC
t_{exit}	0.24 M scans (0.02 ns)	–	12 ns	–	600	–	–
t_{helix}	9.4 M scans (0.9 ns)	68 M scans (6.8 ns)	245 ns	97 ns	44	1900 ns	494
$t_{cluster50}$	1 M scans (0.1 ns)	1.1 M scans (0.11 ns)	10.5 ns	10 ns	98	13 ns	124
t_{acf}	0.034 M scans (3.4 ps)	0.076 M scans (7.6 ps)	0.3 ns	0.2 ns	57	10 ns	1818
dt_{RMSD}	0.01 M scans (0.001 ns)	0.035 M scans (0.0035 ns)	0.016 ns	0.015 ns	10.1	0.08 ns	51

conformational exploration achieved per unit computation time and the pace at which long-term transitional events, such as folding, can be observed.

5.3.5.1   *Relative RMSD Along Trajectories*

A good quantifier of sampling efficiency is the RMSD of backbone atoms at time t with respect to time $t - \Delta t$. If Δt is short, this reveals the short-term local sampling efficiency, such as the fluctuation around a trapped conformation. For larger Δt, the efficiency of long-term major configurational transitions is obtained. When Δt is increased, simulations with many trapped conformations will only exhibit a small increase in average RMSD, since the same conformation is sampled. A large increase indicates frequent transitions to other states, suggesting good sampling. Low values of relative RMSD correspond to stable conformations, whereas large values indicate coil phases with many transitions. Although the spread is large, the average RMSD can be calculated during each phase and the results are plotted in Figure 5.24 as a function of Δt. Better sampling is indicated by a steeper slope. For all systems, the RMSD values, and also the slope of the curve, are largest during the coil phase. Semi-folded conformations, such as partially helical structures, exhibit much slower sampling as only the non-folded part can fluctuate significantly. The lowest values are for completely folded conformations, both the full helix that is ultimately reached by all simulations, and also the β-hairpin conformation. For these conformations, the slope of RMSD versus Δt is virtually zero. In all cases, the implicit membrane simulations reach the same values of RMSD much quicker than the corresponding octane simulations, with an average factor of ~ 10. The least sampling is seen for WALP in the DPPC bilayer, which needs significantly more computational time (> 50 times) to reach the same level of RMSD as the implicit membrane, indicating that the peptide is largely trapped in the explicit bilayer phase, even at a high temperature of 80 °C. The results are independent of starting configuration, showing only minor sampling differences, indicating that simulations are fully converged.

5.3.5.2   *Number of Individual Clusters Along the Trajectories*

The rate of exploration of phase space can also be estimated by the increase in individual conformational clusters with simulation time. For this, all simulations are subjected to a cluster analysis at various points along their trajectories. The clusters can be determined by superimposing structures using backbone least-squares fitting (RMSD), with a similarity cutoff of 1.5 Å. Clustering is usually performed to cover the complete simulation trajectories and the results are shown in Figure 5.25. All curves share a characteristic profile. A large initial rise is observed due to rapid sampling during the coil phase. This is followed by a flat plateau when a stably folded structure (*e.g.* an α-helix) is reached, at which point no further clusters are added. The number of clusters in this phase is variable, depending on how much of phase space was

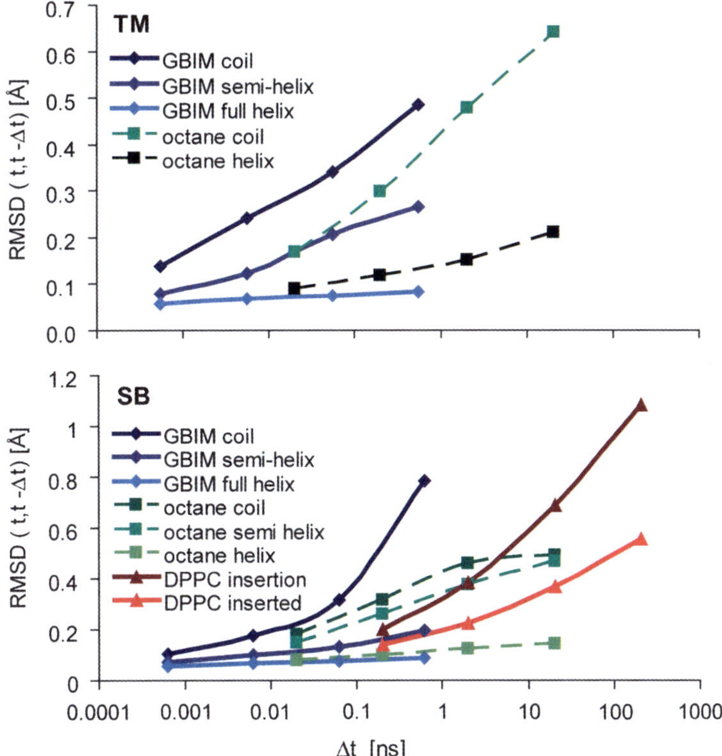

Figure 5.24 RMSD of backbone atoms at time t with respect to time $t - \Delta t$, for the WALP16 simulations starting from transmembrane (TM) and fully solvated (SB) extended configurations. The values are averages and are taken over separate phases of the folding simulations. The slope of the RMSD *versus* Δt is an indicator of sampling efficiency. It is largest during the coil phase, smaller for partially folded conformations and nearly flat for a fully folded helix. With the implicit membrane model, the same RMSD values are reached in much less time than the explicit simulations.

explored in the coil phase. A much faster rise of unique clusters as a function of computational effort is seen with the implicit membrane simulations than for the explicit octane or DPPC simulations, with a factor of ~ 100 in calculation time to reach the same number of clusters.

5.3.5.3 Autocorrelation Functions

Another good indicator of the efficiency to explore configurational space is the correlation time of the torsion angles in the protein backbone. We calculated the autocorrelation functions of the φ, ψ backbone torsion angles for the implicit membrane simulations and compared them with those calculated from the explicit simulations. Figure 5.26 shows the average autocorrelation functions of

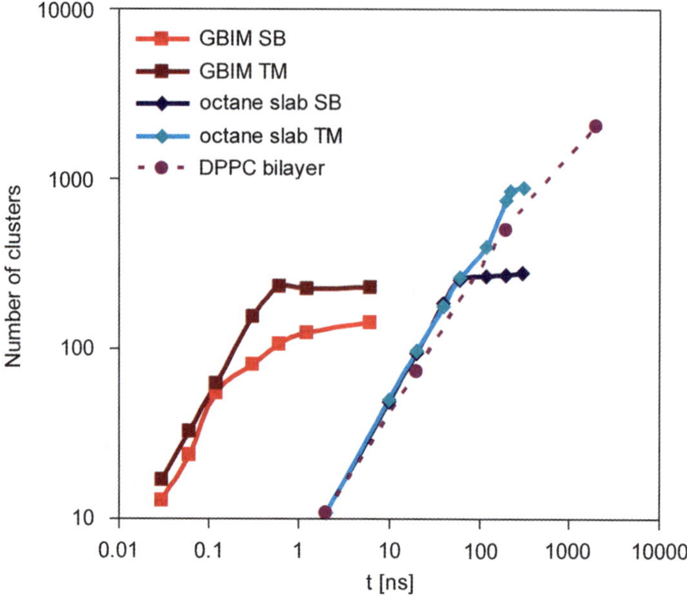

Figure 5.25 Rise in the number of individual clusters as a function of simulation time for the WALP16 peptide. Clustering was performed by pairwise analysis of the backbone RMSD. The graph reveals the rate at which the conformational space is sampled and new structures are encountered. A rapid initial rise is followed by a plateau when the stable helix phase is reached.

all φ/ψ angles. Both the number of Monte Carlo scans and the nanoseconds of molecular dynamics are plotted, with 1×10^{6} MC scans requiring about the same computational effort as 0.1 ns of MD. A striking difference is the rapidly vanishing autocorrelation in the implicit simulations, which is a factor of ~ 57 lower than in the octane runs. Again, there is no dependence of the rate of conformational exploration on the initial starting conformation (TM or SB). The autocorrelation functions were fitted by single exponential functions to quantify the correlation times that are summarized in Table 5.2. Note that in the case of the implicit membrane and the octane layer simulations, the correlation functions are only calculated over the unfolded part of the trajectories. In the case of the DPPC bilayer simulation, the peptide does not fold stably, so the whole $3\,\mu s$ trajectory is used in the analysis. This results in a significantly longer correlation time of the backbone torsions, since the measured correlation is that of the long-term torsional fluctuations in folding/unfolding the α-helix.

5.3.5.4   *Major Conformational Transitions*

A final measure of sampling efficiency is the average simulation time required for the WALP16 peptide to exit the membrane in the TM runs t_{exit} and the average time required for the α-helix to form t_{helix}. In the implicit membrane

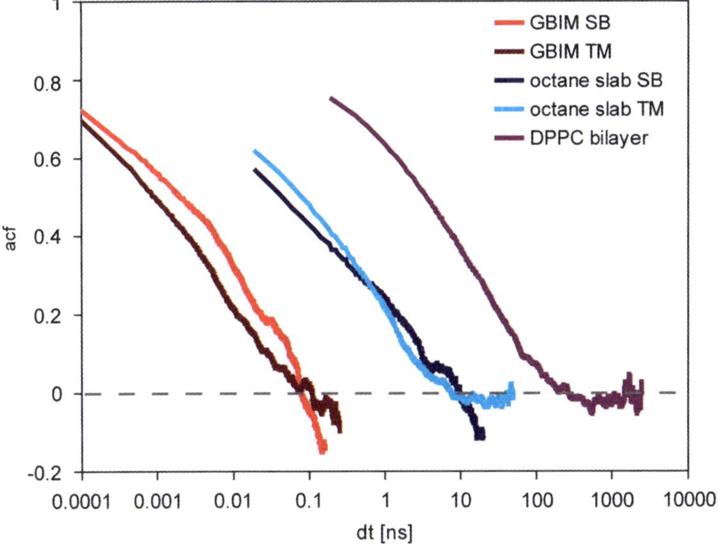

Figure 5.26 Autocorrelation functions of the backbone torsion angles for the simulations of the WALP16 peptide. The autocorrelation was calculated over all φ,ψ backbone torsion angles and averaged, both for the simulations that started from a surface-bound (SB) and a transmembrane (TM) conformation. The implicit simulations reveal ~ 2 orders of magnitude faster decorrelation for the same computational effort than the octane layer simulations. The explicit DPPC bilayer simulation exhibits the longest correlation time.

TM simulation, the peptide exits instantaneously at $\sim 0.24 \times 10^6$ scans, whereas the same process requires ~ 12 ns in the octane simulation. Hence this process is ~ 600 times faster in the implicit model, where no explicit solvent atoms delay an immediate expulsion. To reach a helicity of $>90\%$, the implicit membrane simulations take 9×10^6 scans (TM) and 68×10^6 scans (SB), whereas the corresponding octane simulations require 245 ns (TM) and 97 ns (SB). Hence the stable helix is formed approximately ~ 44 times faster in the implicit model. Helix formation in the DPPC run is much slower, requiring 1900 ns, which is a factor ~ 500 slower than the implicit model.

5.3.5.5 Summary and Conclusions

Modelling a membrane implicitly has clear advantages over fully explicit models. Implicit models promise vastly superior sampling efficiency for two reasons. First, the solvent degrees of freedom (with implicit membranes also the lipids) are removed, thus reducing the number of atoms that have to be explicitly treated by $\sim 80\%$. This will result in a significantly lower cost per time-step as compared with explicit solvent. In addition, the mean-field nature of the solvent allows for substantially accelerated dynamics per time step so that relevant thermodynamic states of biomolecules are sampled much faster than with explicit solvent. This is especially relevant if solvent friction is omitted. With implicit models, solvent viscosity is optional and can be adjusted either to reproduce kinetic properties accurately or switched off if rapid convergence of thermodynamic properties is desired.

For the 16-residue WALP peptide, the implicit membrane algorithm described above was found to be faster by a factor of 50–100 in most examined metrics and the protein displayed similar conformational states, transition events and folding properties. These numbers are with respect to the simple octane/water explicit simulations. Comparison with the fully explicit DPPC bilayer yields even larger numbers, with ~ 5–10 times slower sampling than the octane simulations. The freedom of the peptide to sample efficiently is clearly much more restricted by the charged headgroups (at the interface) and the much stronger order of the lipid tails (if inserted). The overall ranking of sampling efficiency is thus implicit membrane > octane/water > DPPC/water.

The much lower computational effort has been achieved by designing the model to be both simple yet excellent in matching the available experimental data. In addition, the model was optimized to be only ~ 2.5 times slower than vacuum calculations.

5.3.6 Conclusions

Implicit membrane models can now be practically applied to simulate many stages and properties of membrane protein synthesis and function, from the folding of polypeptide fragments to oligomeric helix bundles. To be useful for studying a wide range of peptides, such methods must essentially fulfil several requirements: (a) single transmembrane or surface-bound helices, and also integral membrane proteins, must retain their experimentally observed structure despite being surrounded only by a continuum environment; (b) such stability must not be caused by the use of models and parameter sets that overly bias helical structures or by using artificial constraining potentials; (c) in order to justify the substantial simplifications entailed by an implicit representation of the membrane, the model must be significantly faster than equivalent fully atomistic membrane simulations, permitting extensive conformational sampling that goes beyond simple rigid orientational scans; and (d) a wide range of experimental data must be reproduced, especially the experimentally determined partitioning free energy of polypeptides into both the membrane interfaces and the membrane interior.[34,64,121]

Although the implicit membrane model described above performs well on these points, two key deficiencies have yet to be overcome: the neglect of effects due to the complex lipid headgroup environment and the improper treatment of charged residues. In practical terms, this signifies that the interfacial regions of the membrane are poorly described and some loss of defined secondary structure is observed in the segments of membrane-bound peptides in this region. This will be especially problematic for simulating surface-bound peptides (*e.g.* antimicrobials) and matching experimental partitioning free energy of unfolded polypeptides into membrane interfaces.[141] For charged residues, the generalized Born model predicts a large desolvation penalty on moving into the hydrophobic region. In the biological scale of Hessa *et al.*,[64] the effect of the additional charge is almost non-existent: for example, the apparent free energy of insertion ΔG_{app}^{aa} of an amino acid located at the centre of a 19 residue membrane-spanning helix are roughly equally unfavourable for glutamine (2.36 kcal mol^{-1}) and glutamic acid (2.68 kcal mol^{-1}),[64] whereas the burial penalty due to the additional full charge is large in the implicit model. It would therefore be more appropriate to use variable protonation state models, where residues can be neutralized upon entering the membrane. Alternatively, White and von Heijne suggested that the strong positional dependence of charged residues on the biological scale could be due to distorted bilayer states, where the headgroups are in contact with buried peptide charges and the hydrophobic thickness is significantly reduced.[142] This is currently beyond the limits of the implicit membrane model.

Ultimately, further improvement is required for more accurate modelling of the polar lipid headgroup region of the membrane, which will involve additions to the implicit membrane that go beyond simple dielectric treatment. In addition, the inclusion of variable protonation state models is probably a good idea if sequences with many charged residues are studied.

Acknowledgements

This work was supported by grants from the Wellcome Trust, Deutsche Forschungsgemeinschaft, BIOMS and the Human Frontiers Science Program.

References

1. G. C. Terstappen and A. Reggiani, *In silico* research in drug discovery, *Trends. Pharmacol. Sci.*, 2001, **22**(1), 23–26.
2. R. Grisshammer and C. G. Tate, Overexpression of integral membrane-proteins for structural studies, *Q. Rev. Biophys.*, 1995, **28**(3), 315–422.
3. B. L. de Groot and H. Grubmuller, Water permeation across biological membranes: mechanism and dynamics of aquaporin-1 and GlpF, *Science*, 2001, **294**(5550), 2353–2357.
4. L. R. Forrest and M. S. P. Sansom, Membrane simulations: bigger and better?, *Curr. Opin. Struct. Biol.*, 2000, **10**(2), 174–181.
5. P. C. Biggin and M. S. Sansom, Interactions of alpha-helices with lipid bilayers: a review of simulation studies, *Biophys. Chem.*, 1999, **76**(3), 161–183.
6. C. Domene, P. Bond and M. S. P. Sansom, Membrane protein simulation: ion channels and bacterial outer membrane proteins, *Adv. Protein Chem.*, 2003, **66**, 159–193.

7. R. Braun, D. M. Engelman and K. Schulten, Molecular dynamics simulations of micelle formation around dimeric glycophorin A transmembrane helices, *Biophys. J.*, 2004, **87**(2), 754–763.

8. P. J. Bond, J. M. Cuthbertson, S. S. Deol and M. S. Sansom, MD simulations of spontaneous membrane protein/detergent micelle formation, *J. Am. Chem. Soc.*, 2004, **126**(49), 15948–9.

9. A. Grossfield, S. E. Feller and M. C. Pitman, Convergence of molecular dynamics simulations of membrane proteins, *Proteins*, 2007, **67**(1), 31–40.

10. H. Nymeyer, T. B. Woolf and A. E. Garcia, Folding is not required for bilayer insertion: replica exchange simulations of an alpha-helical peptide with an explicit lipid bilayer, *Proteins Struct. Funct. Bioinf.*, 2005, **59**(4), 783–790.

11. M. B. Ulmschneider and J. P. Ulmschneider, Folding peptides into lipid bilayer membranes, *J. Chem. Theory Comput.*, 2008, **4**, 1807–1809.

12. M. Feig, W. Im and C. L. Brooks, Implicit solvation based on generalized Born theory in different dielectric environments, *J. Chem. Phys.*, 2004, **120**(2), 903–911.

13. T. Lazaridis, Effective energy function for proteins in lipid membranes, *Proteins Struct. Funct. Genet.*, 2003, **52**(2), 176–192.

14. T. Lazaridis, Implicit solvent simulations of peptide interactions with anionic lipid membranes, *Proteins Struct. Funct. Bioinf.*, 2005, **58**(3), 518–527.

15. R. G. Efremov, D. E. Nolde, G. Vergoten and A. S. Arseniev, A solvent model for simulations of peptides in bilayers. I. Membrane-promoting alpha-helix formation, *Biophys. J.*, 1999, **76**(5), 2448–2459.

16. R. G. Efremov, D. E. Nolde, G. Vergoten and A. S. Arseniev, A solvent model for simulations of peptides in bilayers. II. Membrane-spanning alpha-helices, *Biophys. J.*, 1999, **76**(5), 2460–71.

17. W. Im, M. Feig and C. L. Brooks III, An implicit membrane generalized born theory for the study of structure, stability and interactions of membrane proteins, *Biophys. J.*, 2003, **85**, 2900–2918.

18. V. Z. Spassov, L. Yan and S. Szalma, Introducing an implicit membrane in generalized Born/solvent accessibility continuum solvent models, *J. Phys. Chem. B*, 2002, **106**(34), 8726–8738.

19. M. B. Ulmschneider, J. P. Ulmschneider, M. S. P. Sansom and A. Di Nola, A generalized Born implicit membrane representation compared to experimental insertion free energies, *Biophys. J.*, 2007, **92**, 2338–2349.

20. W. C. Still, A. Tempczyk, R. C. Hawley and T. Hendrickson, Semi-analytical treatment of solvation for molecular mechanics and dynamics, *J. Am. Chem. Soc.*, 1990, **112**(16), 6127–6129.

21. S. Tanizaki and M. Feig, A generalized Born formalism for heterogeneous dielectric environments: application to the implicit modeling of biological membranes, *J. Chem. Phys.*, 2005, **122**(12), 124706.

22. S. Tanizaki and M. Feig, Molecular dynamics simulations of large integral membrane proteins with an implicit membrane model, *J. Phys. Chem. B*, 2006, **110**(1), 548–556.

23. W. Im and C. L. Brooks III, De novo folding of membrane proteins: an exploration of the structure and NMR properties of the fd coat protein, *J. Mol. Biol.*, 2004, **337**, 513–519.
24. W. Im and C. L. Brooks, Interfacial folding and membrane insertion of designed peptides studied by molecular dynamics simulations, *Proc. Natl. Acad. Sci. USA*, 2005, **102**(19), 6771–6776.
25. W. Im, J. H. Chen and C. L. Brooks, Peptide and protein folding and conformational equilibria: theoretical treatment of electrostatics and hydrogen bonding with implicit solvent models, in: *Peptide Solvation and H-Bonds,* ed. R. Baldwin and D. Baker, Elsevier Academic Press, San Diego, 2006, **Vol. 72**, Advances in Protein Chemistry, pp. 173–198.
26. J. Lee and W. Im, Implementation and application of helix–helix distance and crossing angle restraint potentials, *J. Comput. Chem.*, 2007, **28**(3), 669–680.
27. J. P. Ulmschneider, M. B. Ulmschneider and A. Di Nola, Monte Carlo folding of transmembrane helical peptides in an implicit generalized Born membrane, *Proteins*, 2007, **69**, 297–308.
28. L. Bu, W. Im and C. L. Brooks III, Membrane Assembly of simple helix homo-oligomers studied via molecular dynamics simulations, *Biophys. J.*, 2007, **92**(3), 854–863.
29. J. P. Ulmschneider and M. B. Ulmschneider, Folding simulations of the transmembrane helix of virus protein U in an implicit membrane model, *J. Chem. Theory Comput.*, 2007, **3**, 2335–2346.
30. T. Carpenter, P. J. Bond, S. Khalid and M. S. P. Sansom, Self-assembly of a simple membrane protein: coarse-grained molecular dynamics simulations of the influenza M2 channel, *Biophys. J.*, 2008, **95**(8), 3790–3801.
31. S. J. Marrink, H. J. Risselada, S. Yefimov, D. P. Tieleman and A. H. de Vries, The MARTINI force field: coarse grained model for biomolecular simulations, *J. Phys. Chem. B*, 2007, **111**(27), 7812–7824.
32. M. Pellegrini-Calace, A. Carotti and D. T. Jones, Folding in lipid membranes (FILM): a novel method for the prediction of small membrane protein 3D structures, *Proteins*, 2003, **50**(4), 537–545.
33. Y. Zhang, M. E. Devries and J. Skolnick, Structure modeling of all identified G protein-coupled receptors in the human genome, *PLoS Comput. Biol.*, 2006, **2**(2), e13.
34. M. B. Ulmschneider, M. S. Sansom and A. Di Nola, Properties of integral membrane protein structures: derivation of an implicit membrane potential, *Proteins Struct. Funct. Genet.*, 2005, **59**(2), 252–265.
35. N. Hurwitz, M. Pellegrini-Calace and D. T. Jones, Towards genome-scale structure prediction for transmembrane proteins, *Philos. Trans. R. Soc. Lond. B Biol. Sci.*, 2006, **361**(1467), 465–475.
36. M. Milik and J. Skolnick, A Monte Carlo model of fd and Pf1 coat proteins in lipid membranes, *Biophys. J.*, 1995, **69**(4), 1382–1386.
37. M. W. Maddox and M. L. Longo, A Monte Carlo study of peptide insertion into lipid bilayers: equilibrium conformations and insertion mechanisms, *Biophys. J.*, 2002, **82**(1), 244–263.

38. M. Milik and J. Skolnick, Insertion of peptide chains into lipid membranes: an off-lattice Monte Carlo dynamics model, *Proteins*, 1993, **15**(1), 10–25.
39. S. J. Marrink, A. H. deVries and A. E. Mark, Coarse grained model for semiquantitative lipid simulations, *J. Phys. Chem. B*, 2004, **108**, 750–760.
40. A. Grossfield, J. Sachs and T. B. Woolf, Dipole lattice membrane model for protein calculations, *Proteins Struct. Funct. Genet.*, 2000, **41**(2), 211–223.
41. R. G. Efremov, D. E. Nolde, G. Vergoten and A. S. Arseniev, A solvent model for simulations of peptides in bilayers. II. Membrane-spanning alpha-helices, *Biophys. J.*, 1999, **76**(5), 2460–2471.
42. D. Sengupta, L. Meinhold, D. Langosch, G. M. Ullmann and J. C. Smith, Understanding the energetics of helical peptide orientation in membranes, *Proteins Struct. Funct. Genet.*, 2005, **58**(4), 913–922.
43. N. Ben-Tal, B. Honig, C. K. Bagdassarian and A. Ben-Shaul, Association entropy in adsorption processes, *Biophys. J.*, 2000, **79**(3), 1180–1187.
44. T. B. Woolf, D. M. Zuckerman, N. D. Lu and H. B. Jang, Tools for channels: moving towards molecular calculations of gating and permeation in ion channel biophysics, *J. Mol. Graphics Modell.*, 2004, **22**(5), 359–368.
45. R. G. Efremov, D. E. Nolde, A. G. Konshina, N. P. Syrtcev and A. S. Arseniev, Peptides and proteins in membranes: what can we learn via computer simulations?, *Curr. Med. Chem.*, 2004, **11**(18), 2421–2442.
46. J. Chen, C. L. Brooks III and J. Khandogin, Recent advances in implicit solvent-based methods for biomolecular simulations, *Curr. Opin. Struct. Biol.*, 2008, **18**(2), 140–148.
47. M. Born, Volumen und Hydratationswärme der Ionen, *Z. Phys.*, 1920, **1**, 45–48.
48. J. P. Ulmschneider and W. L. Jorgensen, Polypeptide folding using Monte Carlo sampling, concerted rotation and continuum solvation, *J. Am. Chem. Soc.*, 2004, **126**(6), 1849–1857.
49. S. Chowdhury, W. Zhang, C. Wu, G. M. Xiong and Y. Duan, Breaking non-native hydrophobic clusters is the rate-limiting step in the folding of an alanine-based peptide, *Biopolymers*, 2003, **68**(1), 63–75.
50. S. Jang, S. Shin and Y. Pak, Molecular dynamics study of peptides in implicit water: ab initio folding of beta-hairpin, beta-sheet and beta beta alpha motif, *J. Am. Chem. Soc.*, 2002, **124**(18), 4976–4977.
51. C. Simmerling, B. Strockbine and A. E. Roitberg, All-atom structure prediction and folding simulations of a stable protein, *J. Am. Chem. Soc.*, 2002, **124**(38), 11258–11259.
52. C. D. Snow, N. Nguyen, V. S. Pande and M. Gruebele, Absolute comparison of simulated and experimental protein-folding dynamics, *Nature*, 2002, **420**(6911), 102–106.
53. D. Qiu, P. S. Shenkin, F. P. Hollinger and W. C. Still, The GB/SA continuum model for solvation. A fast analytical method for the calculation of approximate Born radii, *J. Phys. Chem. A*, 1997, **101**(16), 3005–3014.

54. M. Scarsi, J. Apostolakis and A. Caflisch, Continuum electrostatic energies of macromolecules in aqueous solutions, *J. Phys. Chem. A*, 1997, **101**(43), 8098–8106.
55. W. L. Jorgensen, J. P. Ulmschneider and J. Tirado-Rives, Free energies of hydration from a generalized Born model and an all-atom force field, *J. Phys. Chem. B*, 2004, **108**(41), 16264–16270.
56. A. Parsegian, Energy of an ion crossing a low dielectric membrane: solutions to four relevant electrostatic problems, *Nature*, 1969, **221**(5183), 844–846.
57. M. Schaefer, C. Bartels and M. Karplus, Solution conformations and thermodynamics of structured peptides: molecular dynamics simulation with an implicit solvation model, *J. Mol. Biol.*, 1998, **284**(3), 835–848.
58. J. A. Killian, Synthetic peptides as models for intrinsic membrane proteins, *FEBS Lett.*, 2003, **555**(1), 134–138.
59. M. R. de Planque and J. A. Killian, Protein–lipid interactions studied with designed transmembrane peptides: role of hydrophobic matching and interfacial anchoring, *Mol. Membr. Biol.*, 2003, **20**(4), 271–284.
60. M. C. Wiener and S. H. White, Fluid bilayer structure determination by the combined use of X-ray and neutron diffraction. I. Fluid bilayer models and the limits of resolution, *Biophys. J.*, 1991, **59**(1), 162–173.
61. M. C. Wiener and S. H. White, Fluid bilayer structure determination by the combined use of X-ray and neutron diffraction. II. 'Composition-space' refinement method, *Biophys. J.*, 1991, **59**(1), 174–185.
62. M. C. Wiener and S. H. White, Structure of a fluid dioleoylphosphatidylcholine bilayer determined by joint refinement of -ray and neutron diffraction data. III. Complete structure, *Biophys. J.*, 1992, **61**(2), 437–447.
63. S. H. White, G. I. King and J. I. Cain, Location of hexane in lipid bilayers determined by neutron diffraction, *Nature*, 1981, **290**, 161–163.
64. T. Hessa, H. Kim, K. Bihlmaier, C. Lundin, J. Boekel, H. Andersson, I. Nilsson, S. H. White and G. von Heijne, Recognition of transmembrane helices by the endoplasmic reticulum translocon, *Nature*, 2005, **433**(7024), 377–381.
65. A. Saaf, E. Wallin and G. von Heijne, Stop-transfer function of pseudo-random amino acid segments during translocation across prokaryotic and eukaryotic membranes, *Eur. J. Biochem.*, 1998, **251**(3), 821–829.
66. A. Radzicka and R. Wolfenden, Comparing the polarities of the amino-acids – side-chain distribution coefficients between the vapor-phase, cyclohexane, 1-octanol and neutral aqueous solution, *Biochemistry*, 1988, **27**(5), 1664–1670.
67. S. Yamaguchi, D. Huster, A. Waring, R. I. Lehrer, W. Kearney, B. F. Tack and M. Hong, Orientation and dynamics of an antimicrobial peptide in the lipid bilayer by solid-state NMR spectroscopy, *Biophys. J.*, 2001, **81**(4), 2203–2214.
68. B. Bechinger, M. Zasloff and S. J. Opella, Structure and orientation of the antibiotic peptide magainin in membranes by solid-state nuclear magnetic resonance spectroscopy, *Protein Sci.*, 1993, **2**(12), 2077–2084.

69. A. Ramamoorthy, F. M. Marassi, M. Zasloff and S. J. Opella, Three-dimensional solid-state NMR spectroscopy of a peptide oriented in membrane bilayers, *J. Biomol. NMR*, 1995, **6**(3), 329–334.

70. F. M. Marassi, S. J. Opella, P. Juvvadi and R. B. Merrifield, Orientation of cecropin A helices in phospholipid bilayers determined by solid-state NMR spectroscopy, *Biophys. J.*, 1999, **77**(6), 3152–3155.

71. M. B. Ulmschneider, M. S. P. Sansom and A. Di Nola, Evaluating tilt angles of membrane-associated helices: comparison of computational and NMR techniques, *Biophys. J.*, 2006, **90**, 1650–1660.

72. E. A. Cohen, E. F. Terwilliger, J. G. Sodroski and W. A. Haseltine, Identification of a protein encoded by the Vpu gene of HIV-1, *Nature*, 1988, **334**(6182), 532–534.

73. K. Strebel, T. Klimkait and M. A. Martin, A novel gene of HIV-1, Vpu and its 16-kilodalton product, *Science*, 1988, **241**(870), 1221–1223.

74. F. M. Marassi, C. Ma, H. Gratkowski, S. K. Straus, K. Strebel, M. Oblatt-Montal, M. Montal and S. J. Opella, Correlation of the structural and functional domains in the membrane protein Vpu from HIV-1, *Proc. Natl. Acad. Sci. USA*, 1999, **96**(25), 14336–14341.

75. R. L. Willey, F. Maldarelli, M. A. Martin and K. Strebel, Human-immunodeficiency-virus type-1 Vpu protein regulates the formation of intracellular Gp160–Cd4 complexes, *J. Virol.*, 1992, **66**(1), 226–234.

76. K. Levesque, A. Finzi, J. Binette and E. A. Cohen, Role of CD4 receptor down-regulation during HIV-1 infection, *Curr. HIV Res.*, 2004, **2**(1), 51–59.

77. T. Klimkait, K. Strebel, M. D. Hoggan, M. A. Martin and J. M. Orenstein, The human immunodeficiency virus type 1-specific protein Vpu is required for efficient virus maturation and release, *J. Virol.*, 1990, **64**(2), 621–629.

78. K. Strebel, T. Klimkait, F. Maldarelli and M. A. Martin, Molecular and biochemical analyses of human immunodeficiency virus type-1 Vpu protein, *J. Virol.*, 1989, **63**(9), 3784–3791.

79. F. Maldarelli, M. Y. Chen, R. L. Willey and K. Strebel, Human-immunodeficiency-virus type-1 Vpu protein is an oligomeric type-I integral membrane-protein, *J. Virol.*, 1993, **67**(8), 5056–5061.

80. C. Ma, F. M. Marassi, D. H. Jones, S. K. Straus, S. Bour, K. Strebel, U. Schubert, M. Oblatt-Montal, M. Montal and S. J. Opella, Expression, purification and activities of full-length and truncated versions of the integral membrane protein Vpu from HIV-1, *Protein Sci.*, 2002, **11**(3), 546–557.

81. G. G. Kochendoerfer, D. H. Jones, S. Lee, M. Oblatt-Montal, S. J. Opella and M. Montal, Functional characterization and NMR spectroscopy on full-length Vpu from HIV-1 prepared by total chemical synthesis, *J. Am. Chem. Soc.*, 2004, **126**(8), 2439–2446.

82. U. Schubert, A. V. Ferrer-Montiel, M. Oblatt-Montal, P. Henklein, K. Strebel and M. Montal, Identification of an ion channel activity of the Vpu transmembrane domain and its involvement in the regulation

of virus release from HIV-1-infected cells, *FEBS Lett.*, 1996, **398**(1), 12–18.

83. G. D. Ewart, T. Sutherland, P. W. Gage and G. B. Cox, The Vpu protein of human immunodeficiency virus type 1 forms cation-selective ion channels, *J. Virol.*, 1996, **70**(10), 7108–7115.

84. W. Romer, Y. H. Lam, D. Fischer, A. Watts, W. B. Fischer, P. Goring, R. B. Wehrspohn, U. Gosele and C. Steinem, Channel activity of a viral transmembrane peptide in micro-BLMs: Vpu(1–32) from HIV-1, *J. Am. Chem. Soc.*, 2004, **126**(49), 16267–16274.

85. S. H. Park, A. A. Mrse, A. A. Nevzorov, M. F. Mesleh, M. Oblatt-Montal, M. Montal and S. J. Opella, Three-dimensional structure of the channel-forming transmembrane domain of virus protein 'u' (Vpu) from HIV-1, *J. Mol. Biol.*, 2003, **333**(2), 409–424.

86. S. H. Park, A. A. De Angelis, A. A. Nevzorov, C. H. Wu and S. J. Opella, Three-dimensional structure of the transmembrane domain of Vpu from HIV-1 in aligned phospholipid bicelles, *Biophys. J.*, 2006, **91**, 3032–3042.

87. S. H. Park and S. J. Opella, Tilt angle of a transmembrane helix is determined by hydrophobic mismatch, *J. Mol. Biol.*, 2005, **350**(2), 310–318.

88. S. Sharpe, W. M. Yau and R. Tycko, Structure and dynamics of the HIV-1 Vpu transmembrane domain revealed by solid-state NMR with magic-angle spinning, *Biochemistry*, 2006, **45**(3), 918–933.

89. A. Kukol and I. T. Arkin, Vpu transmembrane peptide structure obtained by site-specific Fourier transform infrared dichroism and global molecular dynamics searching, *Biophys. J.*, 1999, **77**(3), 1594–1601.

90. V. Lemaitre, D. Willbold, A. Watts and W. B. Fischer, Full length Vpu from HIV-1: combining molecular dynamics simulations with NMR spectroscopy, *J. Biomol. Struct. Dyn.*, 2006, **23**(5), 485–496.

91. I. Sramala, V. Lemaitre, J. D. Faraldo-Gomez, S. Vincent, A. Watts and W. B. Fischer, Molecular dynamics simulations on the first two helices of Vpu from HIV-1, *Biophys. J.*, 2003, **84**(5), 3276–3284.

92. F. Sun, Molecular dynamics simulation of human immunodeficiency virus protein U (Vpu) in lipid/water Langmuir monolayer, *J. Mol. Model.*, 2003, **9**(2), 114–123.

93. A. L. Grice, I. D. Kerr and M. S. P. Sansom, Ion channels formed by HIV-1 Vpu: a modelling and simulation study, *FEBS Lett.*, 1997, **405**(3), 299–304.

94. P. B. Moore, Q. F. Zhong, T. Husslein and M. L. Klein, Simulation of the HIV-1 Vpu transmembrane domain as a pentameric bundle, *FEBS Lett.*, 1998, **431**(2), 143–148.

95. F. S. Cordes, A. Kukol, L. R. Forrest, I. T. Arkin, M. S. P. Sansom and W. B. Fischer, The structure of the HIV-1 Vpu ion channel: modelling and simulation studies, *Biochim. Biophys. Acta Biomembr.*, 2001, **1512**(2), 291–298.

96. F. S. Cordes, A. D. Tustian, M. S. P. Sansom, A. Watts and W. B. Fischer, Bundles consisting of extended transmembrane segments of Vpu

from HIV-1: computer simulations and conductance measurements, *Biochemistry*, 2002, **41**(23), 7359–7365.

97. C. F. Lopez, M. Montal, J. K. Blasie, M. L. Klein and P. B. Moore, Molecular dynamics investigation of membrane-bound bundles of the channel-forming transmembrane domain of viral protein U from the human immunodeficiency virus HIV-1, *Biophys. J.*, 2002, **83**(3), 1259–1267.

98. C. G. Kim, V. Lemaitre, A. Watts and W. B. Fischer, Drug–protein interaction with Vpu from HIV-1: proposing binding sites for amiloride and one of its derivatives, *Anal. Bioanal. Chem.*, 2006, **386**(7–8), 2213–2217.

99. V. Lemaitre, R. Ali, C. G. Kim, A. Watts and W. B. Fischer, Interaction of amiloride and one of its derivatives with Vpu from HIV-1: a molecular dynamics simulation, *FEBS Lett.*, 2004, **563**(1–3), 75–81.

100. P. J. Bond, J. Holyoake, A. Ivetac, S. Khalid and M. S. P. Sansom, Coarse-grain molecular dynamics simulations of membrane proteins and peptides, *J. Struct. Biol.*, 2007, **157**(3), 593–605.

101. M. B. Ulmschneider, M. S. Sansom and A. Di Nola, Evaluating tilt angles of membrane-associated helices: comparison of computational and NMR techniques, *Biophys. J.*, 2006, **90**(5), 1650–1660.

102. G. von Heijne, Principles of membrane protein assembly and structure, *Prog. Biophys. Mol. Biol.*, 1997, **66**, 113–139.

103. M. B. Ulmschneider and M. S. P. Sansom, Amino acid distributions in integral membrane protein structures, *Biochim. Biophys. Acta Biomembr.*, 2001, **1512**(1), 1–14.

104. W. M. Yau, W. C. Wimley, K. Gawrisch and S. H. White, The preference of tryptophan for membrane interfaces, *Biochemistry*, 1998, **37**(42), 14713–14718.

105. E. Strandberg, S. Morein, D. T. Rijkers, R. M. Liskamp, P. C. van der Wel and J. A. Killian, Lipid dependence of membrane anchoring properties and snorkeling behavior of aromatic and charged residues in transmembrane peptides, *Biochemistry*, 2002, **41**(23), 7190–7198.

106. M. R. de Planque, J. A. Kruijtzer, R. M. Liskamp, D. Marsh, D. V. Greathouse, R. E. Koeppe II, B. de Kruijff and J. A. Killian, Different membrane anchoring positions of tryptophan and lysine in synthetic transmembrane alpha-helical peptides, *J. Biol. Chem.*, 1999, **274**(30), 20839–20846.

107. S. H. White and G. von Heijne, Transmembrane helices before, during and after insertion, *Curr. Opin. Struct. Biol.*, 2005, **15**(4), 378–386.

108. A. S. Ladokhin and S. H. White, Interfacial folding and membrane insertion of a designed helical peptide, *Biochemistry*, 2004, **43**(19), 5782–5791.

109. T. Hessa, N. M. Meindl-Beinker, A. Bernsel, H. Kim, Y. Sato, M. Lerch-Bader, I. Nilsson, S. H. White and G. von Heijne, Molecular code for transmembrane-helix recognition by the Sec61 translocon, *Nature*, 2007, **450**(7172), 1026–1030.

110. B. Van den Berg, W. M. Clemons Jr, I. Collinson, Y. Modis, E. Hartmann, S. C. Harrison and T. A. Rapoport, X-ray structure of a protein-conducting channel, *Nature*, 2004, **427**(6969), 36–44.

111. J. L. Popot and D. M. Engelman, Membrane-protein folding and oligomerization – the 2-stage model, *Biochemistry*, 1990, **29**(17), 4031–4037.

112. S. H. White, A. S. Ladokhin, S. Jayasinghe and K. Hristova, How membranes shape protein structure, *J. Biol. Chem.*, 2001, **276**(35), 32395–32398.

113. S. H. White, How hydrogen bonds shape membrane protein structure, *Adv. Protein Chem.*, 2006, **72**, 157–172.

114. M. R. R. de Planque and J. A. Killian, Protein–lipid interactions studied with designed transmembrane peptides: role of hydrophobic matching and interfacial anchoring (review), *Mol. Membr. Biol.*, 2003, **20**(4), 271–284.

115. J. P. Ulmschneider, M. B. Ulmschneider and A. Di Nola, Monte Carlo vs molecular dynamics for all-atom polypeptide folding simulations, *J. Phys. Chem. B*, 2006, **110**(33), 16733–16742.

116. D. M. Engelman, Y. Chen, C. N. Chin, A. R. Curran, A. M. Dixon, A. D. Dupuy, A. S. Lee, U. Lehnert, E. E. Matthews, Y. K. Reshetnyak, A. Senes and J. L. Popot, Membrane protein folding: beyond the two stage model, *FEBS Lett.*, 2003, **555**(1), 122–125.

117. R. E. Jacobs and S. H. White, The nature of the hydrophobic binding of small peptides at the bilayer interface: implications for the insertion of transbilayer helices, *Biochemistry*, 1989, **28**(8), 3421–3437.

118. S. H. White and W. C. Wimley, Membrane protein folding and stability: physical principles, *Annu. Rev. Biophys. Biomol. Struct.*, 1999, **28**, 319–365.

119. N. BenTal, D. Sitkoff, I. A. Topol, A. S. Yang, S. K. Burt and B. Honig, Free energy of amide hydrogen bond formation in vacuum, in water and in liquid alkane solution, *J. Phys. Chem. B*, 1997, **101**(3), 450–457.

120. M. A. Roseman, Hydrophobicity of the peptide $C{=}O\cdots H{-}N$ hydrogen-bonded group, *J. Mol. Biol.*, 1988, **201**(3), 621–623.

121. W. C. Wimley, T. P. Creamer and S. H. White, Solvation energies of amino acid side chains and backbone in a family of host–guest penta-peptides, *Biochemistry*, 1996, **35**(16), 5109–5124.

122. A. D. MacKerell, D. Bashford, M. Bellott, R. L. Dunbrack, J. D. Evanseck, M. J. Field, S. Fischer, J. Gao, H. Guo, S. Ha, D. Joseph-McCarthy, L. Kuchnir, K. Kuczera, F. T. K. Lau, C. Mattos, S. Michnick, T. Ngo, D. T. Nguyen, B. Prodhom, W. E. Reiher, B. Roux, M. Schlenkrich, J. C. Smith, R. Stote, J. Straub, M. Watanabe, J. Wiorkiewicz-Kuczera, D. Yin and M. Karplus, All-atom empirical potential for molecular modeling and dynamics studies of proteins, *J. Phys. Chem. B*, 1998, **102**(18), 3586–3616.

123. B. R. Brooks, R. E. Bruccoleri, B. D. Olafson, D. J. States, S. Swaminathan and M. Karplus, Charmm – a program for macromolecular energy, minimization and dynamics calculations, *J. Comput. Chem.*, 1983, **4**(2), 187–217.

124. M. R. R. de Planque, E. Goormaghtigh, D. V. Greathouse, R. E. Koeppe, J. A. W. Kruijtzer, R. M. J. Liskamp, B. de Kruijff and J. A. Killian, Sensitivity of single membrane-spanning alpha-helical peptides to hydrophobic mismatch with a lipid bilayer: effects on backbone structure, orientation and extent of membrane incorporation, *Biochemistry*, 2001, **40**(16), 5000–5010.

125. P. C. A. van der Wel, E. Strandberg, J. A. Killian and R. E. Koeppe, Geometry and intrinsic tilt of a tryptophan-anchored transmembrane alpha-helix determined by H-2 NMR, *Biophys. J.*, 2002, **83**(3), 1479–1488.

126. T. M. Weiss, P. C. A. van der Wel, J. A. Killian, R. E. Koeppe and H. W. Huang, Hydrophobic mismatch between helices and lipid bilayers, *Biophys. J.*, 2003, **84**(1), 379–385.

127. M. R. R. de Planque, D. V. Greathouse, R. E. Koeppe, H. Schafer, D. Marsh and J. A. Killian, Influence of lipid/peptide hydrophobic mismatch on the thickness of diacylphosphatidylcholine bilayers. A H-2 NMR and ESR study using designed transmembrane alpha-helical peptides and gramicidin A, *Biochemistry*, 1998, **37**(26), 9333–9345.

128. H. I. Petrache, D. M. Zuckerman, J. N. Sachs, J. A. Killian, R. E. Koeppe and T. B. Woolf, Hydrophobic matching mechanism investigated by molecular dynamics simulations, *Langmuir*, 2002, **18**(4), 1340–1351.

129. M. R. de Planque, B. B. Bonev, J. A. Demmers, D. V. Greathouse, R. E. Koeppe, F. Separovic, A. Watts and J. A. Killian II, Interfacial anchor properties of tryptophan residues in transmembrane peptides can dominate over hydrophobic matching effects in peptide–lipid interactions, *Biochemistry*, 2003, **42**(18), 5341–5348.

130. H. Nymeyer, T. B. Woolf and A. E. Garcia, Folding is not required for bilayer insertion: replica exchange simulations of an alpha-helical peptide with an explicit lipid bilayer, *Proteins Struct. Funct. Bioinf.*, 2005, **59**(4), 783–790.

131. W. Im, J. Chen and C. L. Brooks III, Peptide and protein folding and conformational equilibria: theoretical treatment of electrostatics and hydrogen bonding with implicit solvent models, *Adv. Protein Chem.*, 2005, **72**, 173–198.

132. M. B. Ulmschneider, D. P. Tieleman and M. S. P. Sansom, Interactions of a transmembrane helix and a membrane: comparative simulations of bacteriorhodopsin helix A, *J. Phys. Chem. B*, 2004, **108**(28), 10149–10159.

133. J. P. Ulmschneider and M. B. Ulmschneider, Sampling efficiency in explicit and implicit membrane environments studied by peptide folding simulations, *Proteins Struct. Funct. Bioinf.*, 2008, **75**, 586–597.

134. R. H. Zhou and B. J. Berne, Can a continuum solvent model reproduce the free energy landscape of a beta-hairpin folding in water?, *Proc. Natl. Acad. Sci. USA*, 2002, **99**(20), 12777–12782.

135. H. Nymeyer and A. E. Garcia, Simulation of the folding equilibrium of a-helical peptides: a comparison of the generalized Born approximation with explicit solvent, *Proc. Natl. Acad. Sci. USA*, 2003, **100**(24), 13934–13939.

136. I. Daidone, M. B. Ulmschneider, A. Di Nola, A. Amadei and J. C. Smith, Dehydration-driven solvent exposure of hydrophobic surfaces as a driving force in peptide folding, *Proc. Natl. Acad. Sci. USA*, 2007, **104**(39), 15230–15235.

137. H. Fan, A. E. Mark, J. Zhu and B. Honig, Chemical theory and computation special feature: comparative study of generalized Born models: protein dynamics, *Proc. Natl. Acad. Sci. USA*, 2005, **102**(19), 6760–6764.

138. V. P. Bojan, Zagrovic solvent viscosity dependence of the folding rate of a small protein: distributed computing study, *J. Comput. Chem.*, 2003, **24**(12), 1432–1436.

139. M. Feig, Kinetics from implicit solvent simulations of biomolecules as a function of viscosity, *J. Chem. Theory Comput.*, 2007, **3**(5), 1734–1748.

140. M. Feig, J. Chocholoušová and S. Tanizaki, Extending the horizon: towards the efficient modeling of large biomolecular complexes in atomic detail, *Theor. Chem. Acc. Theory Comput. Model. (Theor. Chim. Acta)*, 2006, **116**(1), 194–205.

141. K. Hristova and S. H. White, An experiment-based algorithm for predicting the partitioning of unfolded peptides into phosphatidylcholine bilayer interfaces, *Biochemistry*, 2005, **44**(37), 12614–12619.

142. S. H. White and G. von Heijne, Do protein–lipid interactions determine the recognition of transmembrane helices at the ER translocon?, *Biochem. Soc. Trans.*, 2005, **33**(Pt 5), 1012–1015.

Multi-scale Simulations of Membrane Sculpting by N-BAR Domains

YING YIN*, ANTON ARKHIPOV* AND KLAUS SCHULTEN

Department of Physics and Beckman Institute, University of Illinois at Urbana–Champaign, 405 N. Mathews, Urbana, IL 61801, USA

6.1 Introduction

Living cells feature a large variety of lipid membrane shapes, forming barriers, organelles and compartments. Numerous essential cellular processes, including growth, division and movement, rely on the dynamic modification of the membrane shape, which is carried out by proteins that employ a variety of mechanisms.[1–14] Due to technological advances, observational membrane morphogenesis is currently progressing rapidly and a number of exciting discoveries have been achieved.[4,9,15–18] Theoretical and computational studies have complemented experiments through important insights.[8,11,12,19–21]

Proteins of the BAR domain superfamily drive the formation of tubular and vesicular membrane structures in cells.[15,22–24] *In vitro*, BAR domains are found to bind to liposomes and reshape low-curvature spheres into high-curvature tubes and vesicles.[25] The crystallographically resolved structures of BAR domains[15,16,24,26–31] exhibit crescent-shaped dimers with a high density of

---

*These authors contributed equally to this work.

RSC Biomolecular Sciences No. 20
Molecular Simulations and Biomembranes: From Biophysics to Function
Edited by Mark S.P. Sansom and Philip C. Biggin

Published by the Royal Society of Chemistry, www.rsc.org

positively charged residues on one side of their curved surface. Extensive studies suggest that BAR domains induce membrane curvature by scaffolding a negatively charged membrane with the protein's curved and positively charged surface. To curve large pieces of membrane, multiple BAR domains have to work together, but it has been unclear how this happens. A breakthrough came when Frost *et al.*[24] demonstrated through cryo-electron microscopy (cryo-EM) reconstructions that a type of BAR domain, F-BAR domain, forms well-organized rows or spirals on the surface of membrane tubes. Other members of the BAR domain family, such as amphiphysin BAR domains, were found to form similar arrangements,[24,25] although a high-resolution visualization has not been obtained yet for these proteins. Hence the experimental evidence suggests that BAR domains act concertedly in tight, well-ordered formation, forming longitudinal lines around membrane tubes. Differences in this formation may lead to the observed variability in tube geometry.

Despite extensive studies, many questions regarding the molecular mechanism underlying membrane scaffolding by BAR domains remain unanswered. How much curvature does a single BAR domain induce? How does the arrangement of BAR domains in rows determine the tube diameter? Furthermore, do the fascinating cryo-EM images of F-BAR domain-induced tubes,[24] obtained after multiple cycles of annealing and freezing, correspond to the tubulation mechanism at physiological temperature? These images furnish static pictures of membrane tubes and the question is what the dynamic picture of tube formation looks like. The difficulty of addressing the stated questions is due to the heterogeneity of the membrane bending process, where the constituent lipids exist in partially disordered phases and where the participating proteins are distributed non-uniformly, restricting the application of many experimental imaging methods.

The problems faced by experimental techniques can be overcome using molecular dynamics (MD) simulations,[32–34] which offer a highly detailed description (in both space and time) of molecular systems, as demonstrated by recent computational studies.[8,11,14] Unfortunately, with current computing capabilities, MD simulations are limited to system sizes and time scales that are too small and too short. In order to overcome the limitations of MD simulations, one can utilize a multi-scale approach, which permits one to reach desired length and time scales, albeit at the expense of dispensing with atomic level detail.

A four-scale computational approach, applied to one type of BAR domain, is described below. This multi-scale treatment allows one to observe membrane sculpting by a single BAR domain and also by multiple BAR domains. One specific BAR domain type, namely the amphiphysin N-BAR domain from *Drosophila*, is considered. The name 'N-BAR' refers here to the fact that its structure includes N-terminal helices, which may contribute to membrane bending and protein anchoring to the membrane. Various lattices formed by multiple copies of N-BAR domains arranged on membrane surface are found to induce a wide range of membrane curvatures. Detailed, dynamic pictures of the formation of membrane tubes by N-BAR domain lattices over time scales of $\sim 200\,\mu s$ are obtained by the computational approach.

The multi-scale approach employs computations at different resolution levels, namely (i) all-atom MD, (ii) residue-based coarse-grained MD,[35–38] which resolves single amino acids and lipid molecules, (iii) shape-based coarse-grained MD,[38–40] which resolves overall protein and membrane shapes, and (iv) a mesoscopic continuum description of membrane and proteins through a model based on elasticity theory.[41] All-atom MD provides single-atom resolution for the process of interest, but falls short of describing overall membrane remodeling due to limited size and time scales. Here it is applied in the extreme to a system of 2.3 million atoms for 0.3 µs. Coarse-grained (CG) MD techniques permit one to reach larger system sizes and longer time scales by employing a simplified representation of the participating molecules, but at the price of losing accuracy. CG MD approaches, under development in recent years, have usually been derived for specific systems, such as lipid membranes,[35,42–53] specific proteins[54–56] or protein–membrane systems and protein assemblies.[20,36–40,57–65] At an even coarser level of description, continuum models were used to study elastic properties, remodeling and fusion of membranes.[41,66–70] The disadvantage of continuum models is that interactions between individual proteins and between proteins and membrane are not resolved, but the models permit fast computation that guides more detailed CG MD simulations, which in turn guide all-atom MD simulations. We show below that combining all-atom MD with two CG MD techniques and a continuum description opens up the opportunity of moving systematically between multiple levels of resolution.

6.2 Methods

Interactions of a membrane-sculpting protein, N-BAR domain, with negatively charged lipid bilayer membranes are studied at four scales: the atomic level, the residue-based coarse-grained (RBCG) level with ~10 atoms represented by a CG bead and single-residue resolution (see Figure 6.1A), the shape-based coarse-grained (SBCG) level with ~150 atoms per CG bead and a group of beads per protein (see Figure 6.1B) and the continuum level. The RBCG and SBCG descriptions are described and reviewed elsewhere.[38] Each description level is parameterized based on the more detailed description level. All-atom MD uses the CHARMM force-field,[71,72] derived from quantum chemistry calculations. The next level of coarseness is furnished by RBCG MD,[35–37] which describes single residues by a few CG beads (*e.g.* an amino acid is represented by two beads, one for the backbone and the other for the side chain). The RBCG model is parameterized using experimental data (as done for lipids in the Marrink model[35]), structural considerations[37] and all-atom simulations.[11] The SBCG model[39,40] uses CG beads to represent protein segments, whose dynamics are described as in MD simulations. SBCG parameterization is based on results of all-atom simulations and on experimental data, such as area per lipid for membrane bilayers.[11] For all three levels, MD simulations are performed using NAMD,[73] and analysis is carried out with

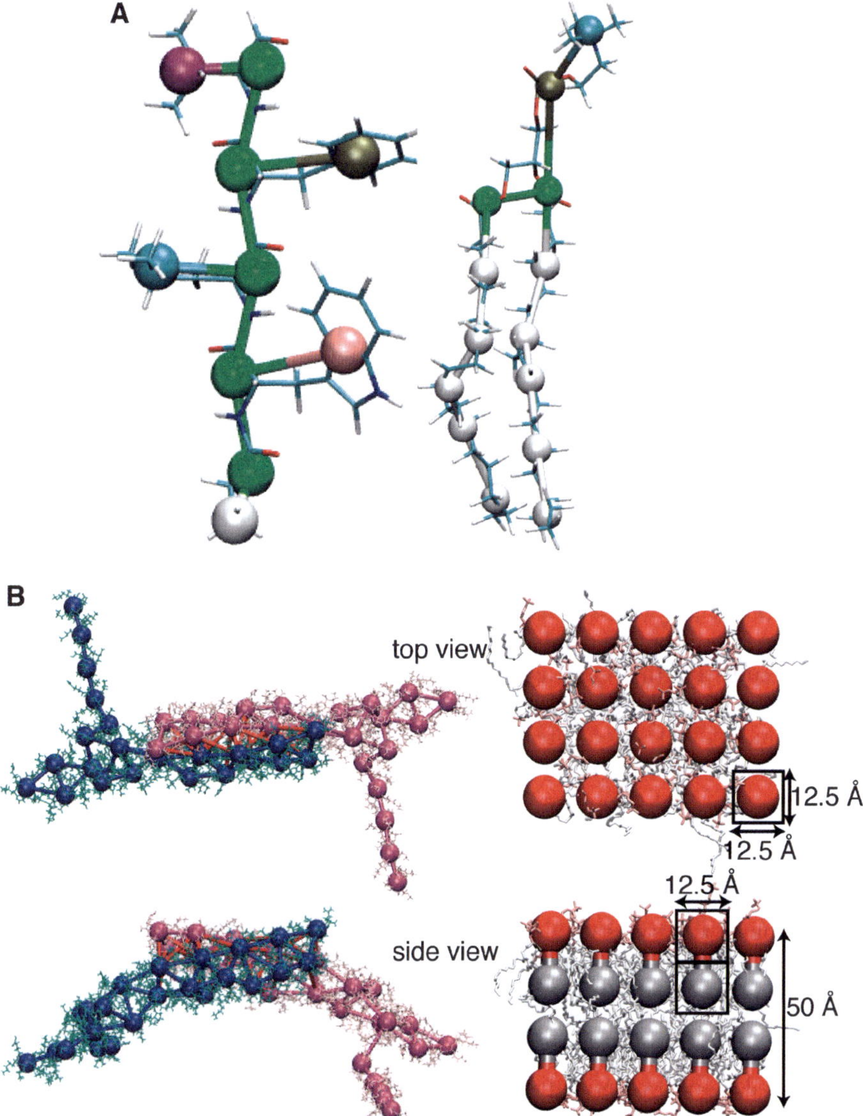

Figure 6.1 Residue-based coarse-grained (RBCG) and shape-based coarse-grained (SBCG) models. (A) Overlap of all-atom (in licorice representation) and RBCG models (the latter shown through CG beads) for a short peptide (sequence AWLFV) and for a DOPC lipid. RBCG uses ∼10 atoms per CG bead; an amino acid is represented by one bead for the backbone and another one for the side chain. (B) Overlap of all-atom and SBCG models of N-BAR domains (dimer, different color per monomer) and lipids viewed from top and side. In the SBCG model used here, each SBCG bead represents ∼150 atoms. The N-BAR domain dimer is represented by 50 SBCG beads, the two monomers (in blue and purple) being connected by bonds (red) to preserve the integrity of the dimer. Each SBCG lipid 'molecule' represents ∼2.2 lipids; the SBCG 'head' beads are shown in red and the 'tail' beads in gray.

VMD.[74] The elastic membrane model at the continuum level is parameterized using SBCG simulations and experimental data (*e.g.* with regard to the Young modulus for membrane extensibility).[11]

The atomic coordinates of *Drosophila melanogaster* amphiphysin BAR domain were taken from the Protein Data Bank (pdb code: 1URU[15]); 25 residues missing at the N-terminus were modeled as a short helix and a flexible link.[11] An N-BAR domain homodimer (see Figure 6.1B) was used as a unit protein in the simulations, since the homodimer is expected to be an active form of the protein in *in vitro* experiments. Both RBCG and SBCG models of the N-BAR domain were based on the atomic model.

A lipid membrane studied was composed of 70% DOPC lipids (neutral) and 30% DOPS lipids (each DOPS lipid carries a charge of $-1|e|$). Lipid bilayers in all simulations were initially flat, with one or more N-BAR domains placed on top of the membrane. Periodic boundary conditions used in the simulations are illustrated in Figure 6.2. The bilayers were discontinuous in their long dimension to avoid the influence of periodic boundary conditions on membrane bending and complete tubulation. During simulations, the discontinuous membrane edges rearranged to adopt a semi-micellar structure at a time scale of a few nanoseconds. In experiments, BAR domains pull tubes out of liposomes, instead of bending planar, discontinuous membrane patches, but simulating a liposome tubulation over relevant time scales would be impossible even with CG methods. Furthermore, conditions of our simulations and of experiments differ in the respect that the simulations employ periodic boundary conditions, which may affect curvature formation and they cover only short times (microseconds, *versus* minutes in experiments). Despite these limitations, employing discontinuous membrane patches allows one to probe various arrangements of N-BAR domains in a straightforward manner. As a result, a broad range of N-BAR lattice types is sampled here to investigate physical principles behind the concerted action of N-BAR domains.

A significant speed-up in computations is achieved when the RBCG and SBCG models are used. The RBCG model reduces the number of particles in the system ten-fold, permitting an integration time step of 20 fs, *versus* 1–2 fs common for all-atom MD. The SBCG model with ~ 150 atoms per CG bead and describing the solvent implicitly, results in an overall reduction of system size of ~ 750-fold, permitting an integration time step of 100 fs. For example, simulations that would encompass up to 100 million atoms in an all-atom description were performed in an SBCG representation with slightly over 10 000 CG beads on 48 processors, with the average performance of $\sim 1\,\mu s$ per day.

6.3 All-atom Simulations

All-atom simulations of a single N-BAR domain were used to parameterize the RBCG and SBCG models.[11] Such simulations described also a 2.3-million atom water–lipid–protein system of a lattice formed by eight N-BAR domains

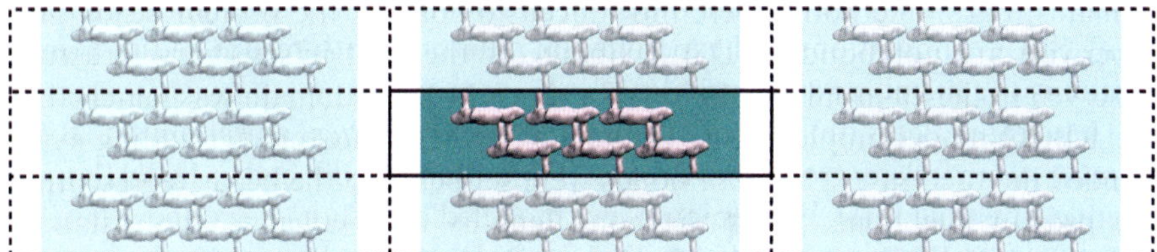

Figure 6.2 Example of a system setup with periodic boundary conditions for simulations of an N-BAR domain lattice on a membrane patch. A single periodic cell is highlighted as a solid square and a few periodic images are shown in dimmed color; boundaries of periodic cells are marked by dotted lines. The view is from the top on the surface of the membrane bilayer. A lattice of six BAR domains is placed on the membrane surface. In the horizontal dimension, the membrane size is smaller than the periodic cell size, to allow for easier bending of the membrane.

to complement and validate SBCG simulations. A membrane patch with dimensions of $46 \times 10\,\mathrm{nm}^2$ was employed for all-atom MD simulation of one N-BAR domain and patch dimensions of $64 \times 8\,\mathrm{nm}^2$ were used for the eight N-BAR domains simulation. After adding water and ions, the simulated systems consisted of 671 331 and 2 304 973 atoms, respectively. Simulation details can be found elsewhere.[11,14] Briefly, the simulations employed the CHARMM[71,72] force field and were performed in the NpT ensemble (temperature 310 K and pressure 1 atm). An integration time step of 2 fs was used. The simulations covered 25 ns for the single N-BAR domain and 0.3 µs for the eight N-BAR domain case.

6.4   Residue-based Coarse-grained Simulations

The RBCG method is based on the Marrink CG model,[35,46] the recent releases of which are called MARTINI.[52] The MARTINI model or its modifications have been extensively used to study lipid–protein systems, such as high-density lipoproteins (HDLs)[36,37,60,61] and integral membrane proteins.[57–59,65,75,76] For studying N-BAR domains, we employed the RBCG model where ~ 10 atoms (including hydrogens) are represented by a single CG bead (Figure 6.1A). For example, a DOPC lipid is represented by 14 CG beads: one for the choline group, one for the phosphate group, two for each of the glycerol groups and 10 to represent the two hydrocarbon tails. Four water molecules are represented by a single CG bead; an ion together with its first hydration shell (six water molecules) is represented by one CG bead; each amino acid is represented by two CG beads, one for the backbone and one for the side chain (glycine is represented by a single backbone CG bead). Effective interaction potentials between CG beads are assigned to reproduce their hydrophobic/hydrophilic properties,[35] and time evolution is described using classical MD.

Originally, this RBCG model was parameterized[37,60] to maintain only secondary structure of a protein. To preserve the tertiary structure of N-BAR domains, harmonic bonds were introduced to connect the protein beads that otherwise are not bonded. The flexibility of the protein's tertiary structure observed in the all-atom simulation of a single N-BAR domain was matched in the RBCG model (implemented through a NAMD[73] feature for adding extra bonded interactions). The extra bonds were set between the beads representing the protein backbone,[11] and the bond lengths were equal to the distances between the beads as found in the crystal structure.

Several distances and angles characterizing the overall structure of N-BAR domain, as illustrated in Figure 6.3A, were monitored in the all-atom simulation of a single N-BAR domain and in equivalent RBCG simulations with different strength of the extra bonds (Figure 6.3B–G). Figure 6.3 shows that the strength for extra bonds that gives the best agreement with all-atom simulations is $K = 5\,\text{kcal}\,\text{mol}^{-1}\,\text{Å}^{-2}$). Simulations with weaker extra bonds resulted in a significant change of the structure, while stronger extra bonds made the protein stiffer than it is in the all-atom simulation [see, *e.g.*, the root mean square deviation (RMSD) values in Figure 6.3B].

All-atom simulations of an N-BAR domain[8,11,13,14] suggest that membrane bending is due to electrostatic interactions. Unfortunately, the RBCG model does not provide a description to reproduce fine-scale electrostatic effects. A medium with uniform dielectric constant is used instead to model electrostatic interactions. In order to parameterize this approach for the N-BAR domain system, several RBCG simulations of a single N-BAR domain interacting with a membrane were performed assuming dielectric constants $\varepsilon = 1$, 2, 10 and 20. Among these simulations, only the one with $\varepsilon = 1$ exhibited significant membrane bending, comparable to that observed in all-atom simulations. Therefore, Coulomb interactions in the RBCG model were modeled using $\varepsilon = 1$.

RBCG simulations were performed for a single N-BAR domain dimer ($47 \times 10\,\text{nm}^2$ membrane patch, the total system consisted of 74 916 beads) and for two lattices of six N-BAR domains (110×12 and $74 \times 18\,\text{nm}^2$ for the membrane patches, the fully solvated and ionized system comprising 223 017 and 181 873 CG beads). RBCG simulations were performed in the NpT ensemble (temperature 310 K and pressure 1 atm). A time step of 20 fs was used. Three single N-BAR domain simulations were performed, each covering 50 ns and one simulation was performed for each of the six N-BAR domain systems, covering 20 and 50 ns for the larger and smaller systems, respectively.

6.5 Shape-based Coarse-grained Simulations

The SBCG method was developed to simulate proteins and their assemblies[11,14,39,40] using a ratio of 150 or 500 atoms per CG bead. For the study of N-BAR domains, 150 atoms per bead were used. Each SBCG bead describes a chunk of a protein or a Voronoi cell (which for every bead is comprised from atoms that are closer to that bead than to any other bead). The mass of the Voronoi cell and its charge are assigned to the bead and time evolution of the whole system is described using classical MD. Interactions between beads are

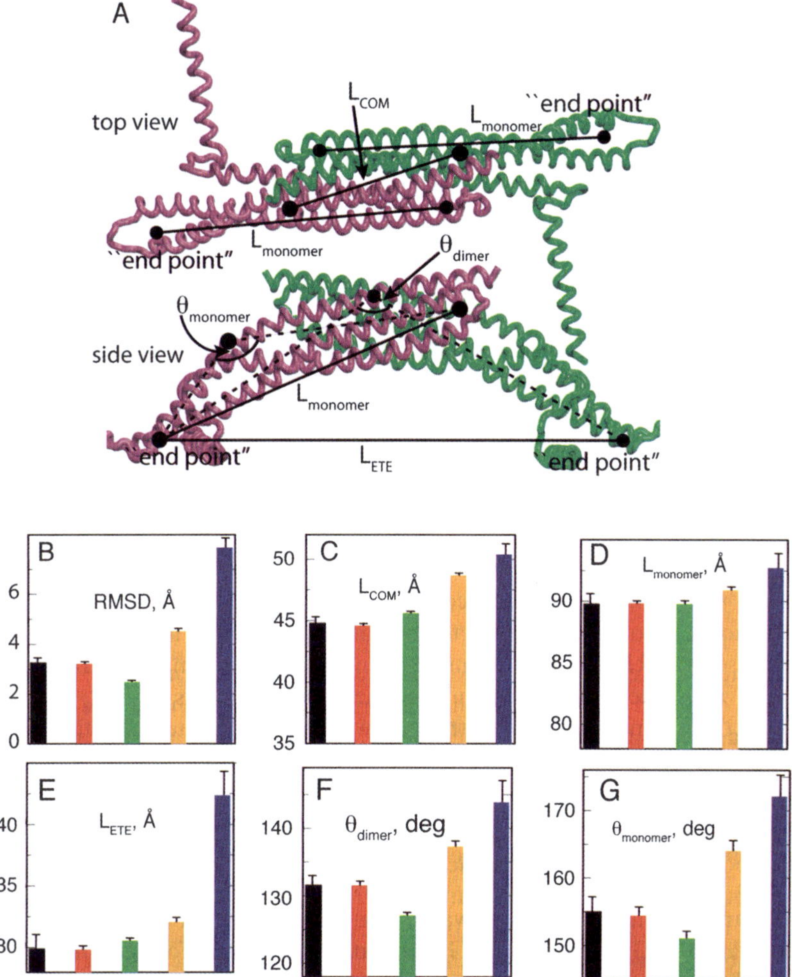

Figure 6.3 Constraints used to maintain the tertiary structure and inter-domain arrangement for the RBCG model of N-BAR domain. (A) The N-BAR domain dimer (one monomer in purple, the other in green) is shown from the top and from the side. Several distances and angles are chosen that characterize the overall structure of the protein, namely, the distance between the centers of mass of the two monomers (L_{COM}), end-to-end distance for one monomer ($L_{monomer}$), end-to-end distance for the dimer (L_{ETE}) and the opening angles for the dimer and for the monomer (θ_{dimer} and $\theta_{monomer}$). Averages of these values, and also of C_a-RMSD for the dimer, are shown (B–G) for, from left to right, the all-atom simulation (black) and RBCG simulations with $K = 5$ (red), 25 (green), 0.5 (orange) and $0\,\text{kcal}\,\text{mol}^{-1}\,\text{Å}^{-2}$ (blue). $K = 5\,\text{kcal}\,\text{mol}^{-1}\,\text{Å}^{-2}$ provides the best agreement between the all-atom and RBCG simulations.

described by a CHARMM-like force field,[71] *i.e.* bonded interactions are represented by harmonic bond and angle potentials (but no dihedral potentials) and the non-bonded potentials include 6–12 Lennard-Jones (LJ) and Coulomb terms.

The solvent is modeled implicitly, by using the Langevin equation to simulate the movements of CG beads.[11] The Langevin equation contains terms corresponding to fluctuating and frictional forces acting on each bead of the protein or membrane; these terms are tuned to represent water viscosity and thermal motion. The solvent also screens electrostatic interactions and this effect is modeled simply by using the uniform dielectric constant ε. For the same reasons as in case of the RBCG model, we set $\varepsilon = 1$.

The SBCG approach assumes a simple implicit solvent characterized, besides ε, through frictional and fluctuating forces, the last two accounted for by a single parameter, the damping constant γ.[39] To select an appropriate γ, we ran simulations of a single free-floating N-BAR domain dimer with $\gamma = 0.5$, 2 and $5\,\mathrm{ps}^{-1}$. Ten 1 µs long simulations were carried out for each value of γ and the diffusion constant D of the N-BAR domain was computed applying the relation to the protein's center of mass $\vec{r}_{\mathrm{c}}$:

$$\langle [\vec{r}_{\mathrm{c}}(t) - \vec{r}_{\mathrm{c}}(0)]^2 \rangle = 6Dt \tag{6.1}$$

where the average $\langle \ldots \rangle$ was taken over all simulations for given γ. We found that the diffusion constant obtained from simulations with $\gamma = 2\,\mathrm{ps}^{-1}$ is in the closest agreement with the experimental value[77,78] for a protein of a size comparable to an N-BAR domain (55 000 amu). Accordingly, $\gamma = 2\,\mathrm{ps}^{-1}$ was used for our further SBCG simulations.

The SBCG model represents the shape of a protein through several CG beads whose positions are assigned as described elsewhere;[79] the beads are connected by harmonic bonds to maintain protein shape. The amphiphysin N-BAR domain dimer was represented through 50 CG beads, 25 for each monomer (Figure 6.1B). Two CG beads within the dimer were connected by a bond if the distance between these beads in the original SBCG structure (derived from the all-atom crystal structure) was below 18 Å.

Interaction parameters were extracted from the all-atom structure and MD simulation of a single N-BAR domain system. The non-bonded interaction strength ε of the LJ interaction between beads i and j was computed as $\varepsilon_{ij} = \sqrt{\varepsilon_i \varepsilon_j}$, where ε_i and ε_j are the strengths for each bead. The value of ε_i was assigned for each bead i based on the hydrophobic solvent accessible surface area (SASA) for the Voronoi cell represented by the bead:[11]

$$\varepsilon_i = \varepsilon_{\max} \left(\frac{\mathrm{SASA}_i^{\mathrm{hphob}}}{\mathrm{SASA}_i^{\mathrm{tot}}} \right)^2 \tag{6.2}$$

where $\mathrm{SASA}^{\mathrm{hphob}}$ and $\mathrm{SASA}_i^{\mathrm{tot}}$ are the hydrophobic and total SASA of the Voronoi cell of bead i, respectively, and $\varepsilon_{\max} = 10\,\mathrm{kcal\,mol}^{-1}$. This assignment serves the purpose of allowing hydrophobic beads to aggregate and hydrophilic beads to dissolve in the 'solvent'. For a pair of completely hydrophilic beads holds $\varepsilon_{ij} = 0$, in which case the two beads are free to dissociate unless they are bound to other particles; ε_{ij} for two completely hydrophobic beads is 10 kcal mol^{-1}, which is significantly higher than the thermal energy ($k_{\mathrm{B}}T = 0.6$ kcal mol^{-1} at 300 K), but still permits thermal fluctuations.

The terms for bonded interactions in the SBCG method[39] are described through potentials $V_{\mathrm{bond}}(r) = K_{\mathrm{b}}(r - r_0)^2$ and $V_{\mathrm{a}}(\theta) = K_{\mathrm{a}}(\theta - \theta_0)^2$ for bond length r and angle θ, where K_{b}, r_0, K_{a} and θ_0 are the force-field parameters. The values for these parameters are derived from all-atom simulations in an iterative procedure aimed at reproducing the stiffness of the protein as observed in the all-atom simulation. The derivation proceeds as follows.

First, the Boltzmann inversion method is used to obtain initial values for the force-field parameters. The method requires one to obtain for each variable x (such as ith bond length r_i) the distribution $\rho(x)$ from an all-atom simulation. One uses then the Boltzmann relation $\rho(x) = \rho_0 \exp[-V(x)/k_{\mathrm{B}}T]$ to obtain the potential $V(x)$. Since a harmonic potential is assumed for bonds and angles, obtaining the complete $\rho(x)$ is not necessary, but only the average $\langle x \rangle$ and RMSD $\sqrt{\langle x^2 \rangle - \langle x \rangle^2}$ need to be obtained to determine an initial estimate for K_{b}, r_0, K_{a} and θ_0.[11]

Force-field parameters for the equilibrium bond lengths r and angles θ_0 obtained with the Boltzmann inversion usually serve well for reproducing the results of all-atom simulations by an SBCG simulation. Bond and angle strengths K_{b} and K_{a}, however, usually require further refinement. Figure 6.4A shows that when values of K_{b} obtained with the Boltzmann inversion are used, the actual bond stiffness recorded in the SBCG simulation (dotted red line) is much higher than that seen in an all-atom simulation (solid black line). This deficiency arises since the Boltzmann inversion applied holds only if x is an independent variable, whereas for a network of bonds as in the case of an SBCG protein model, all bonds and angles are coupled. To rectify this problem, force constants K_{b} and K_{a} obtained from the single variable Boltzmann inversion are calculated and scaled repeatedly as shown in Figure 6.4A and as explained elsewhere,[11] until the stiffness of the SBCG model agrees with that of the all-atom model (compare the black and blue curves in Figure 6.4A).

For SBCG modeling of lipids, each leaflet of a lipid bilayer is represented by two layers of CG beads: one for the upper half of the leaflet (which is mostly hydrophilic) and one for the lower part (which is hydrophobic), as shown in Figure 6.1B. The head–tail pairs of CG beads are connected by harmonic bonds. The leaflet thickness is ~ 25 Å and, therefore, each bead accounts for the part of the leaflet that is 12.5 Å in height or for the volume $12.5 \times 12.5 \times 12.5$ Å (since the bead is symmetric in each dimension). Thus, a two-bead CG 'molecule' (Figure 6.1B) stretching across the leaflet accounts for the area 12.5×12.5 Å and occupies a volume of $12.5 \times 12.5 \times 25$ Å. With the DOPC area per lipid being ~ 70 Å^2, each two-bead DOPC 'molecule' represents 2.2 DOPC lipids on average or ~ 300 atoms, *i.e.* 150 atoms per CG bead.

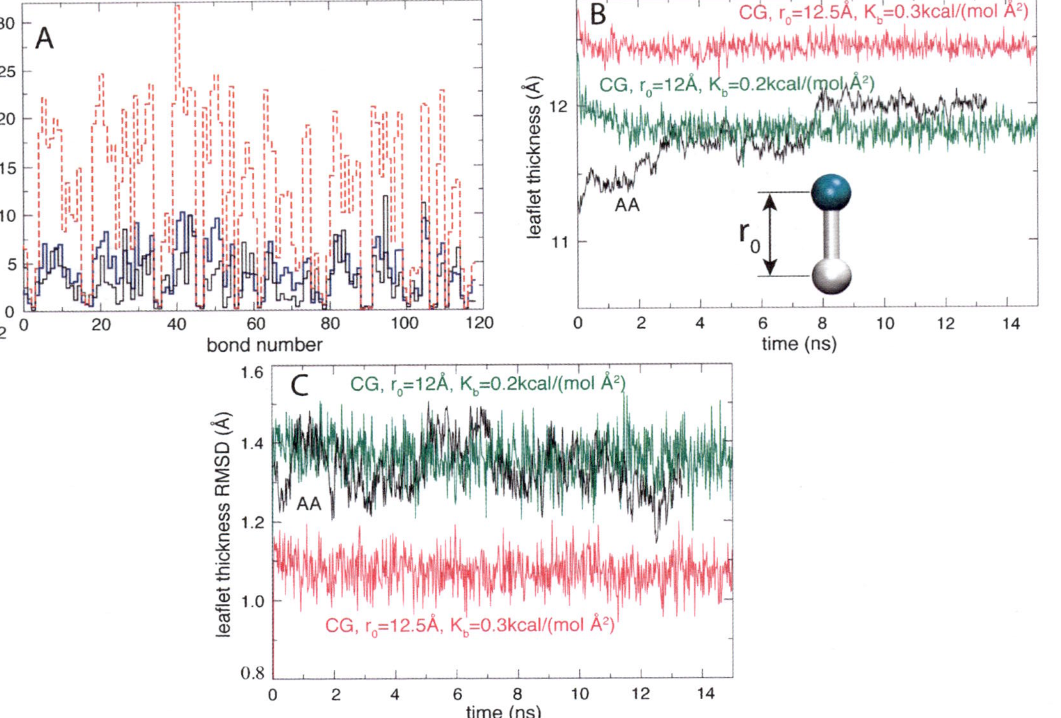

Figure 6.4 Determining bond parameters for SBCG models. (A) Bond constants K_b for all bonds in the SBCG model of the N-BAR domain. Using single variable Boltzmann inversion, K_b values are extracted from an all-atom simulation (black). An SBCG simulation is then performed using these values of K_b and the single variable Boltzmann inversion on the resulting SBCG trajectory gives now K_b values shown by the red dotted line. The K_b values used for the SBCG simulation are then scaled (multiplied by 0.3 in this example) and a new SBCG simulation is performed using the scaled K_b. Boltzmann inversion on the new SBCG trajectory is shown in blue. (B, C) DOPC leaflet thickness (defined as the distance between the centers of mass of the upper and lower parts of a lipid, averaged over the leaflet) and its RMSD recorded in all-atom (black) and SBCG (green and purple) simulations of a patch of DOPC bilayer. The all-atom and SBCG simulations agree when the SBCG bond parameters are $r_0 = 12.0\,\text{Å}$ and $K_b = 0.2\,\text{kcal}\,\text{mol}^{-1}\text{Å}^{-2}$ (green curves), while using, *e.g.* $r_0 = 12.5\,\text{Å}$ and $K_b = 0.3\,\text{kcal}\,\text{mol}^{-1}\text{Å}^{-2}$ (purple curves) results in a noticeable deviation.

The lipids were parameterized based on simulations of a DOPC membrane. The length of the bond between head and tail beads, r_0, and the bond strength, K_b, control the average membrane leaflet (or bilayer) thickness and its RMSD. Figure 6.4B and C show that using $r_0 = 12\,\text{Å}$ and $K_b = 0.2\,\text{kcal}\,\text{Å}^{-2}\,\text{mol}^{-1}$ yields good agreement with the all-atom simulation; these parameters were used for the SBCG simulations. The LJ parameters for lipid beads were chosen to reproduce the area per lipid ($\sim 70\,\text{Å}^2$). The LJ energy ε_i for a tail bead was set to $10\,\text{kcal}\,\text{mol}^{-1}$ and that for a head bead to $0.1\,\text{kcal}\,\text{mol}^{-1}$, and the LJ radii were set to $6.8\,\text{Å}$ for either one.[11] The area per lipid was computed from the area per two-bead SBCG 'molecule' assuming that such a molecule represents 2.2 lipids on average.

Each bead in the DOPC model had zero charge. To model negatively charged DOPS lipids, a charge of $-2.2|e|$ was assigned to the head bead. The masses were $864.75\,\text{amu}$ for the DOPC head and tail beads and $866.76\,\text{amu}$ for the DOPS head bead. To match the charge of the DOPS beads, 'ions' with the charge of $\pm 2.2|e|$ and mass of $1000\,\text{amu}$ were introduced, roughly corresponding to eight ions of mixed nature (such as both Na^+ and Cl^-) with their hydration shells.

In test simulations of pure lipid systems, the SBCG model qualitatively captured the self-assembly properties of real lipid molecules, with correct phases such as micellar, lamellar and inverted hexagonal forming, depending on the lipid concentration.[11] Furthermore, the SBCG membrane possessed the bending rigidity $k_c \approx (20 \pm 10)k_B T$ (T = 300 K); experimentally, for a biological membrane one finds $k_c = 10\text{--}20 k_B T$.[80]

SBCG simulations were performed for a single N-BAR domain dimer and a number of lattices composed of multiple N-BAR domains, sampling lattice densities from 6 to 35 N-BAR dimers per $1000\,\text{nm}^2$ of the membrane surface. Most of these simulations involved a membrane patch of dimension $64\times16\,\text{nm}^2$. Several simulations were performed with a longer, $\sim 200\times16\,\text{nm}^2$, membrane patch, which is suitable for observation of complete membrane tubulation by BAR domains, as the length of the patch allows it to be wrapped into a tubular segment with a curvature that is naturally induced by N-BAR domains (see below). Such simulations involved $\sim 11\,000$ CG beads. SBCG simulations were carried out with a time step of 100 fs. The longest simulations of the $\sim 200\times16\,\text{nm}^2$ membrane patch with 24 N-BAR dimers covered $200\,\mu\text{s}$, whereas the accumulated simulation time for all SBCG simulations of N-BARs with the membrane was over 1 ms.

6.6 Continuum Elastic Membrane Model

To estimate overall properties of the membrane bending by N-BAR domains, we employed a continuum elastic membrane (EMe) model as introduced recently.[11] A 'string' of linked elements in two dimensions (2D) was used to represent the profile of the membrane along its center line; the membrane was

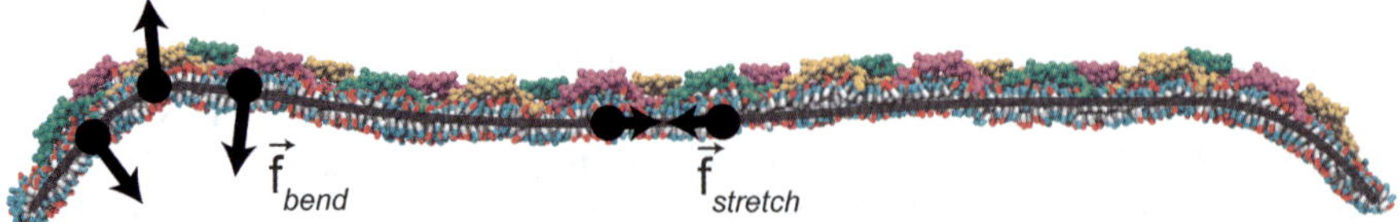

Figure 6.5   Continuum elastic membrane model. A membrane patch (shown in white, red and cyan) bent by multiple N-BAR domains (colored in green, purple and yellow) in an SBCG simulation is shown from the side. The profile (black solid line) of the membrane is reproduced by a 2D continuum elastic membrane model, which consists of elements on a string. Five elements are indicated by black disks. In actual calculations of the type shown, 200 elements, evenly spaced along the profile, were employed.

assumed to be fully periodic in the third dimension (Figure 6.5). The string dynamics were described by classical mechanics and calculated numerically.

The string of length L was represented as a chain of point-like elements $i = 1$, ..., N. Positions of each element at each time moment t were described by vectors $\vec{r}_i(t) = [x_i(t), y_i(t)]$. The forces acting on each element arise due to three causes: membrane stretching ($\vec{f}_{\text{stretch}}$), membrane bending ($\vec{f}_{\text{bend}}$) and damping from the environment. Thus, the dynamics of each element were described as

$$\ddot{\vec{r}}_i(s, t) = \sum \vec{f}_{\text{stretch}}(s, t) + \sum_i \vec{f}_{\text{bend}}(s, t) - \gamma_{\text{EM}} \dot{\vec{r}}(s, t) \tag{6.3}$$

where $\vec{f}_{\text{stretch}}$ and $\vec{f}_{\text{bend}}$ are proportional to actual forces (with the proportionality being defined by the linear density of the string, which does not enter the final equations).

The stretching term was described by a harmonic spring acting between any two adjacent elements in the chain:

$$\vec{f}_{\text{stretch}} = \pm A_{\text{stretch}}(|\Delta\vec{r}| - \Delta s)\frac{\Delta\vec{r}}{|\Delta\vec{r}|} \tag{6.4}$$

where $\Delta\vec{r}$ is the vector connecting the two elements, $\Delta s = L/(N-1)$ is the distance between the elements along the straight string at rest and A_{stretch} is a constant. The number N was chosen to obtain $\Delta s \approx 1\,\text{nm}$, which permitted a smooth representation of the membrane segments that were considered, typically with overall lengths $L = 50 - 200\,\text{nm}$.

The bending term was designed to maintain a straight membrane if no N-BAR domains are present or sculpt the membrane to acquire a certain curvature, if N-BAR domains are present. The bending term is defined for each triple of consecutive elements. For each triple, the absolute values of $\vec{f}_{\text{bend}}$ acting on the two edge elements are equal to each other and are given by

$$|\vec{f}_{\text{bend}}| = A_{\text{bend}}|K(s, t) - K_0| \tag{6.5}$$

where K_0 is the assigned intrinsic curvature ($K_0 = 0$ for no N-BAR domains), A_{bend} is a constant and

$$K(s,t) = \frac{|x'(s,t)y''(s,t) - y'(s,t)x''(s,t)|}{[x'(s,t)^2 + y'(s,t)^2]^{\frac{3}{2}}} \tag{6.6}$$

is the local curvature (x' and x'' being the first and second derivatives with respect to the element's position s along the string length, $s \in [0,L]$). Directions of $\vec{f}_{bend}$ applied to the edge elements of the triple were defined identically with the case of a bonded angle potential in an MD force field, *i.e.* $\vec{f}_{bend}$ acted to close the angle formed by the triple if $K(s,t) < K_0$ and to open the angle if $K(s,t) > K_0$. The term $\vec{f}_{bend}$ for the central element was a negative of the sum of $\vec{f}_{bend}$ from the two edges.

The EMe model used four parameters: K_0, $A_{stretch}$, A_{bend} and γ_{EM}. K_0 was set to a uniform (over s) value chosen to reproduce the final curvatures of the membrane obtained in RBCG and SBCG simulations of six N-BAR domains. $A_{stretch}$ was fixed[11] at $\sim 10^4\,ns^{-2}$ by comparing the stretching of the model membrane with that of real membranes, for which the Young's modulus for the extensibility is available from experiment.[80] A_{bend} and γ_{EM} were tuned to reproduce SBCG simulations, using the final membrane curvature value and the time-dependence of the membrane end-to-end distance for reference. The best-matching values, used for further EMe computations, were $A_{bend} \approx 7$ $nm^2\,ns^{-2}$ and $\gamma_{EM} \approx 0.06\,ns^{-1}$.

6.7   Results and Discussion

We have investigated membrane sculpting by N-BAR domains using simulations in all-atom, reside-based coarse grained (RBCG), shape-based coarse-grained (SBCG) and continuum representations. The simulations describe how single N-BAR domain dimers induce local membrane curvature and how lattices of multiple N-BAR domains acting in concert generate global membrane curvature. Different N-BAR domain lattices are investigated through extensive sampling in a large number of SBCG simulations. An all-atom simulation of 2.3 million atoms, covering $0.3\,\mu s$, corroborates the results. Complete membrane tubulation, on time scales up to $200\,\mu s$, is simulated in both SBCG and EMe descriptions.

6.8   Simulations of a Single N-BAR Domain

Simulations of a single N-BAR domain dimer on top of a $46 \times 10\,nm^2$ planar membrane were performed in three representations, all-atom, RBCG and SBCG MD.[11] Figure 6.6 shows that in all three representations, local membrane curvature developed beneath the N-BAR domain within tens of nanoseconds, in agreement with all-atom simulations reported elsewhere.[8] Three independent RBCG simulations of a single N-BAR domain dimer were carried

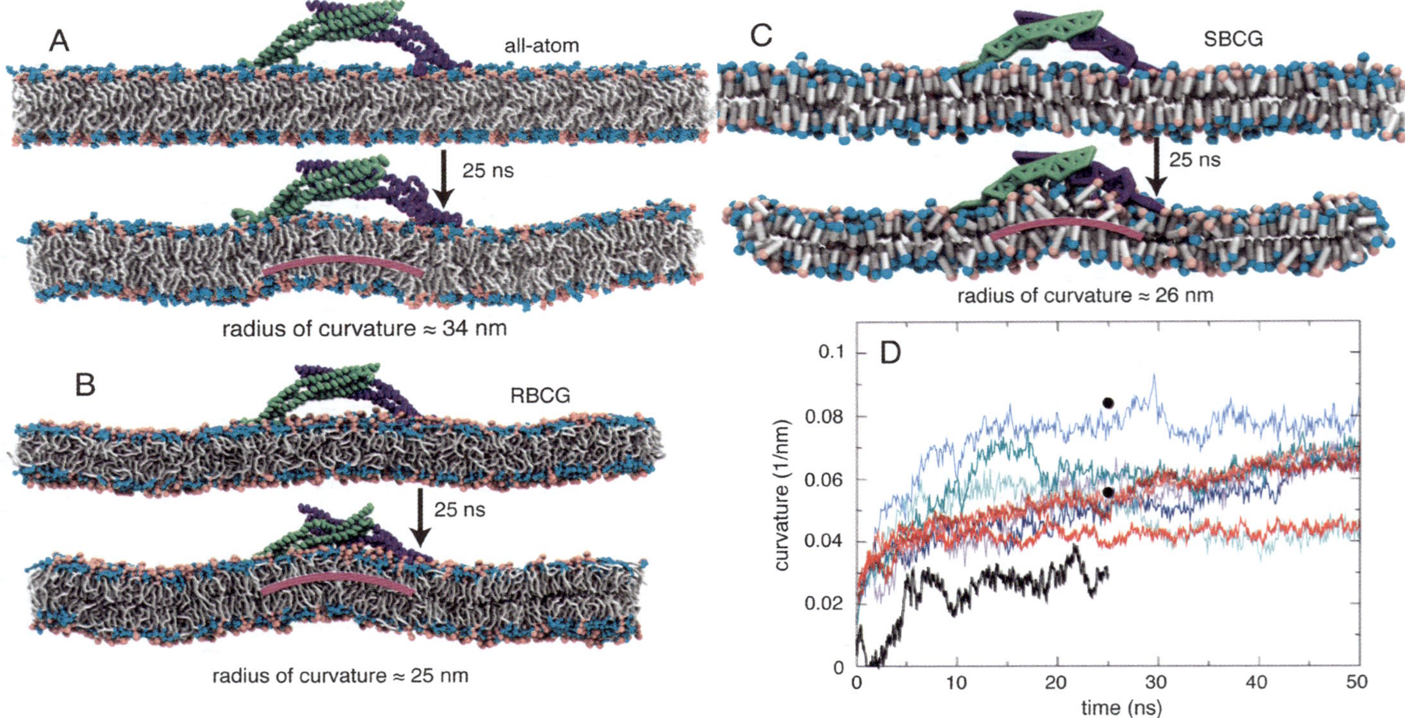

Figure 6.6 Single N-BAR domain bending a membrane. (A–C) Snapshots from all-atom, RBCG and SBCG simulations. Membranes consist of 30% negatively charged DOPS (PS head groups are shown in pink) and 70% DOPC (PC head groups are shown cyan). Monomers of the N-BAR domain dimer are shown in green and purple. (D) Membrane curvature generated by single BAR domain dimers. The two black dots are the curvatures obtained from two all-atom simulations reported elsewhere.[8] The black curve is from our all-atom simulation; deep red, pink and red are from RBCG simulations; the remaining curves are from SBCG simulations.

out, the N-BAR domain inducing local curvature each time. The same behavior was observed in five independent SBCG simulations (Figure 6.6C). In each simulation, membrane curvature developed within ~25 ns and stabilized afterwards. Figure 6.6D shows the time evolution of the membrane curvatures in all single N-BAR domain simulations, the curvatures being obtained by fitting the profile of the corresponding membrane section to a circle in the plane perpendicular to the initial plane of the membrane.[11] In our all-atom simulations, the radius of curvature produced by a single N-BAR domain reached ~34 nm (black curve in Figure 6.6D). This membrane curvature is somewhat lower than that in all-atom simulations performed by Blood *et al.*[8] (two black circles in Figure 6.6D). However, in other all-atom simulations,[81] Blood observed membrane curvatures lower than those reported previously,[8] suggesting that single BAR domains generate a range of membrane curvatures. Indeed, the curvature radius observed in our all-atom, RBCG and SBCG simulations falls in the range of 10–50 nm and varies from simulation to simulation.

6.9 Comparison of RBCG and SBCG Simulations for Systems with Six N-BAR Domains

Using RBCG and SBCG models, we simulated systems of six N-BAR domains on top of an initially planar membrane in a non-staggered arrangement (Figure 6.7A and C) and in a staggered arrangement (Figure 6.7B and D), as reported elsewhere.[11] The RBCG systems contained ~200 000 beads. The simulations were stopped at 20 ns for the non-staggered arrangement and at 50 ns for the staggered arrangement. In the SBCG description, dynamics of these systems was simulated for up to 5 μs (Figure 6.8).

We found that N-BAR domains bend the membrane in a substantially different manner for the non-staggered and staggered arrangements. Within tens of nanoseconds, N-BAR domains in the non-staggered arrangement induced local membrane curvatures, but not a global one (Figure 6.7A and C). The membrane directly beneath each N-BAR domain became bent by the concave charged surface of a protein, whereas the membrane between two adjacent N-BAR domains remained fairly flat. Thus, a ripple-like overall membrane structure (Figure 6.7A and C) developed in about 15 ns. In the case of a staggered arrangement, a global bending developed, with the radius of curvature reaching ~40 nm within tens of nanoseconds. The membrane beneath each N-BAR domain also became bent locally, but because the membrane between two adjacent N-BAR domains in one row is covered by another N-BAR domain from the next row, a collective bending resulted (Figure 6.7B and D). The local curvatures observed for both arrangements are close to those in simulations of single N-BAR domains (Figure 6.6D) and exhibit a similar scatter.

The SBCG simulations reproduce the results of RBCG simulations over tens of nanoseconds ideally, but reach further in time. Over a long time scale, global

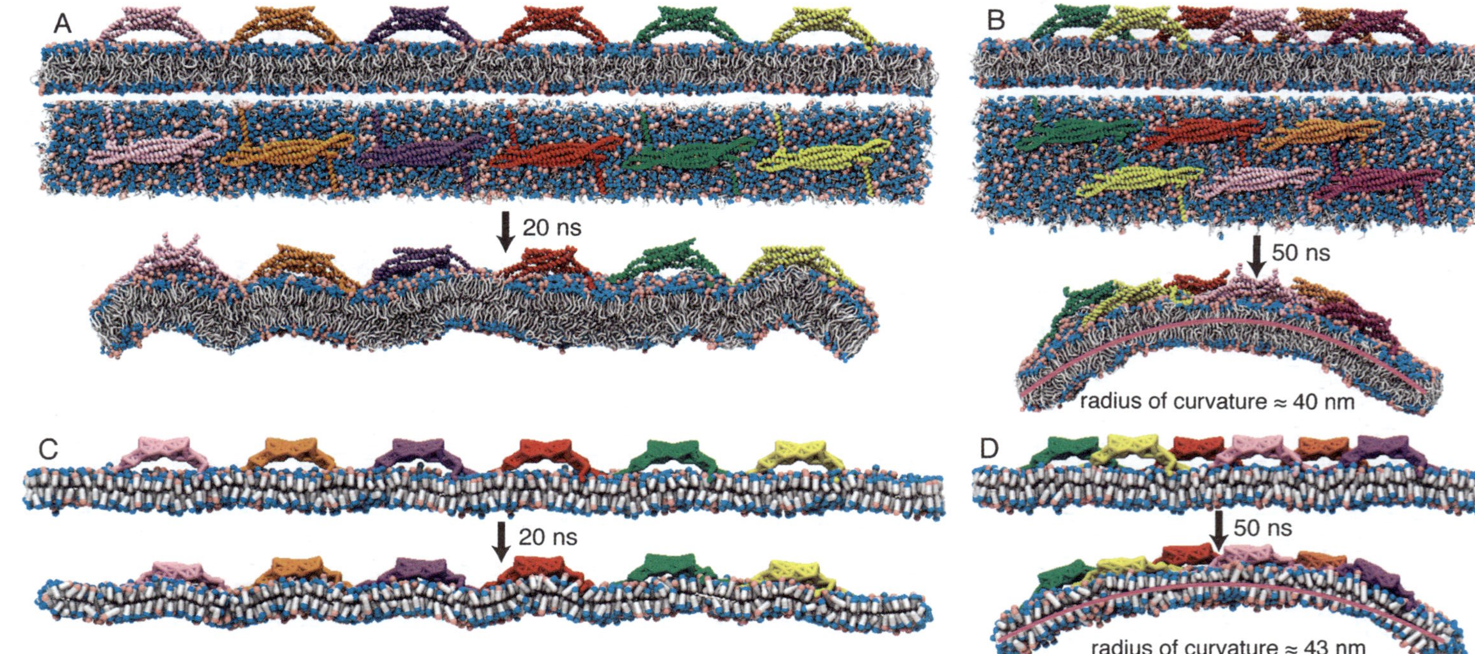

Figure 6.7 RBCG and SBCG simulations of multiple N-BAR domains inducing membrane curvatures. (A, B) RBCG simulations of six N-BAR domains in a non-staggered arrangement (A) and in a staggered arrangement (B). Upper and middle panels in (A) and (B) show side and top views of the initial setup. Lower panels are snapshots after 20 and 50 ns. (C, D) SBCG simulations of N-BAR domains in non-staggered (C) and staggered arrangements (D).

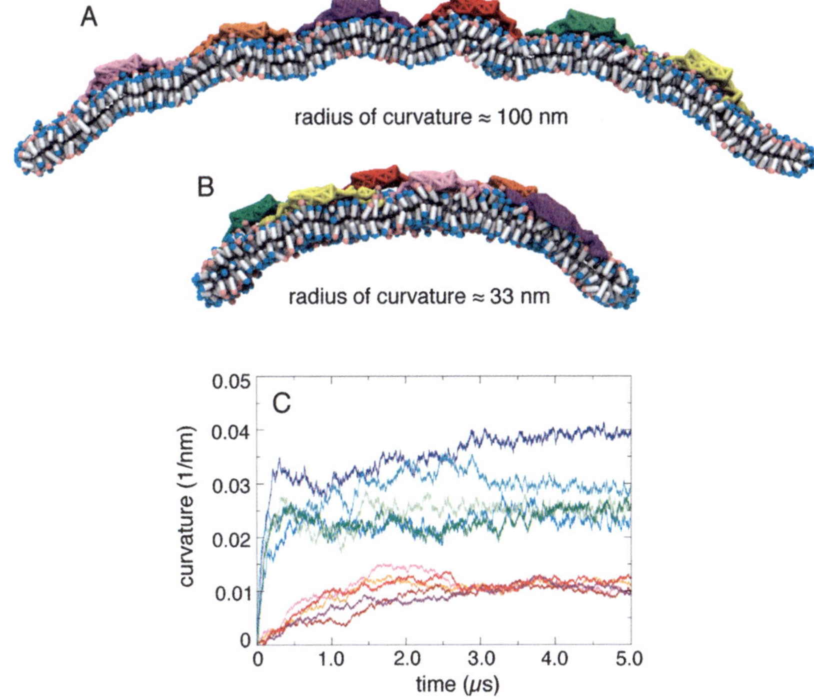

Figure 6.8 Microsecond SBCG simulations of six N-BAR domain systems. (A, B) Snapshots at 5 µs for non-staggered N-BAR domain arrangement (A) and staggered N-BAR domain arrangement (B). (C) Time evolution of global membrane curvatures generated by the non-staggered arrangement (bottom five curves) and the staggered arrangement (top five curves).

curvature arises also in the case of the non-staggered N-BAR domains arrangement (Figure 6.8): for up to 100–400 ns, the membrane remained overall flat, but then started to bend globally (Figure 6.8C). The bending reached its maximum at 1.5 µs, with a radius of global curvature of ∼70–120 nm. The radius converged to a value of about 100 nm at ∼2.7 µs when the membrane conformation in all five independent SBCG simulations stabilized. In contrast, the membrane in the staggered system reached a global curvature radius of 30–40 nm during the first 300 ns of simulation and did not exhibit significant changes thereafter.

Simulations of the six N-BAR domain systems exhibit remarkable agreement between the RBCG and SBCG models. The differences in the curvatures reached and the dynamics of bending between simulations with non-staggered and staggered arrangements of six N-BAR domains indicate that the arrangement of N-BAR domains on the membrane is crucial for the efficiency of bending. With the significant computational gain of SBCG simulations, one can explore further how various N-BAR domain arrangements affect membrane sculpting.

6.10 Effect of Different N-BAR Domain Lattices on Membrane Curvatures

With the SBCG model, we investigated membrane bending by various arrangements, *i.e.* various lattices of N-BAR domains, sampling the action of more than 20 different lattice types on the same membrane patch of dimension $64 \times 16 \, \text{nm}^2 \approx 1000 \, \text{nm}^2$.[14] Several examples are shown in Figure 6.9. In the simulations we sampled N-BAR domain densities from 6 to 35 dimers per $1000 \, \text{nm}^2$, distances S between rows of N-BAR domains from 1.9 to 8 nm, spiraling angles θ from 0 to 25° and connections between N-BAR domains within one row from 'no contact' to 'center-to-center' contact (Figure 6.9).

The highest observed membrane curvatures in all the simulations corresponded to radii in the range $R = 13$–20 nm, while the curvature radius of the protein itself is 11 nm. Such high curvatures are generated by lattices with approximately 10–20 dimers per $1000 \, \text{nm}^2$, $S = 3$–6 nm and $\theta = 0$–5°. These features of 'optimal' lattices for membrane bending are in qualitative agreement with cryo-EM images of membrane tubes formed by amphiphysin N-BAR domains,[15,25] showing striations with $S \approx 5$–10 nm. High-resolution cryo-EM reconstructions for F-BAR domains,[24] which are larger and feature lower intrinsic curvature than amphiphysin N-BAR domains, correspond to $S \approx 6$–8 nm and $\theta = 10$–15°.

A high membrane curvature was generated by lattices with an intermediate density of N-BAR domains (10–20 dimers per $1000 \, \text{nm}^2$). Lattices with lower densities produced a relatively low membrane curvature due to insufficient bending forces (*e.g.* 6BARs in Figure 6.9). Surprisingly, densities higher than approximately 20 dimers per $1000 \, \text{nm}^2$ also lead to a low curvature (*e.g.* 32BARs in Figure 6.9). Analysis of the interactions between the membrane and the protein's concave surface for dense lattices showed[14] that such interactions are occluded by the tips and N-terminal helices of the neighboring N-BAR domains. Thus, interdigitation of neighboring N-BAR domains prevents them from establishing a strong contact with the membrane and as a result efficient scaffolding of the membrane by the proteins' charged concave surface cannot proceed.

To characterize membrane bending further, we estimated[14] the elastic bending energy of the membrane E_{bend} per one N-BAR domain for each simulated lattice using the Helfrich elastic membrane theory.[41] For lattices producing significantly high curvature ($R < 30$ nm), values of E_{bend} are 1–$3k_B T$, whereas for lattices with low curvature we found $E_{\text{bend}} < 0.5 \, k_B T$. Similarly, one can estimate,[14] based on all-atom and CG simulations of a single N-BAR domain,[8,11,13] that a single N-BAR domain imposes a bending energy in a broad range from 0.6 to $15 k_B T$. The actual value in each case depends on how well the contact between protein and membrane is established. These numbers

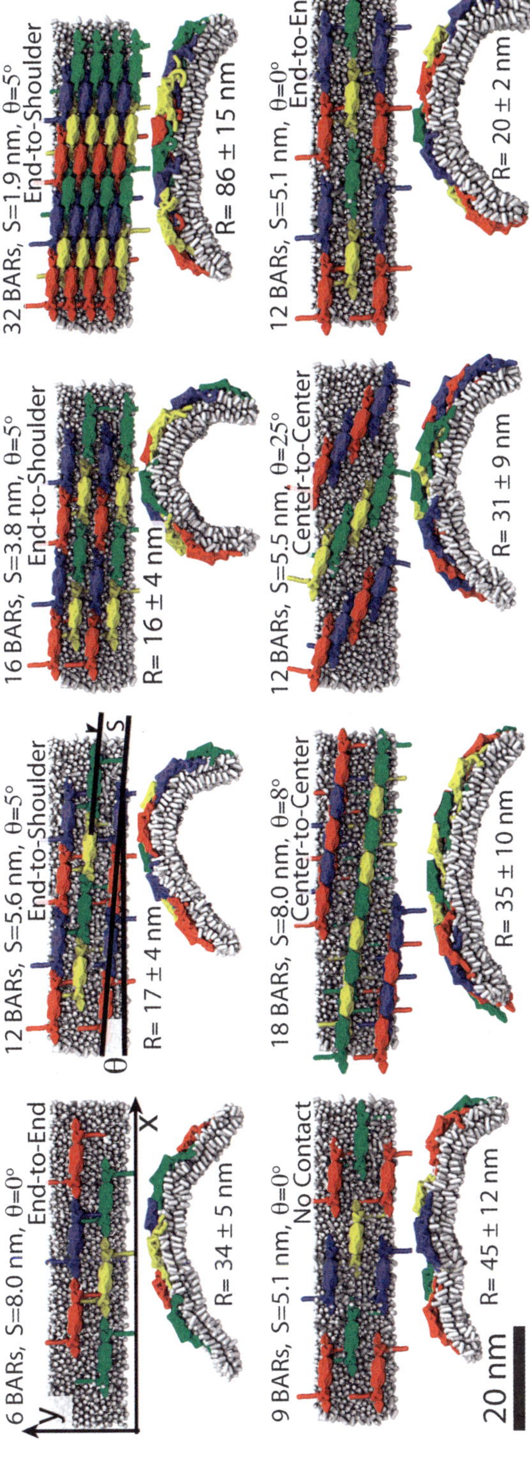

Figure 6.9 Membrane curvatures generated by various lattices of N-BAR domains. Variations in the number of N-BAR domain dimers are denoted '6BARs', '12BARs', *etc.* Shown are distance S between rows, spiraling angle θ (between rows and the *x*-axis) and the contact between N-BAR domains within one row ('end-to-end', 'end-to-shoulder', *etc.*). For each lattice, the top view of the initial configuration and the side view at $t = 1.5$ μs are shown. The R values are the curvature radii at $t = 1.5$ μs, averaged over five independent simulations and shown with their respective standard deviations.

show that even in the case of significant membrane bending, only a relatively moderate amount of bending is produced by each individual N-BAR domain ($1-3k_BT$ *versus* the maximally possible $15k_BT$), highlighting the efficiency of organizing multiple N-BAR domains into lattices with an 'optimal' geometry.

The arrangement of dimers within one row of a lattice was also found to be decisive for the induced curvature. For F-BAR domains, whose membrane tubulation activity was recently studied by high-resolution cryo-EM tomography,[24] end-to-end contacts within the rows of the protein lattice were suggested initially based on the crystal structure,[16] but the cryo-EM reconstructions showed[24] that F-BAR domains engaged instead in end-to-center contacts. Due to the high intrinsic curvature of amphiphysin BAR domains, a lattice with end-to-center contacts could not be realized as an initial arrangement in a simulation starting from a planar membrane,[14] but a similar lattice with center-to-center contacts could be studied (Figure 6.10). Crystal packing of amphiphysin BAR domains[15] involves extensive contacts between residues 87 and 116 of a given monomer and between residues 161 and 185 of its neighbor, *i.e.* spanning from the BAR domain center to its tip, which is remotely similar to center-to-center contact, although the crystal packing involves a 90° tilt of monomers with respect to each other, in contrast to the uniform orientation of N-BAR domains in the simulated lattices. However, lattices with center-to-center contacts produced relatively low curvature in simulation (Figure 6.9). These results suggest that different contacts in the crystal and on the sculpted membrane tube should arise for amphiphysin N-BAR domains as they do for F-BAR domains.

The highest membrane curvature (Figure 6.9) was observed when N-BAR domains were arranged 'end-to-end' or 'end-to-shoulder' (the 'shoulder' refers

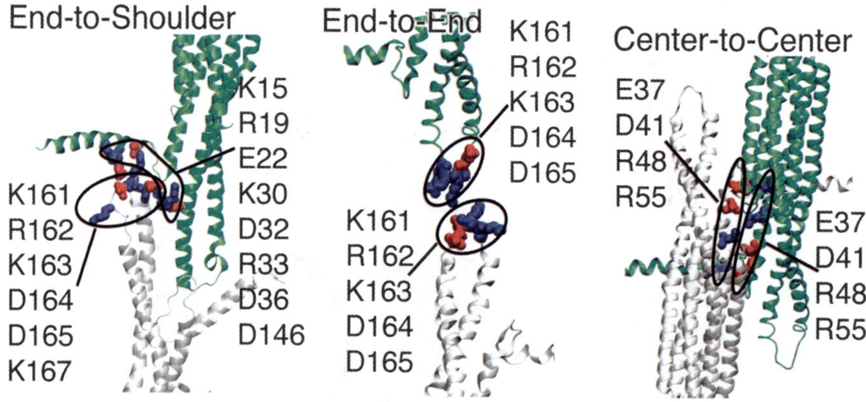

Figure 6.10 Contacts between N-BAR domain dimers within various lattices probed in simulations (see Figure 6.9). Two contacting dimers in all-atom representation are shown in white and green. Amino acids that are involved in the contacts are highlighted in blue (positively charged) and red (negatively charged).

to the place where the main body of the dimer connects to the N-terminal helix). All contacts discussed here were stabilized by electrostatic interactions between CG beads, which at the atomic level would correspond to interactions between charged residues. The 'end-to-shoulder' arrangement involves a particularly high concentration of charged residues at the contact point, *e.g.* Lys161, Arg162, Lys163, Asp164, Asp165 and Lys167 at the tip of the N-BAR domain and Lys30, Asp32 and Asp146 at the 'shoulder' (Figure 6.10), most of which are highly conserved.[15] Interestingly, it was found experimentally[15] that mutations of Lys161 and Lys163 into glutamates inhibits membrane tubulation.

6.11 Comparing All-atom and SBCG Simulations of an N-BAR Domain Lattice

Due to computational limitations, exploring various N-BAR domain lattices on the microsecond time scale can only be achieved through the SBCG approach. However, for one lattice we carried out a 0.3 μs all-atom simulation to test the SBCG model.[14] The all-atom system was composed of eight N-BAR domains on top of a $68 \times 8 \, nm^2$ membrane patch, containing ~ 2.3 million atoms (Figure 6.11). An analogous SBCG simulation was also carried out.

The N-BAR lattice probed in the all-atom simulation is the closest to the 16 BARs lattice explored in SBCG simulations of various different lattices (Figure 6.9, third panel from the left in the top row). Similarly to SBCG simulations (Figure 6.11A), the all-atom simulation showed that the assumed N-BAR domain lattice remains stable and generates global membrane curvature (Figure 6.11B), which continues to develop throughout the 300 ns simulation. The concave surfaces of seven out of the eight N-BAR domains established close contact with the membrane; the eighth N-BAR domain did not yet form such contact. Differences in membrane contacts between different N-BAR domains within a lattice were also observed in the SBCG simulations. Membrane bending in the all-atom simulation developed slower than in SBCG simulations: at 300 ns, the radius of curvature was $R \approx 59 \, nm$, whereas for the SBCG simulation of the same lattice this value of R was reached within 50 ns, and at 300 ns one finds in the SBCG case $R \approx 36 \, nm$. A likely reason for the speed discrepancy is that the all-atom representation contains many more degrees of freedom than the SBCG model, which results in stronger friction being manifested through slower bending. Another reason is that at the beginning of the all-atom simulation a longer time than in the SBCG simulation is required for the N-BAR domains to form proper contacts with the membrane and with each other.

Despite the difference in bending speed between the all-atom and SBCG simulations, the N-BAR domain lattices generated global membrane curvature in a similar fashion, by scaffolding the membrane with the proteins' charged concave surfaces. The protein lattice was stable in either representation, maintaining the end-to-shoulder contacts (as shown in Figure 6.10) that were

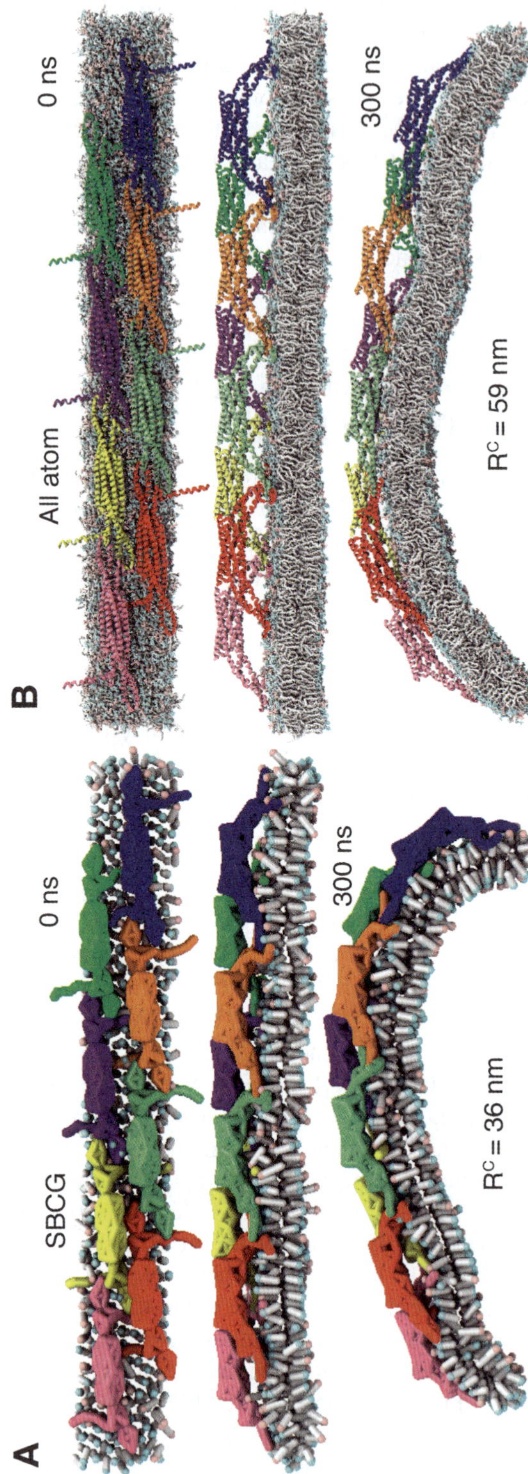

Figure 6.11 All-atom and SBCG simulations of a lattice of N-BAR domains. (A, B) Snapshots of the simulations in SBCG and all-atom representations, respectively. Eight individual N-BAR domain dimers are shown in different colors.

established in the initial structure from which the simulations started. The N-terminal helices of N-BAR domains were fairly flexible[14] in both all-atom and SBCG simulations, in agreement with a recent experimental–computational study that characterized the N-helices as being flexible and relatively disordered.[82]

6.12 Complete Membrane Tubulation by Lattices of BAR Domains

SBCG simulations were further used to study complete tubulation of a $200 \times 16\,\mathrm{nm}^2$ membrane patch by lattices of N-BAR domains (Figure 6.12). Two lattices were studied, with 43 and 24 N-BAR domains corresponding to the '6BARs' and '12BARs, end-to-shoulder' lattices in Figure 6.9, respectively. As shown in Figure 6.12, in each case the bending was initiated at the edges of the membrane within a few hundred nanoseconds. The bending then propagated towards the center, rounding the entire membrane on a time scale of 30–200 µs. Eventually, the membrane was driven into a near-tubular state, in which the edges can come into contact and fuse, producing a complete, stable tube. Figure 6.12 shows snapshots from two simulations in which complete tubulation is achieved within 35 and 200 µs.

The radii of membrane tubes formed by either 43 or 24 N-BAR domains were approximately the same, *i.e.* $\sim 25\,\mathrm{nm}$. These radii were mainly determined by the original length of the membrane, which was chosen based on a tube with a size typically observed in *in vitro* experiments for amphiphysin N-BAR domains.[15] However, experimentally, tubes with a wide range of radii can form and our simulations with various lattice types showed that many different radii are viable indeed, depending on the lattice (Figure 6.9). The lattice corresponding to 24BARs (Figure 6.9, lattice '6BARs') was found to bend the membrane to the curvature radius of $R = 34 \pm 5\,\mathrm{nm}$, whereas that corresponding to 43BARs (Figure 6.9, lattice '12BARs, end-to-shoulder') resulted in $R = 17 \pm 4\,\mathrm{nm}$, thus $R = 25\,\mathrm{nm}$ is in the middle of the range of radii produced by these two lattices. The lattice in 43BARs system favored a smaller radius than that in the 24BARs system and, due to the stronger bending, the 43BARs lattice (Figure 6.12B) achieved the tubulation about six times faster than the 24BARs lattice (Figure 6.12C).

6.13 Elastic Membrane Computations

The EMe model describes the dynamics of the membrane profile in 2D and serves to estimate the time scale of membrane bending. Typical results of EMe computations are shown in Figure 6.13. EMe computations for a membrane patch 64 nm in length (Figure 6.13A) are compared with SBCG simulations of six N-BAR domains on the $64 \times 16\,\mathrm{nm}^2$ membrane patch (Figure 6.8B). End-to-end distances recorded in several SBCG simulations of this system (the simulations diverged to some extent due to the thermal noise) are shown together

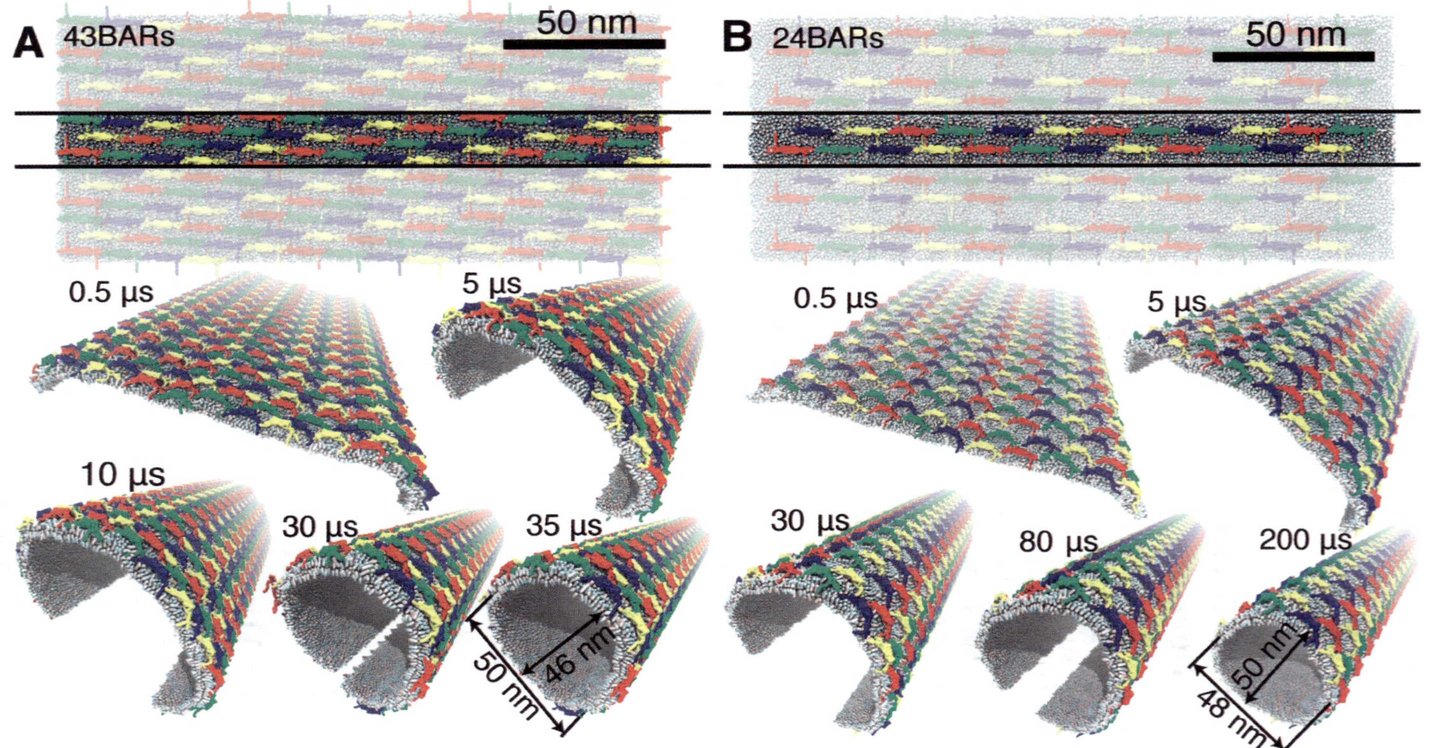

Figure 6.12 Complete membrane tubulation by N-BAR domain lattices. (A, B) Tubulation of planar membranes by lattices containing 43 and 24 N-BAR domains (colored in blue, yellow, green and red) per unit cell, respectively. For each of the two simulations, the top view of the initial setup is provided with one unit cell highlighted within two lines. Snapshots during the simulated tubulation processes are shown, with several periodic images included to demonstrate formation of a tube.

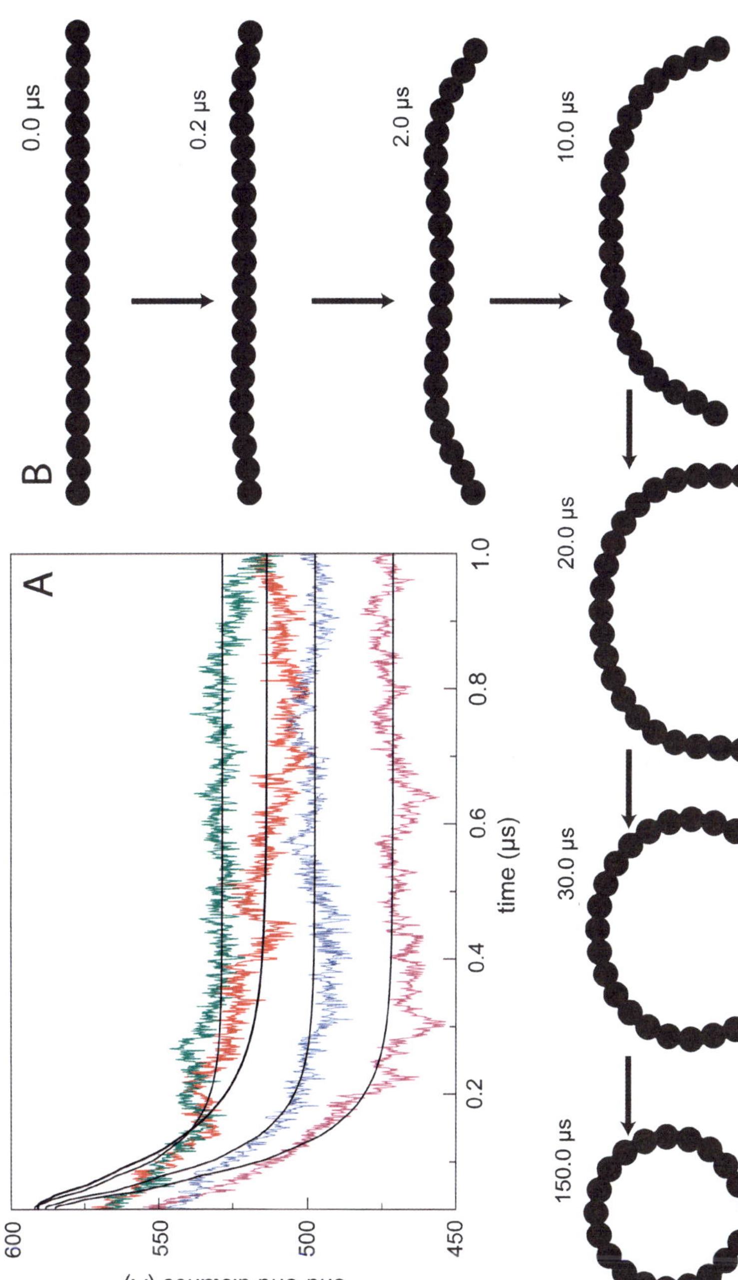

Figure 6.13 Membrane dynamics in computations using the EMe membrane model. (A) End-to-end distance of the membrane over time. The fluctuating curves are taken from SBCG simulations of the membrane with six N-BAR domains and smooth curves are from analogous EMe computations. (B) Snapshots of the membrane shape in a typical EMe computation of membrane tubulation. For clarity, only a small subset of string elements along the membrane profile is marked by circles.

with those obtained from EMe computations. Of the four parameters defining the EMe model (see Methods), three were held constant, namely A_{stretch}, A_{bend} and γ_{EM} [see equations (6.3)–(6.5)]. The EMe intrinsic curvature K_0 [see equation (6.5)] was varied to match the final membrane curvature observed in each SBCG simulation. Once an appropriate value of K_0 was chosen, the EMe computation faithfully reproduced the time dependence of the membrane end-to-end distance observed in the respective SBCG simulation, as shown in Figure 6.13A.

EMe computations were also performed to survey time scales of membrane tubulation by N-BAR domains. A membrane patch 200 nm long was considered as shown in Figure 6.13B, in preparation for the SBCG simulations of membrane tubulation (see Figure 6.12). The N-BAR domain density chosen for the EMe computation illustrated in Figure 6.13B is about six dimers per $1000 \, \text{nm}^2$, *i.e.* the same as in the 24BARs simulation described above (Figure 6.12B). The EMe computations showed that the 200 nm membrane with such a density of N-BAR domains undergoes tubulation within $\sim 150 \, \mu\text{s}$ and, indeed, SBCG simulations reached tubulation within $\sim 200 \, \mu\text{s}$. Of course, the intermediate membrane shapes arising in SBCG simulations and in the EMe computations were also fairly similar. The SBCG approach provides much more details about the system than the EMe model, but only at a significantly higher computational cost. Hence further large-scale studies of membrane remodeling by proteins will benefit from use of the EMe model guiding SBCG and all-atom simulations.

6.14   Conclusion

Many processes in cells, such as endocytosis, synaptic vesicle fusion and cell division, involve an interplay between protein assemblies and membranes and arise, at least partially, at the scales covered in this study, namely hundreds of nanometers and hundreds of microseconds. The methods employed here for membrane-sculpting BAR domains, a combination of atomic resolution and CG simulations, can be applied to elucidate the dynamics of such processes as the underlying structural features of systems become known.

Most fascinating in the study of membrane sculpting and fusion is that the proteins involved work in teams, naturally requiring larger and longer simulations than needed for conventional single protein investigations. The proteins pose the challenge of identifying their team organization. In the case of BAR domains, the proteins team up into lattice-like scaffolds, at least in the case of the experimentally studied *in vitro* systems. Hundreds of simulations were needed to elucidate the relationship between the lattice chosen and membrane geometry generated. The results are rewarding as they show that the organization into specific lattices matters for the overall function. Surely, other types of protein teams will be discovered in their involvement in various types of membrane morphogenesis in cells and tissues. The beautiful lattices of BAR domains offer a glimpse of a new science of molecular systems biology.

Our approach bridges the size (100 nm) and time (100 µs) scales needed to describe membrane sculpting by introducing a hierarchical reduction in resolution: at the 1 Å resolution level we employ all-atom MD simulations, at the 5 Å level RBCG MD, at the 25 Å level SBCG MD, and finally at the 125 Å level a continuum description. The higher resolution descriptions determine and test the lower resolution descriptions such that membrane sculpting can be extended to 200 µs and sampled overall for 1 ms. However, the ultimate test for the multi-scale approach are the all-atom simulations and indeed key to our findings is one such simulation on a 2.3 million atom system lasting 0.3 µs. This simulation shows membrane bending in agreement with a shape-based CG MD simulation, resolving in the act essential details of the interactions among BAR domains arranged in a lattice. All-atom and SBCG simulations have also been recently combined[83] to discern how membrane sculpting is affected by insertion of the N-terminal helical amphipathic segments of N-BAR domains into the membrane, as well as by scaffolding. All-atom simulations of 2.3 million atoms over hundreds of nanoseconds and multiple SBCG simulations covering tens of microseconds showed that complete N-BAR domains and BAR domains without the N-terminal amphipathic helices bend the membrane, whereas the helices alone do not.

With the current increase in processor numbers available for scientific computing, with the ongoing improvement of MD codes to take advantage of the new computer power and with the advent of improved force fields that include atomic polarizability, one anticipates that our study, at present still at the frontier of technology, will soon become common place. However, even with the expected technological advances researchers will benefit more from multi-scale strategies in computational modeling than relying just on huge all-atom simulations. It is hoped that our multi-scale approach will be a guidepost for future modeling studies of membrane molding at the molecular systems level. Such studies will undoubtedly open up an exciting new chapter in the biology of living cells, aimed at an understanding of protein team work in molecular detail.

Acknowledgements

This work was supported through National Institutes of Health grants R0-GM067887 and P41-RR05969; A.A. was supported by the L. S. Edelheit Fellowship. The authors acknowledge supercomputer time provided by NSF (Large Resources Allocation Committee grant MCA93S028) and through the University of Illinois. This research also used resources of the Argonne Leadership Computing Facility at Argonne National Laboratory, which is supported by the Office of Science of the US Department of Energy under contract DE-AC02-06CH11357.

References

1. M. Marsh and H. T. McMahon, *Science*, 1999, **285**, 215.
2. T. Kirchhausen, *Annu. Rev. Biochem.*, 2000, **69**, 699.
3. H. T. McMahon and I. G. Hills, *Curr. Opin. Cell Biol.*, 2004, **16**, 379.
4. H. T. McMahon and J. L. Gallop, *Nature*, 2005, **438**, 590.

5. T. Itoh, K. S. Erdmann, A. Roux, B. Habermann, H. Werner and P. De Camilli, *Dev. Cell*, 2005, **9**, 791.

6. W. Cho and R. V. Stahelin, *Annu. Rev. Biophys. Biomol. Struct.*, 2005, **296**, 153.

7. C. Gurkan, S. M. Stagg, P. LaPointe and W. E. Balch, *Nat. Rev. Mol. Cell Biol.*, 2006, **7**, 727.

8. P. D. Blood and G. A. Voth, *Proc. Natl. Acad. Sci. USA*, 2006, **103**, 15068.

9. W. Römer, L. Berland, V. C. K. Gaus, B. Windschieg, D. Tenza, M. R. E. Aly, V. Fraisier, J.-C. Florent, D. Perrais, C. Lamaze, G. Raposo, C. Steinem, P. Sens, P. Bassereau and L. Johannes, *Nature*, 2007, **450**, 670.

10. M. A. Lemmon, *Nat. Rev. Mol. Cell Biol.*, 2008, **9**, 99.

11. A. Arkhipov, Y. Yin and K. Schulten, *Biophys. J.*, 2008, **95**, 2806.

12. D. Chandler, J. Hsin, C. B. Harrison, J. Gumbart and K. Schulten, *Biophys. J.*, 2008, **95**, 2822.

13. P. D. Blood, R. D. Swenson and G. A. Voth, *Biophys. J.*, 2008, **95**, 1866.

14. Y. Yin, A. Arkhipov and K. Schulten, *Structure*, 2009, **17**, 882–892.

15. B. J. Peter, H. M. Kent, I. G. Mills, Y. Vallis, P. J. G. Butler, P. R. Evans and H. T. McMahon, *Science*, 2004, **303**, 495.

16. A. Shimada, H. Niwa, K. Tsujita, S. Suetsugu, K. Nitta, K. Hanawa-Suetsugu, R. Akasaka, Y. Nishino, M. Toyama, L. Chen, Z. Liu, B. Wang, M. Yamamoto, T. Terada, A. Miyazawa, A. Tanaka, S. Sugano, M. Shirouzu, K. Nagayama, T. Takenawa and S. Yokoyama, *Cell*, 2007, **129**, 761.

17. O. Daumke, R. Lundmark, Y. Vallis, S. Martens, P. J. G. Butler and H. T. McMahon, *Nature*, 2007, **449**, 923.

18. J. Hu, Y. Shibata, C. Voss, T. Shemesh, Z. Li, M. Coughlin, M. M. Kozlov, T. A. Rapoport and W. A. Prinz, *Science*, 2008, **319**, 1247.

19. J. Zimmerberg and M. M. Kozlov, *Nat. Rev. Mol. Cell Biol.*, 2006, **7**, 9.

20. B. J. Reynwar, G. Illya, V. A. Harmandaris, M. M. Müller, K. Kremer and M. Deserno, *Nature*, 2007, **447**, 461.

21. H. T. M. Sascha Martens and M. M. Kozlov, *Science*, 2007, **316**, 1205.

22. J. Zimmerberg and S. McLaughlin, *Curr. Biol.*, 2004, **14**, R250.

23. G. Ren, P. Vajjhala, J. S. Lee, B. Winsor and A. L. Munn, *Microbiol. Mol. Biol. Rev.*, 2006, **70**, 37.

24. A. Frost, R. Perera, A. Roux, K. Spasov, O. Destaing, E. H. Egelman, P. De Camilli and V. M. Unger, *Cell*, 2008, **132**, 807.

25. K. Takei, V. I. Slepnev, V. Haucke and P. De Camilli, *Nat. Cell Biol.*, 1999, **1**, 33.

26. W. Weissenhorn, *J. Mol. Biol.*, 2005, **351**, 653.

27. T. H. Millard, G. Bompard, M. Y. Heung, T. R. Dafforn, D. J. Scott, L. M. Machesky and K. T. Futterer, *EMBO J.*, 2005, **24**, 240.

28. M. Masuda, S. Takeda, M. Sone, T. Ohki, H. Mori, Y. Kamioka and N. Mochizuki, *EMBO J.*, 2006, **25**, 2889.

29. J. L. Gallop, C. C. Jao, H. M. Kent, P. J. Butler, P. R. Evans, R. Langen and H. T. McMahon, *EMBO J.*, 2006, **25**, 2898.

30. W. M. Henne, H. M. Kent, M. G. J. Ford, B. G. Hegde, O. Daumke, P. J. G. Butler, R. Mittal, R. Langen, P. R. Evans and H. T. McMahon, *Structure*, 2007, **15**, 1.

31. P. K. Mattila, A. Pykäläinen, J. Saarikangas, V. O. Paavilainen, H. Vihinen, E. Jokitalo and P. Lappalainen, *J. Cell Biol.*, 2007, **176**, 953.
32. A. R. Leach, *Molecular Modelling: Principles and Applications*, 2nd edn, PrenticeHall, Upper Saddle River, NJ, 2001.
33. M. Karplus and J. McCammon, *Nat. Struct. Biol.*, 2002, **265**, 654.
34. K. Y. Sanbonmatsu and C.-S. Tung, *J. Struct. Biol.*, 2007, **157**, 470.
35. S. J. Marrink, A. H. de Vries and A. E. Mark, *J. Phys. Chem. B*, 2004, **108**, 750.
36. A. Y. Shih, A. Arkhipov, P. L. Freddolino and K. Schulten, *J. Phys. Chem. B*, 2006, **110**, 3674.
37. A. Y. Shih, P. L. Freddolino, A. Arkhipov and K. Schulten, *J. Struct. Biol.*, 2007, **157**, 579.
38. P. L. Freddolino, A. Arkhipov, A. Y. Shih, Y. Yin, Z. Chen and K. Schulten, *Coarse-Graining of Condensed Phase and Biomolecular Systems*, ed. G. A. Voth, Chapman and Hall/CRC Press, Taylor and Francis Group, Boca Raton, FL, 2008, p. 299.
39. A. Arkhipov, P. L. Freddolino and K. Schulten, *Structure*, 2006, **14**, 1767.
40. A. Arkhipov, P. L. Freddolino, K. Imada, K. Namba and K. Schulten, *Biophys. J.*, 2006, **91**, 4589.
41. W. Helfrich, *Z. Naturforsch.*, 1973, **28**, 693.
42. B. Smit, P. A. J. Hilbers, K. Esselink, L. A. M. Rupert, N. M. van Os and A. G. Schlijper, *Nature*, 1990, **348**, 624.
43. J. C. Shelley, M. Y. Shelley, R. C. Reeder, S. Bandyopadhyay and M. L. Klein, *J. Phys. Chem. B*, 2001, **105**, 4464.
44. M. J. Stevens, *J. Chem. Phys.*, 2004, **121**, 11942.
45. G. S. Ayton and G. A. Voth, *Biophys. J.*, 2004, **87**, 3299.
46. S. J. Marrink, J. Risselada and A. E. Mark, *Chem. Phys. Lipids*, 2005, **135**(2), 223.
47. S. Izvekov and G. A. Voth, *J. Chem. Phys.*, 2005, **123**, 134105.
48. J. C. Shillcock and R. Lipowsky, *Nat. Mater.*, 2005, **4**, 225.
49. Q. Shi, S. Izvekov and G. A. Voth, *J. Phys. Chem. B*, 2006, **110**(31), 15045.
50. P. M. Kasson, N. W. Kelley, N. Singha, M. Vrljic, A. T. Brunger and V. S. Pande, *Proc. Natl. Acad. Sci. USA*, 2006, **103**, 11916.
51. V. A. Harmandaris and M. Deserno, *J. Chem. Phys.*, 2006, **125**, 204905.
52. S. J. Marrink, H. J. Risselada, S. Yefimov, D. P. Tieleman and A. H. de Vries, *J. Phys. Chem. B*, 2007, **111**, 7812.
53. S. J. Marrink, A. H. de Vries, T. A. Harroun, J. Katsaras and S. R. Wassall, *J. Am. Chem. Soc.*, 2008, **130**, 10.
54. F. Tama, O. Miyashita and C. L. Brooks III, *J. Mol. Biol.*, 2004, **337**, 985.
55. V. Tozzini, *Curr. Opin. Struct. Biol.*, 2005, **15**, 144.
56. J.-W. Chu and G. A. Voth, *Proc. Natl. Acad. Sci. USA*, 2005, **102**, 13111.
57. P. J. Bond and M. S. P. Sansom, *J. Am. Chem. Soc.*, 2006, **128**, 2697.
58. P. J. Bond and M. Sansom, *Proc. Natl. Acad. Sci. USA*, 2007, **104**, 2631.
59. P. J. Bond, J. Holyoake, A. Ivetac, S. Khalid and M. Sansom, *J. Struct. Biol.*, 2007, **157**, 593.

60. A. Y. Shih, A. Arkhipov, P. L. Freddolino, S. G. Sligar and K. Schulten, *J. Phys. Chem. B*, 2007, **111**, 11095.
61. A. Y. Shih, P. L. Freddolino, S. G. Sligar and K. Schulten, *Nano Lett.*, 2007, **7**, 1692.
62. D. T. Mirijanian, J. W. Chu, G. S. Ayton and G. A. Voth, *J. Mol. Biol.*, 2007, **365**, 523.
63. J. Zhou, I. F. Thorpe, S. Izvekov and G. A. Voth, *Biophys. J.*, 2007, **92**(12), 4289.
64. M. Rueda, P. Chacon and M. Orozco, *Structure*, 2007, **15**, 565.
65. X. Periole, T. Huber, S.-J. Marrink and T. P. Sakmar, *J. Am. Chem. Soc.*, 2007, **129**, 10126.
66. Y. Kozlovsky, L. V. Chernomordik and M. M. Kozlov, *Biophys. J.*, 2002, **83**, 2634.
67. L. V. Chernomordik and M. M. Kozlov, *Annu. Rev. Biochem.*, 2003, **72**, 175.
68. K. Katsov, M. Müller and M. Schick, *Biophys. J.*, 2004, **87**, 3277.
69. K. Katsov, M. Müller and M. Schick, *Biophys. J.*, 2006, **90**, 915.
70. G. S. Ayton, P. D. Blood and G. A. Voth, *Biophys. J.*, 2007, **92**, 3595.
71. A. MacKerell Jr, D. Bashford, M. Bellott, R. L. Dunbrack Jr, J. Evanseck, M. J. Field, S. Fischer, J. Gao, H. Guo, S. Ha, D. Joseph, L. Kuchnir, K. Kuczera, F. T. K. Lau, C. Mattos, S. Michnick, T. Ngo, D. T. Nguyen, B. Prodhom, I. W. E. Reiher, B. Roux, M. Schlenkrich, J. Smith, R. Stote, J. Straub, M. Watanabe, J. Wiorkiewicz-Kuczera, D. Yin and M. Karplus, *J. Phys. Chem. B*, 1998, **102**, 3586.
72. S. E. Feller, *Curr. Opin. Colloid Interface Sci.*, 2000, **5**(3–4), 217.
73. J. C. Phillips, R. Braun, W. Wang, J. Gumbart, E. Tajkhorshid, E. Villa, C. Chipot, R. D. Skeel, L. Kale and K. Schulten, *J. Comput. Chem.*, 2005, **26**, 1781.
74. W. Humphrey, A. Dalke and K. Schulten, *J. Mol. Graphics*, 1996, **14**, 33.
75. S. Yefimov, E. van der Giessen, P. R. Onck and S. J. Marrink, *Biophys. J.*, 2008, **94**, 2994.
76. W. Treptow, S.-J. Marrink and M. Tarek, *J Phys Chem B*, 2008, **112**, 3277.
77. G. D. Fasman and H. A. Sober, *Handbook of Biochemistry and Molecular Biology*, CRC Press, Cleveland, OH, 1976.
78. M. Liu, P. Li and J. C. Giddings, *Protein Sci.*, 1993, **2**, 1520.
79. H. Ritter, T. Martinetz and K. Schulten, *Neural Computation and Self-Organizing Maps: an Introduction*, Addison-Wesley, New York, 1992.
80. E. Evans and W. Rawicz, *Phys. Rev. Lett.*, 1990, **64**, 2094.
81. P. D. Blood. *Computer Simulation Studies of Membrane Remodeling by Shear Flow and N-BAR Domain Protein Modules*, PhD thesis, University of Utah, Salt Lake City, UT, 2007.
82. C. Löw, U. Weininger, H. Lee, K. Schweimer, I. Neundorf, A. G. Beck-Sickinger, R. W. Pastor and J. Balbach, *Biophys. J.*, 2008, **95**, 4315.
83. A. Arkhipov, Y. Yin and K. Schulten, *Biophys. J.*, 2009, **97**, 2727.

Continuum Electrostatics and Modeling of K^+ Channels

JANICE L. ROBERTSON,[a, b] VISHWANATH JOGINI[b] AND BENOÎT ROUX[b]

[a] Department of Physiology, Biophysics and Systems Biology, Weill Graduate School of Cornell University, New York, NY, 10021, USA; [b] Department of Biochemistry and Molecular Biology, The University of Chicago, Chicago, IL, 60637, USA

7.1 Introduction

The movement of ions across biological membranes is one of the most fundamental processes occurring in living cells. By virtue of the long-range nature of Coulombic interactions, the electric potential generated by the movements of small metal ions across the cell membrane enables the synchronization of specific molecular processes taking place at distances far larger than typical short-range intermolecular forces. It constitutes a unified communication system for physiological information throughout the cell. The flow of ions and its ability to generate membrane voltages underlies numerous important physiological processes in living systems, ranging from propagation of the nerve impulse, cell excitability, volume regulation, excitation-secretion coupling, cellular motility, transduction by sensory receptors, the development of the embryo and many more. Specific macromolecular transport systems, ion channels and pumps, provide mechanisms to facilitate and control the passage of ions across the lipid membrane. Ion channels are necessary for the excitability of biological membranes in that they provide selective and energetically

RSC Biomolecular Sciences No. 20
Molecular Simulations and Biomembranes: From Biophysics to Function
Edited by Mark S.P. Sansom and Philip C. Biggin
© Royal Society of Chemistry 2010
Published by the Royal Society of Chemistry, www.rsc.org

favorable passage for ions to diffuse rapidly according to their electrochemical potential. Without ion channels, the lipid membrane would present a prohibitively high-energy barrier to the passage of any ion due to the unfavorable differences in the solvation free energy of the ion in the membrane relative to the aqueous solution. The problem reduces largely to differences in electrostatic stabilization in the two environments. In the solvent, the water molecules are capable of orienting their dipoles around the charge, forming an electrostatically stable hydration structure; water has a high dielectric value and can easily be polarized to offer electrostatic stabilization of the ion in solution whereas the hydrocarbon of the lipid bilayer has a small dielectric value and does not offer such stabilization. For ions to cross a membrane efficiently, there must be some molecular structure that can reduce the electrostatic energy barrier. Accordingly, ion channels can be considered catalysts of ion permeation.

Initial breakthroughs came when, high-resolution structures of many different types of ion channels were determined through X-ray crystallography,[1–3] revealing the diversity of structures employed to facilitate ion passage. Channels such as bacterial porins and mechanosensitive channels form wide, water-filled holes that open up the membrane and provide an environment for non-selective ion flux.[1,2] On the other hand, potassium channels present a set of structural features that support highly selective, diffusion-limited flux of K^+ across the membrane. The determination of the structure of the prokaryotic K^+ channel KcsA showed that the pore is formed by a narrow, multi-ion binding selectivity filter in the external half of the membrane and a large aqueous vestibule in the lower half.[3] Further K^+ channel structures show modulations to the canonical KcsA-like pore. For instance, inwardly rectifying K^+ channels possess a cytoplasmic domain that extends the ion conduction pathway more than twice the length of the KcsA pore[4] and voltage-gated K^+ channels have an additional four transmembrane helices that form a voltage-sensing domain that regulates pore gating[5] (Figure 7.1). In voltage-gated potassium channels, the voltage across the bilayer is responsible for the protein rearrangement within the voltage-sensing domain that leads to opening of the ion conduction pathway.[6] There are many questions underlying ion behavior and voltage modulation within all of these K^+ channels that remain to be addressed. With so much structural information becoming available over time, the field of ion channels demands higher level analysis with physical models in order to assess the role and functional behavior of ions in these channels. Also, these structures can be used to assess the voltage dependence of ions in the pore and the influence of external voltage on the charged entity of the voltage sensor domain. Since the problems of ion permeation and voltage sensing are largely electrostatics issues, we turn to the study of continuum electrostatics as a first-line approach to examine robust physical features of K^+ channel structures.

Continuum electrostatics approximations are based on the fundamental Poisson equation, in which the surrounding solvent is represented as a featureless dielectric medium.[7,8] In 1920, Born pioneered the application of continuum electrostatics for calculating the solvation free energy of spherical ions.[9] The approach was later extended by Kirkwood[10] and Onsager[11] to treat

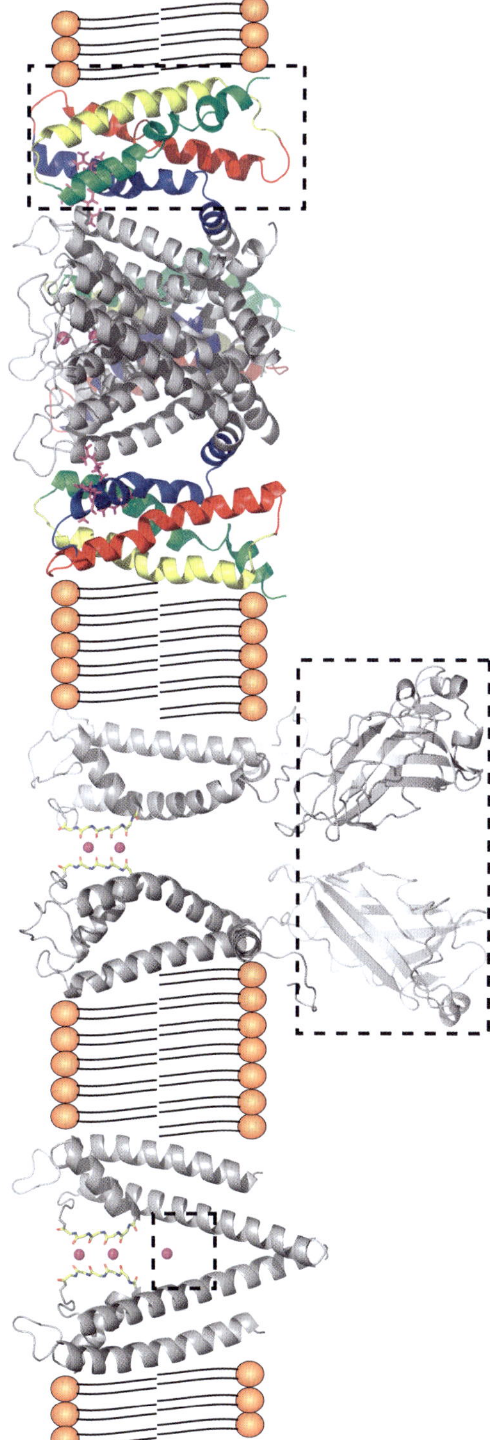

Figure 7.1 Potassium channels. Left: KcsA K$^+$ channel in the closed state. The box highlights the intracellular vestibule/cavity that provides a stabilizing environment for an ion in the center of the low dielectric membrane. Middle: inward rectifier K$^+$ channel KirBac1.1 in the putative open state. The cytoplasmic domain that extends the ion conducting pore is highlighted by the dashed lines. Right: voltage-gated K$^+$ channel Kv1.2 in the open-activated or -inactivated state. The box highlights the S1–S4 of the voltage sensor (S1, green; S2, yellow; S3, red; S4, blue). Shown in magenta are the charged residues that sense the external voltage. The fourth voltage sensor domain in front of the channel has been removed for clarity.

arbitrary charge distributions inside a spherical cavity. The effect of ionic screening arising from the presence of mobile ions in the solvent was first incorporated into Debye–Hückel and Gouy–Chapman theories, which ultimately led to the Poisson–Boltzmann (PB) equation. The treatment of complex molecular systems of arbitrary geometries and charge distribution is made possible by mapping the problem onto a discrete grid and using finite-difference relaxation numerical algorithms to solve the PB equation.[7,12,13] Consideration of the transmembrane potential led to the modified PB-voltage (PB-V) equation.[14] Although continuum electrostatics is an approximation by definition, it originates directly from fundamental laws of electrostatics and is theoretically justified. For this reason, it is often a fruitful first approach to begin to analyze protein structures, especially in the case of ion channels whose function is so directly dependent on electrostatic behavior of the system.

In this chapter, we review the application of continuum electrostatics to the investigation of K^+ channels. In Section 7.2, we begin by giving an overview of the PB and PB-V equations and how different solutions of these equations can yield key information about the structures. In particular, we discuss how to calculate the total electrostatic free energy and decompose it into physically relevant components such as the reaction field and static field energies. In Section 7.3, we discuss applications of such calculations to investigate three structural features that are present in K^+ channels: (i) the intracellular vestibule/cavity in different conformational states of prokaryotic KcsA and KirBac1.1, (ii) long-pore electrostatics in inward rectifier K^+ channels and the role of the large cytoplasmic domain and (iii) the significance of the transmembrane potential in K^+ channels and its correspondence to charges in the voltage sensing domain of Kv1.2 (Figure 7.1).

7.2 Theory and Methods

7.2.1 The Poisson–Boltzmann (PB) Equation

The PB equation:

$$\nabla \cdot [\varepsilon(\mathbf{r})\nabla\phi(\mathbf{r})] + \bar{\kappa}^2(\mathbf{r})\phi(\mathbf{r}) = -4\pi\rho(\mathbf{r}) \tag{7.1}$$

describes the electrostatic potential $\phi(\mathbf{r})$ created by a distribution of charges, $\rho(\mathbf{r})$, with Debye–Hückel ionic screening, $\bar{\kappa}(\mathbf{r})$, in a dielectric continuum, $\varepsilon(\mathbf{r})$. For complex irregular geometries, a numerical solution of the potential field can be determined by mapping the system on to a 3D grid and using a finite-difference algorithm to obtain the solution to the differential equation. Details about these methods can be found in numerous papers.[7,12,13] Briefly, the PB equation can be formally rewritten as a linear algebra problem:

$$\mathbf{M}^{-1} \cdot \phi = -4\pi\rho \Rightarrow \phi = -4\pi(\mathbf{M} \cdot \rho) \tag{7.2}$$

where the matrix $\mathbf{M}^{-1}$ represents the finite-difference form of the Green's function for the differential equation (7.1) and ϕ is the numerical solution obtained on a discrete grid across the entire system (see reference 13 for further details).

7.2.2 Calculation of Electrostatic Free Energies and Decomposition

For a system of n charges ($q_1, \ldots, q_n$), the electrostatic potential energy is the sum of each charge multiplied by the potential at the position of that charge, $\phi(\mathbf{r}_i)$:

$$\Delta G_{\text{elec}} = \frac{1}{2} \sum_i^n q_i \phi(\mathbf{r}_i) \tag{7.3}$$

where the $\frac{1}{2}$ term in front accounts for the linear response in the thermodynamic charging process.[15] In general, this equation can be formally expressed using Green's function:

$$\Delta G_{\text{elec}} = \frac{1}{2} \sum_{i,j=1}^n q_i \mathbf{M}_{ij} q_j \tag{7.4}$$

such that $\mathbf{M}_{ij}q_j$ is the electrostatic potential created by charge q_j at the position of the charge q_i.

With these definitions, the electrostatic free energy of interaction corresponding to the energy of transferring an ion from the bulk into the channel pore can be determined by finding the difference in electrostatic energy between three separate systems:

$$\Delta\Delta G_{\text{int}} = \Delta G_{\text{IC}} - \Delta G_{\text{C}} - \Delta G_{\text{I}} \tag{7.5}$$

where 'IC' represents the ion-channel complex, 'C' represents the channel alone and 'I' represents the ion alone in bulk. For simplicity, let us assume a system of one free ion, q_1, and the channel charges ($q_2, \ldots, q_n$). Each of the individual terms above can be expanded as follows:

$$\Delta G_{\text{IC}} = \frac{1}{2} \sum_{i,j=1}^n q_i M_{ij}^{\text{IC}} q_j = \frac{1}{2} q_1 M_{1,1}^{\text{IC}} q_1 + \sum_{j=2}^n q_1 M_{1,j}^{\text{IC}} q_j + \frac{1}{2} \sum_{i,j=2}^n q_i M_{ij}^{\text{IC}} q_j \tag{7.6}$$

$$\Delta G_{\text{C}} = \frac{1}{2} \sum_{i,j=2}^n q_i M_{ij}^{\text{C}} q_j \tag{7.7}$$

$$\Delta G_{\text{I}} = \frac{1}{2} q_1 M_{1,1}^{\text{I}} q_1 \tag{7.8}$$

Each of these terms can be calculated directly by solving the PB equation for the three different system definitions. Additionally, inserting each of the above expansions back into the total electrostatic free energy [equation (7.5)] shows that the relationship can be rearranged as a function that is dependent on the ion charge, q_1:

$$\Delta\Delta G_{\text{int}} = \left(\frac{1}{2} q_1 M_{1,1}^{\text{IC}} q_1 + \sum_{j=2}^{n} q_1 M_{1,j}^{\text{IC}} q_j + \frac{1}{2} \sum_{i,j=2}^{n} q_i M_{ij}^{\text{IC}} q_j \right)$$
$$- \left(\frac{1}{2} \sum_{i,j=2}^{n} q_i M_{ij}^{\text{C}} q_j \right) - \left(\frac{1}{2} q_1 M_{1,1}^{\text{I}} q_1 \right) \tag{7.9}$$

$$\Delta\Delta G_{\text{int}} = \frac{1}{2} q_1 \left(M_{1,1}^{\text{IC}} - M_{1,1}^{\text{I}} \right) q_1 + \sum_{j=2}^{n} q_1 M_{1,j}^{\text{IC}} q_j$$
$$+ \frac{1}{2} \sum_{i,j=2}^{n} q_i \left(M_{ij}^{\text{IC}} - M_{ij}^{\text{C}} \right) q_j \tag{7.10}$$

which demonstrates the parabolic dependence of the electrostatic free energy on the free ion charge $q_1 = Q_{\text{ion}}$:

$$\Delta\Delta G_{\text{int}} = \frac{1}{2} A Q_{\text{ion}}^2 + B Q_{\text{ion}} + C$$
$$A = M_{1,1}^{\text{IC}} - M_{1,1}^{\text{I}}$$
$$B = \sum_{j=2}^{n} M_{1j}^{\text{IC}} q_j \tag{7.11}$$
$$C = \frac{1}{2} \sum_{i,j=2}^{n} q_i \left(M_{ij}^{\text{IC}} - M_{ij}^{\text{C}} \right) q_j$$

where the first energetic term, $\frac{1}{2} A Q_{\text{ion}}^2$, is the energy that an ion experiences due to the reaction field created by the protein dielectric in response to the ion charge ($\Delta G_{\text{RF-prot}}$). The second term, $B Q_{\text{ion}}$, is the energy of the ion in the static field established by the protein charges (ΔG_{SF}). The last term C, reflects the energy due to the reaction field created by the ion dielectric in response to the protein charges ($\Delta G_{\text{RF-ion}}$). The complete energetic balance of an ion in the channel depends on the sum of these three terms and follows a quadratic dependency on the ion charge.[16–18]

7.2.3 The Modified PB Equation for Treatment of Transmembrane Voltage

The transmembrane potential acts as a driving force on the translocation of permeating ions and on the opening and closing transitions of voltage-gated channels.[19] Microscopically, it arises from charge separation at the membrane interfaces, resulting in an electrostatic potential across the system. In the simplest case of a planar membrane, the potential is linear across the width of the membrane. For more complicated structures, such as that created by an ion channel in the membrane, the transmembrane potential is difficult to predict and must be calculated by finite difference solution of a modified PB equation that includes the contribution from the membrane potential:[14]

$$\nabla[\varepsilon(\mathbf{r})\nabla\phi(\mathbf{r})] - \bar{\kappa}^2(\mathbf{r})[\phi(\mathbf{r}) - V_{\mathrm{mp}}\Theta(\mathbf{r})] = -4\pi\rho(\mathbf{r}) \tag{7.12}$$

where $\Theta(r)$ is a Heaviside step-function set to 0 on one side of the membrane and 1 on the other. The additional term, $V_{\mathrm{mp}}\Theta(\mathbf{r})$, accounts for the externally imposed transmembrane potential, V_{mp}. Accordingly, the total electrostatic free energy of a membrane protein submitted to a transmembrane potential is

$$\Delta G_{\mathrm{elec}} = \frac{1}{2}CV_{\mathrm{mp}} + \frac{1}{2}\sum_i^n q_i\phi_{\mathrm{rf}}(\mathbf{r}_i) + \sum_i^n q_i\phi_{\mathrm{mp}}(\mathbf{r}_i) \tag{7.13}$$

where C is the capacitance of the system and $\phi_{\mathrm{mp}}(\mathbf{r})$ is the transmembrane field determined by solving equation (7.12) with the charge distribution set to zero, $\rho(\mathbf{r}) = 0$. $\phi_{\mathrm{rf}}(\mathbf{r})$ is the reaction field obtained by solving the system with V_{mp} set to zero; the standard PB equation is then recovered and the associated free energy reduces back to equation (7.3). The 'gating charge', Q_{gating}, represents the cumulative charge that is moved across the membrane electric field during a conformational change of the channel. Alternatively, the gating charge can be related to the reversible electric work, $Q_{\mathrm{gating}}V_{\mathrm{mp}}$,[19] which is associated to the voltage-dependent component of the relative free energy of the open and closed state of the channel, *i.e.* $Q_{\mathrm{gating}} = \partial[G_{\mathrm{elec}}(\mathrm{open}) - G_{\mathrm{elec}}(\mathrm{closed})]/\partial V_{\mathrm{mp}}$, which leads to

$$Q_{\mathrm{gating}} = \frac{1}{V_{\mathrm{mp}}}\left[\sum_i q_i\phi_{\mathrm{mp}}\left(r_i^{(\mathrm{o})}\right) - \sum_i q_i\phi_{\mathrm{mp}}\left(r_i^{(\mathrm{c})}\right)\right] \tag{7.14}$$

To evaluate Q_{gating} from equation (7.14), the PB-V equation (7.12) must be solved twice, once for the open state and once for the closed state. Equation 7.14 also shows that an important quantity is the dimensionless factor $\delta = \phi_{\mathrm{mp}}(\mathbf{r})/V_{\mathrm{mp}}$, representing the 'fraction' of the transmembrane potential acting on a charge at the point $\mathbf{r}$. While it provides an important measure of the voltage coupling of charged residues of a voltage-gated channel, it is used to characterize the locations where various blockers and permeant ions[20–23] can bind along the pore.

7.3   Applications

7.3.1   Electrostatics in the Intracellular Vestibule of K^+ Channels

The general template for the K^+ channel transmembrane pore was determined from numerous high-resolution structures of prokaryotic K^+ channels such as KcsA and KirBac (Figure 7.1).[3,4,24] The extracellular facing half of the channel contains the selectivity filter, a series of binding sites formed by backbone carbonyls of the 'TVGYG' signature sequence. In these sites, dehydrated K^+ ions bind in close proximity to one another, allowing for strong selectivity while maintaining high-throughput conductance through multi-ion repulsion. In the lower half of the membrane, the channel forms an aqueous cavity or intracellular vestibule that has been shown to stabilize a hydrated ion in the closed state.[3,4,25] This was a significant finding, demonstrating that the K^+ channel is capable of removing the energetic barrier of the membrane, and allows the stabilization of K^+ at the center of the hydrophobic core. Moreover, since the ion remains hydrated by water and is not in direct coordination with any protein residues, it appears that this stabilization occurs through long-range electrostatic interactions from the protein.

The electrostatic environment inside the potassium channel was investigated by calculating the total free energy (electrostatic and van der Waals energies) of transferring a K^+ ion from the bulk into the pore of KcsA[18] and KirBac1.1[26] and is shown in Figure 7.2. The closed-state energetics were calculated for the 2 Å resolution structure of KcsA (pdb: 1K4C)[25] and the 3.6 Å resolution structure of KirBac1.1 (pdb: 1P7B).[4] For KirBac1.1, two missing portions of the cytoplasmic domain were built by homology based on the complete cytoplasmic structure of Kir3.1/GIRK (pdb: 1N9P).[27] In addition, the charged states of several residues were modified in both channels, E71 in KcsA and E106/E130 in KirBac1.1, in order to fit with experimental and pK_a calculation predictions.[18,28] Based on these structures, the calculations demonstrate that in the closed states of K^+ channels, there is a large barrier throughout the lower region of the transmembrane pore. As the ion approaches the cavity at $Z = 0$ Å, the interaction energy inside KcsA becomes favorable whereas an ion inside KirBac1.1 continues to experience > 10 kcal mol^{-1} destabilization. Even though both channels are in the closed state, there are differences in the stabilization energy of the cavity ion because the KirBac1.1 calculations are carried out in the presence of the cytoplasmic domain (see Section 7.3.2). The structures of numerous K^+ channels have also been determined with the pore in the open states: MthK (pdb: 1LNQ),[29] KvAP (pdb: 1ORQ)[30] and Kv1.2 (pdb: 2A79),[5] allowing for homology modeling of open KcsA.[18] In addition, 9 Å resolution 2D electron crystallography data for KirBac3.1 were determined with a putative open pore,[31] allowing for the extrapolation of a homology model for an open state of KirBac1.1.[39] In these open structures, the overall barrier goes to zero at ion positions $Z < -10$ Å and the cavity shows a moderate repulsive energy throughout the transmembrane region of the pore in

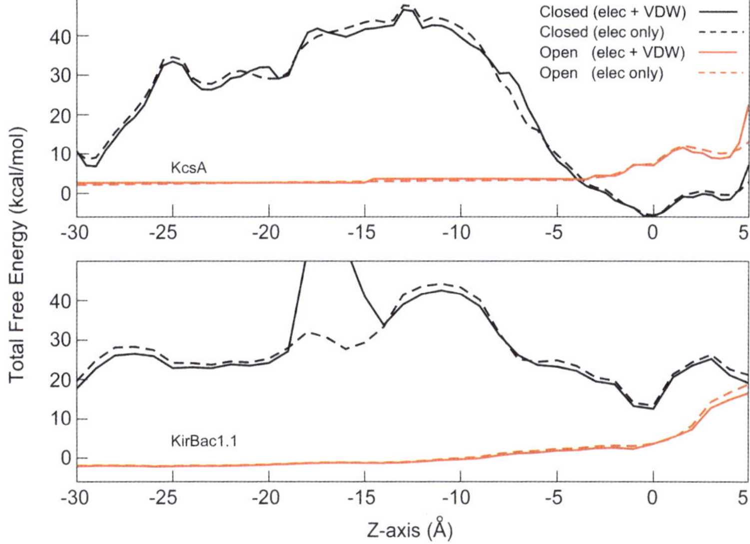

Figure 7.2 Total free energy in the closed and open states of KcsA and KirBac1.1. The free energy of transferring an ion from the bulk into the central pore is shown for KcsA (top) and KirBac1.1 (bottom) for closed (black) and open (red) state models. The total free energy (solid line) includes the electrostatic free energy and the van der Waals interaction of the ion with the channel. The isolated electrostatic component is represented by the dashed lines. The membrane spans from $-12.5\,\text{Å} < Z < 12.5\,\text{Å}$ and the intracellular vestibule/cavity is centered at $Z = 0\,\text{Å}$.

both channels. For additional calculations based on different open/closed structures and models of K^+ channels, we refer the reader to work by Jogini and Roux.[18]

The dominant contribution to the large barrier to K^+ permeation in the closed state of both channels comes from the high reaction field energy along the pore (Figure 7.3). A reaction field occurs when a charge polarizes a high dielectric medium, thereby establishing a repulsive electric field at nearby low dielectric interfaces, and is a particular consequence in the narrowed closed state. This effect is conserved between the two channels even though the details of their closed-state structures are different. In KirBac1.1, the lower transmembrane helices do not physically cross, although they do narrow, allowing for a bulky hydrophobic residue, F139, to occlude the pore sterically, as demonstrated by the abrupt peak in the van der Waals energy around $Z \approx -16.5\,\text{Å}$ (Figure 7.2). On the other hand, KcsA forms a bundle crossing, acting like an aperture restricting the pore size, yet even in the closed state it is still wide enough to allow for the passage of a dehydrated K^+ ion without significant van der Waals clashes. As expected, the reaction field contribution to

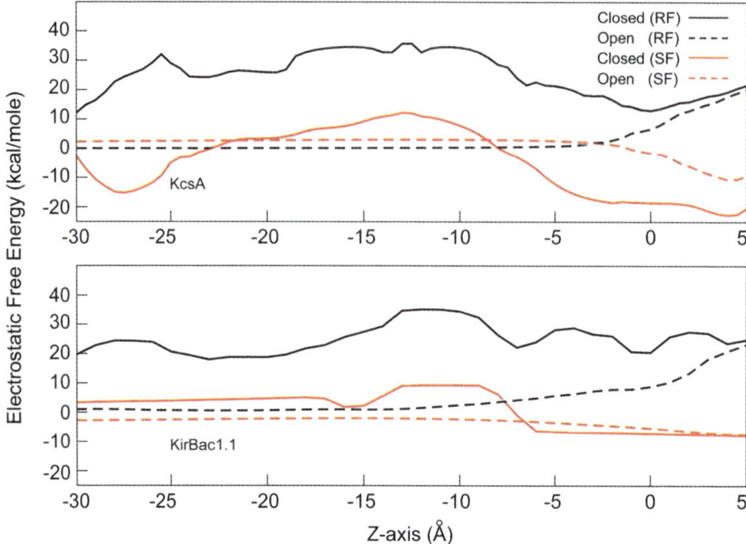

Figure 7.3 Reaction field and static field energies. The reaction field energy profiles (black) and static field profiles (red) for KcsA (top) and KirBac1.1 (bottom). The membrane spans from $-12.5\,\text{Å} < Z < 12.5\,\text{Å}$ and the cavity is centered at $Z = 0\,\text{Å}$.

the free energy is reduced in the open states relative to the closed states as the dielectric interfaces differ extensively. In the open state of both channels, the diameter of the pore on the inner side is larger and the energy barrier at the intracellular entrance is reduced to $\sim 0\,\text{kcal}\,\text{mol}^{-1}$. This demonstrates that a channel modulates the electrostatic barrier of the membrane by opening the lower transmembrane pore and that physical occlusion is not necessary to have a physical closed state.

For an ion at the center of a membrane represented by a $30\,\text{Å}$ slab with a dielectric constant of 2, the electrostatic free energy is extremely unfavorable, on the order of $40\,\text{kcal}\,\text{mol}^{-1}$.[15] As demonstrated by the total free energy profiles in Figure 7.2, an ion at the center of the membrane inside the K^+ channel intracellular vestibule is substantially more stable, even in the buried low dielectric of the closed state structures. However, the repulsion of the reaction field is never completely removed, even in the open state, and so the protein must confer an additional favorable energy term in order to achieve this stability. This comes from the static field, established by the protein charge distribution shielded by the dielectric environment of the channel and membrane. The total static field was calculated along both channels and is shown in Figure 7.3. These profiles demonstrate that the protein charges create a generally favorable component within the membrane region, which is strongest in the closed states. The molecular composition of the static field at the cavity was determined by dissecting the contribution from each individual residue (Figure 7.4). This analysis demonstrates that the backbone of the pore helix contributes a significant amount to the cavity stabilization. Often referred to as the pore–helix dipole,[16] this contribution appears to be a common feature of both channels, although it is more significant in the structure of KcsA. These differences correspond to deviations in the pore–helix alignment with respect to the center of the pore; while the helices of KcsA are all pointed directly at the cavity, they are slightly rotated outward in the case of KirBac1.1 and thereby the field is not as focused.[4] In KirBac1.1, the strongest electrostatic contribution comes from residue T110 at the position where the helix turns to form the selectivity filter. Note that since the cavity is embedded within the low dielectric, the static field at this position is particularly sensitive to changes in charge distributions throughout the channel. Previous results have shown that the cavity energy for KirBac1.1 becomes favorable if E106/E130 are in their native charged state.[26,33] Also, calculations of the isolated TM domain show the cavity energy as favorable;[18] however, this becomes destabilized on adding the cytoplasmic domain due to positive charges near the membrane interface (Figure 7.2). The original crystal structure of KirBac1.1 does not show a cavity ion.[4] However, a higher resolution structure of a chimaera of the TM pore of KirBac3.1 and cytoplasmic domain of Kir3.1/GIRK does show a hydrated ion density,[25] although this may come about due to long-range electrostatic effects of the different cytoplasmic domains or from charged lipids that were bound to the channel exterior. Further functional information on such structures is necessary for defining the appropriate structural features for electrostatic analysis.

A counterintuitive finding that is common between both KcsA and Kir-Bac1.1 is that the total electrostatic free energy for an ion at the cavity is generally unfavorable in the open channels. In fact, for KcsA, the ion is actually more stable in the closed structure ($-5.6\,\mathrm{kcal\,mol^{-1}}$) than in a widely open state ($7.6\,\mathrm{kcal\,mol^{-1}}$). It might be expected that the opposite would be true, since the reaction field energy is typically on the order of 11–$13\,\mathrm{kcal\,mol^{-1}}$ for the closed state and about 2–$3\,\mathrm{kcal\,mol^{-1}}$ for the open channels. Nonetheless, the overall reduction in reaction field energy does not necessarily lead to a net increase in stability for a cation in the cavity because the static field contribution is also reduced. When the channel is open, the cavity becomes more exposed to bulk water, resulting in a high dielectric environment in the vicinity of the cavity ion and the pore helices. As a consequence, the contribution of the pore helices in stabilizing an ion in the cavity is much smaller in the open state than in the closed state (Figure 7.4). In fact, the static field from all residues is smaller in the open state relative to the closed state. Therefore, despite a decrease in the unfavorable reaction field contribution in the open state, a cation in the cavity is significantly less stable in the open state than the closed state because the static field is extensively reduced. In the open state, strong interactions inside the cavity may slow conductance, since this site is not as strongly coupled to the multi-ion repulsion in the filter. Therefore, the unfavorable cavity energy in the open state may be indicative of an optimization for

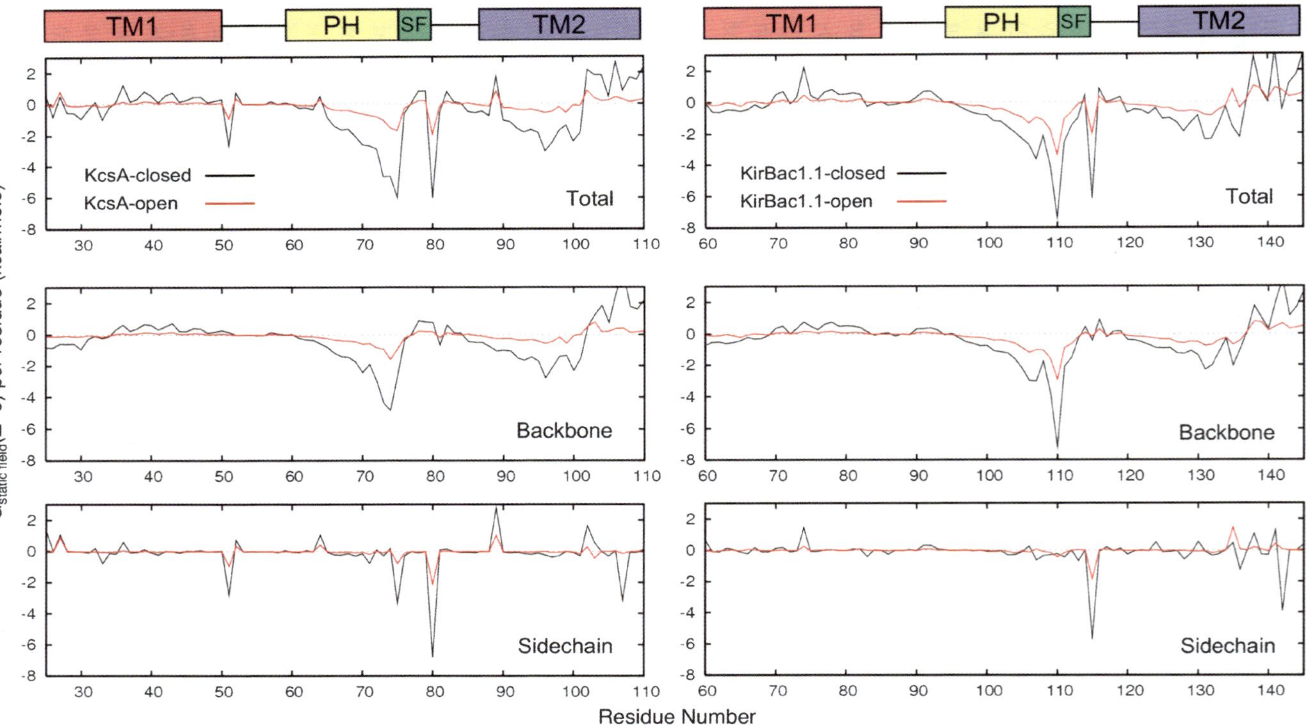

Figure 7.4 Static field decomposition by residue inside the transmembrane cavity at $Z = 0$ Å. The individual residue contributions of the static field energy at the cavity are shown for KcsA (left) and KirBac1.1 (right). The top panel shows the total residue contribution and the middle and bottom panels show the isolated backbone and side-chain energies, respectively. Residues E71 in KcsA, E106 and E130 in KirBac1.1 have been protonated in accord with pK_a calculations from Jogini and Roux.[18] A schematic of the protein topology is shown at the top of the plots (TM1, the first transmembrane helix; PH, pore helix; SF, selectivity filter; TM2, the second transmembrane helix).

fast ion flux.[18] However, in the presence of negatively charged residues in the vestibule region (for example, A108D in KcsA), the cavity ion is stabilized even in the open state due to static field effects. Although the ion is stabilized, an increased local concentration of K^+ due to negative electrostatic potential can easily knock off the cavity ion, resulting in large conductance as seen in Ca^{2+} and voltage-gated BK channels.[34] It has also been shown that the presence of charged residues in the extracellular region can modulate the flux of the channel using a similar mechanism.[35]

A consequence that arises out of the strong dependence at the cavity site on both the static and reaction field is that it is also a site of valence selectivity for the ion charge. Described in Section 7.2, the electrostatic interaction energy of a single ion is a quadratic function of the ion charge, $\Delta\Delta G_{int} = (1/2) AQ_{ion}^2 + BQ_{ion} + C$ [equation (7.11)], where A is the reaction field created by the ion charge on the protein dielectric, B is the static field created by the protein charges and C is a charge-independent reaction field energy from the protein charges on the ion dielectric. In the case of KcsA and KirBac1.1, this final term is negligible at the cavity. The function creates a parabolic dependence on the ion charge, with the minimum yielding the optimal (best) charge at that position, $Q_{best} = -B/A$. These charge dependence curves are shown for the cavity site for KcsA and KirBac1.1 in the closed and open states in Figure 7.5A. Both channels show optimal charges of $\sim +0.5e$ to $+1.4e$, indicating a preference for monovalent cations such as K^+. This does not change from the closed to the open state; rather, the conformational change affects the magnitude of the stabilization. For comparison, we also examine the cavity charge relationships for two inward rectifiers, a subfamily of K^+ channels that will be discussed in more detail in the next section. Strong inward rectifier channels, such as Kir2.1/IRK, selectively restrict outward current through cytoplasmic block by Mg^{2+} and polyamines, with the presumed site for rectification block near aspartic acid residues within the cavity. Weak rectifiers such as Kir1.1/ROMK do not contain these cavity charges and in turn do not exhibit the strong rectification behavior. Figure 7.5B shows the charge dependence at the cavity in open-state models of these two channels, showing that the weak rectifier has an optimal charge of $\sim +1e$, whereas the strong rectifier prefers $> +2e$, in agreement with this site favoring stronger charge species and particularly divalent Mg^{2+}. This type of analysis has also been used to clarify experimental results of mutations in the pore–helix and cavity of inward rectifiers.[36] As mentioned previously, residues at the base of the pore–helix contribute significant electrostatic stabilization inside the pore. However, the corresponding mutation T141K in Kir2.1/IRK channels does not affect K^+ conductance but does specifically reduce the affinity of Ba^{2+} blockade.[36] Electrostatic calculations introducing these mutations into KirBac1.1, at the S4 binding site of the selectivity filter, the proposed location of Ba^{2+} blockade, show that the charge preference shifted from $+1.6e$ to $+0.8e$ in the mutant channel. The electrostatics predict that these mutations affect the divalent energetics much more than the monovalent energies, in agreement with the observations in the experimental studies.

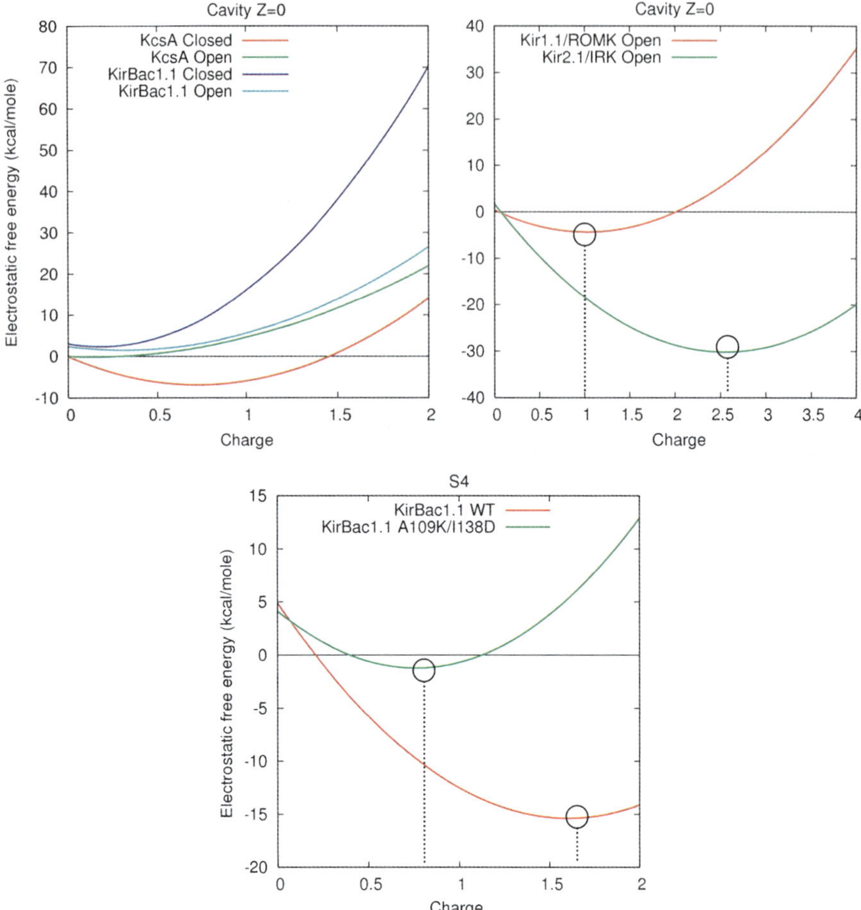

Figure 7.5 Charge dependence in the cavity and S4 site. Parabolic curves following the charge dependence equation $\Delta\Delta G_{\mathrm{int}} = (1/2)AQ^2_{\mathrm{ion}} + BQ_{\mathrm{ion}} + C$ for different channels and states. The optimal charge is given by the minimum of the parabola $Q_{\mathrm{best}} = -B/A$ and is designated by a circle. Top left: energetics at the cavity position, $Z = 0\,\text{Å}$, for KcsA and KirBac1.1 in the closed and open states. The optimal charge inside the cavity is $\sim +0.5e$ to $+1e$ for both channels. Top right: the charge dependence inside the cavity of mammalian inward rectifier K^+ channels, Kir1.1/ROMK (weakly rectifying) and Kir2.1/IRK (strongly rectifying). The optimal charges are $+1e$ and $+2.5e$, respectively, supporting preferential binding of stronger charge ions over monovalents in the strong rectifier channels. Bottom: energetics at the S4 binding site in the selectivity filter of KirBac1.1. The mutation A109K/I138D corresponding to T141K/D172 in Kir2.1/IRK shifts the optimal charge from $+1.6e$ to $+0.8e$, supporting the observed reduction in Ba^{2+} block affinity but not K^+ conductance in experiments. The values for this plot were obtained from the supplementary table in Chatelain *et al.*[36]

7.3.2 Long-pore Electrostatics in K^+ Channels

So far, we have focused on the K^+ channel pore and the influence of the cavity vestibule in modulating the electrostatic barrier of the membrane. We now change direction and look at a different structural feature of K^+ channels that is specific to the subfamily of inward rectifiers. These channels possess a large cytoplasmic domain that extends the ion conducting pore up to $\sim 85\,\text{Å}$ (Figure 7.1), more than twice the length of KcsA. Although the cytoplasmic domain is far from the membrane, it is strongly involved in the process of ion permeation by regulating conduction, rectification affinity, rectification kinetics and gating. Here we examine the electrostatic consequences of the long Kir channel pore through continuum electrostatics analysis. For this analysis, we use the low-resolution 3D structure of the KirBac1.1 channel obtained by electron microscopy[31] as a template to construct putative open-state models of additional Kir channels by exploiting their sequence similarity with KirBac1.1.

The electrostatic free energy of transferring a K^+ ion from the bulk into the center of the pore was calculated along the putative open state models of the prokaryotic inward rectifier homologue KirBac1.1, and also two mammalian Kir channels, the weakly rectifying Kir1.1/ROMK and the strong rectifier Kir2.1/IRK (Figure 7.6).[33] The cytoplasmic domain, located between $-60\,\text{Å}$ $<Z<-30\,\text{Å}$, provides a favorable environment for the stabilization of cations in mammalian channels but not in the case of the KirBac1.1. Since the cytoplasmic pore is fairly wide, $>12\,\text{Å}$ throughout, and also far from the low-dielectric membrane, there is little repulsive contribution of the reaction field. Instead, it is the static field that accounts for the majority of the total electrostatic free energy inside the cytoplasmic domain.

The static field decomposition of the contribution from each residue within the channel reveals a striking observation about the electrostatic nature inside the long Kir pore (Figure 7.7). The static field from certain residues is significantly long ranged, even up to $40\,\text{Å}$ away from their Z position along the pore (Figure 7.7, black line). This effect persists even when the protein dielectric is increased to 10, albeit with reduced magnitude. In the cytoplasmic domain, these long-range residues are either pore lining or in the vicinity of the pore and are exposed to solvent high dielectric. However, the long pore of the Kir channel extends the low dielectric into the cytoplasm, potentiating the static field even though there is high-dielectric screening inside the pore. It is well known that long-ranged electrostatics is a key factor governing the behavior of ions inside channels;[16,33,37,38] however, they are often difficult to grasp intuitively except in the case of the simplest geometries, and for this reason many aspects are not readily understood. Most experiments involving mutations of charged residues are interpreted by assuming that this type of perturbation is local, but our results imply that this assumption is not generally valid, especially within Kir channels. This suggests that cytoplasmic residues can affect ion energetics throughout the pore and possibly blocker affinities, even at sites that are far away such as inside the transmembrane cavity. This is an important consideration when consolidating the results of mutagenesis studies with the long-pore structure of Kir channels.

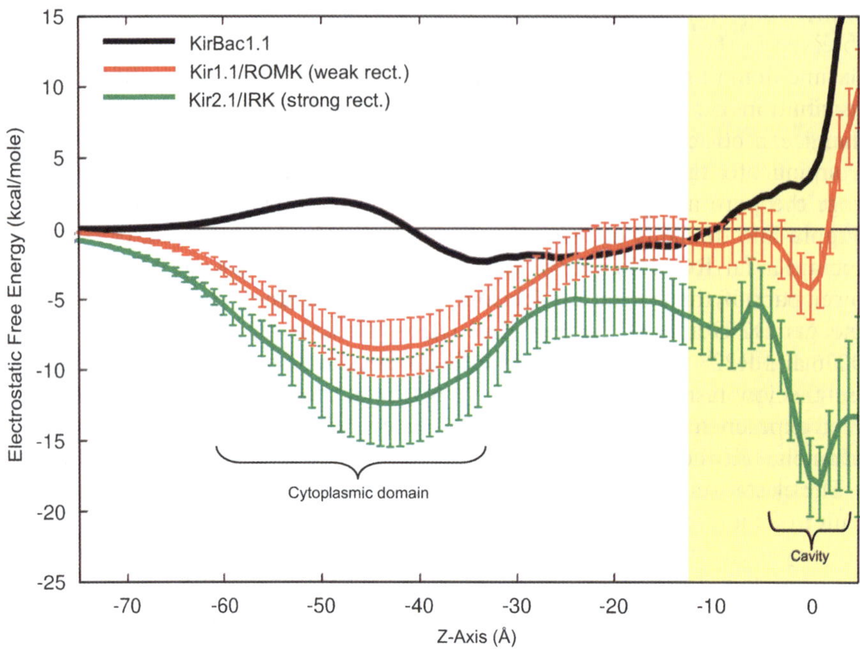

Figure 7.6   Electrostatic free energy inside open-state inward rectifier K^+ channels. The total electrostatic free energy is shown for a cation along the putative open-state models of KirBac1.1 (black), Kir1.1/ROMK (red) and Kir2.1/ IRK (green). In the KirBac1.1 calculations, residues E106 and E130 are in the protonated form in accord with pK_a calculations from Jogini and Roux.[18] Kir1.1/ROMK residue K80 is neutralized corresponding to putative conductive electronic state of this channel.[33] The two areas that are highlighted are the cytoplasmic domain, from $-60\,\text{Å} < Z < -30\,\text{Å}$, and the cavity at $Z = 0\,\text{Å}$. The low dielectric membrane slab is represented in yellow. The error bars represent the standard deviation in calculation values based on 12 different homology models for each of the mammalian channels. Adapted from Robertson *et al.*, 2008.[33]

The static field decomposition in Figure 7.7 also demonstrates the degree of variability between the charge distributions of the different cytoplasmic domains. In a study of five inward rectifiers (KirBac1.1, Kir1.1/ROMK, Kir2.1/IRK, Kir3.1/GIRK and Kir6.2/KATP), it was found that four charged residues are conserved and together create a common background of favorable static field for cations throughout the entire length of the pore (Figure 7.8A). In turn, 39 other positions were identified as variable sites for electrostatic modulation, responsible for defining the specific electrostatic profile particular to the cytoplasmic domain of each channel.[33] For instance, KirBac1.1 contains an excess of positively charged residues, accounting for the overall repulsive energy in Figures 7.6 and 7.8B. Alternatively, the mammalian Kir channels consistently have an excess of negative static field in the cytoplasmic domain. In fact, for Kir1.1/ROMK and Kir2.1/IRK, the sum profiles of the strongly contributing residues are nearly identical (Figure 7.8B). The differences observed in the total electrostatic free energy profiles between these two cytoplasmic domains in Figure 7.6 comes from long-range influences from charge distributions outside this region. In particular, the slight excess negative field in Kir2.1 can be accounted by the contribution from the cavity aspartates and a grouping of negative charges at the cytoplasmic entrance that are far away from the pore axis. Although the core electrostatics of the two mammalian cytoplasmic domains are similar, their molecular compositions are very different. Kir1.1/ROMK contains two negative charges, D254 and E258, near the lower part of the central cytoplasmic pore, which are also present in Kir2.1/IRK as D255 and D259. Yet Kir2.1/IRK also possesses five others: D249, D276 and R228 located at the intracellular entrance, and E224 and R260 in the central cytoplasmic pore (Figure 7.8). Many of these residues have been identified experimentally as having some effect on the functional behavior of these channels, but their exact roles are not yet well understood.

Electrostatic analysis has assisted in clarifying some of the key considerations in interpreting mutagenesis and chimaera studies between these channels. For instance, the mutation E224G reduces the strength of rectification in Kir2.1/IRK, whereas the corresponding mutation in Kir1.1/ROMK, G223E, does not confer the expected increased rectification behavior.[39] Since the profiles of the two cytoplasmic domains are similar, the exchange of charged residues between these channels does not equate to exchanging the electrostatic features. Neutralization of E224 significantly destabilizes ion energetics along the entire pore of Kir2.1/IRK, whereas introducing a glutamate at the same position in Kir1.1/ROMK does not confer favorable electrostatics; instead, it increases the field in an already favorable domain. Another point that arises from the analysis is the demonstration of how residues that are apparently close to one another might affect ion behavior differently. For instance, residues D255 and D259 in Kir2.1/IRK are fairly close to each other in the pore; however, D259 has a direct effect on the strength and affinity of rectification, whereas D255 only affects rectification kinetics.[40,41] In fact, the static field contributions of the two residues are different, D259 being more long-range and having a stronger contribution towards the transmembrane pore (Figure 7.7). This comes about through the different dielectric environment for each residue, in which D259 is located centrally inside the core of the cytoplasmic domain, whereas D255 is more exposed to the high dielectric of the bulk solution at the entrance. These results are in line with the suggestion that D255 plays a role in concentrating rectification blockers and thereby is capable of modulating kinetics of blockade.[41] A final remark can be made regarding the original chimaera studies, which showed that exchanging the cytoplasmic domain of Kir1.1/ROMK with that of Kir2.1/IRK confers strong rectification even in the absence of the cavity aspartates in the pore.[39,42] Since the static fields from the cytoplasmic domains of Kir1.1/ROMK and Kir2.1/IRK are very similar, a general electrostatic explanation for change in rectification behavior does not hold. The specific differences between these charge distributions and how these affect multi-ion configurations or direct interactions

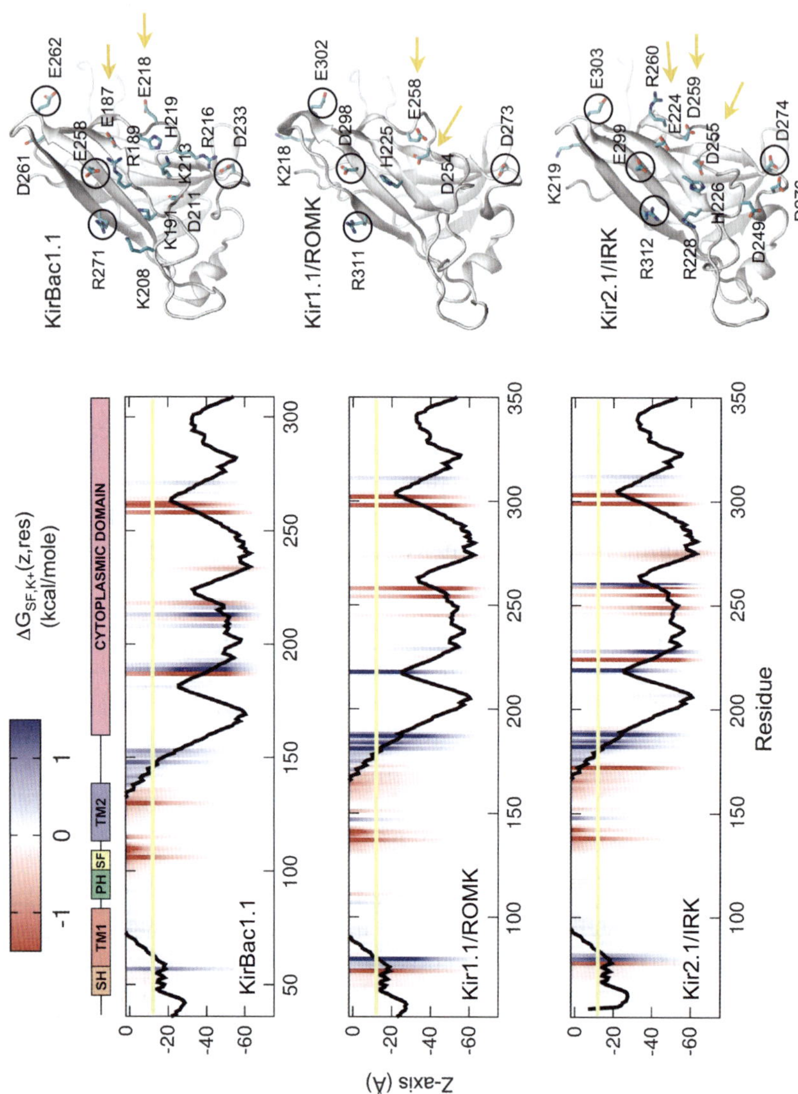

KirBac1.1
E262
D261
E258
R271
K208
E187
R189
K191
D211
H219
K213
R216
E218
D233

Kir1.1/ROMK
E302
D298
R311
K218
H225
E258
D254
D273

Kir2.1/IRK
E303
E299
R312
K219
R228
H226
D249
E224
R260
D259
D255
D274
D276

ΔG_SF,K+(z,res)
(kcal/mole)
1
0
-1
SH
TM1
PH
SF
TM2
CYTOPLASMIC DOMAIN

KirBac1.1
Kir1.1/ROMK
Kir2.1/IRK

Z-axis (Å)
0
-20
-40
-60

Residue
50
100
150
200
250
300
350

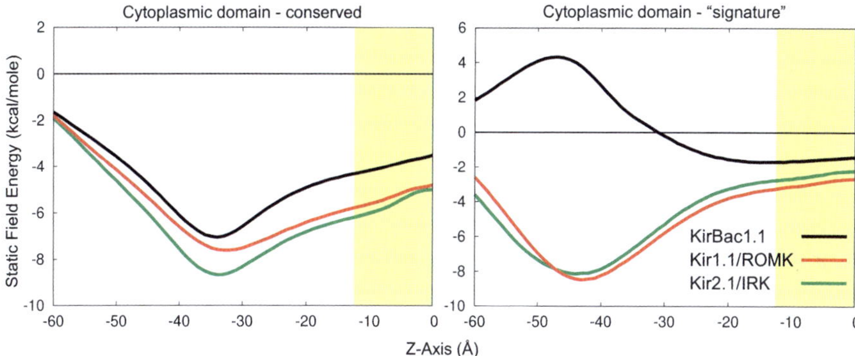

Figure 7.8 Static field contributions from the cytoplasmic domain. Left: sum contribution from the four conserved electrostatic residues in the cytoplasmic domain. Right: the profile from the remaining residues individual to each channel, otherwise referred to as the cytoplasmic 'signature' profiles. The yellow slab represents the position of the low dielectric membrane. Adapted from Robertson *et al.*, 2008.[33]

with blockers are likely important for the understanding of the mechanisms underlying ion conduction and rectification.

7.3.3 K^+ Channels and the Transmembrane Potential

The transmembrane potential across an ion channel system can be calculated with continuum electrostatics by solving the PB-V equation (7.12) described in Section 7.2.14. To illustrate the dependence of the voltage profile on the channel structure, the potential was calculated along the pore of KcsA in a low-dielectric membrane slab. The voltage profile was calculated for different channel conformations, going from the closed to the open state. The closed state is based on the X-ray structure of KcsA and the open state is a model

Figure 7.7 Decomposition of the static field energy per residue along the Kir channel pore. The left panels show the static field decomposition by residue for KirBac1.1 (top), Kir1.1/ROMK (middle) and Kir2.1/IRK (bottom), along the full length of the pore. Each streak represents the individual contribution by a particular residue towards the ion interaction at the central pore axis. A schematic of the channel topology is shown at the top of the plots depicting the residue alignment with each channel domain. The solid black line shows the Z position of the C_α atom of each residue, and the yellow line represents the bottom of the membrane. A single subunit of each cytoplasmic domain is shown at right. The residues contributing an absolute value greater than $1.0\,\text{kcal}\,\text{mol}^{-1}$ are represented explicitly. An exception is the histidine residues, which are neutral in the static field calculations but are shown because they are in position to provide significant contributions if protonated. The circled residues depict the conserved electrostatic group in the cytoplasmic domain and the yellow markers highlight the strongest negative residues at the central core of the cytoplasmic pore. Adapted from Robertson *et al.*, 2008.[33]

based on the X-ray structure of MthK. The intermediate conformations were generated by linear interpolation of dihedral angles between the closed and open confirmations (Figure 7.9).[15,18] In the closed state, the potential is distributed almost evenly across the pore with a 50% drop at the selectivity filter located in the upper half of the membrane. In contrast, the voltage drop is focused almost entirely across the selectivity filter in the open state.

Major aspects of the nature of the transmembrane potential inside the K^+ channel pore have been predicted before such theoretical calculations were made. In 1988, Neyton and Miller[20,21] investigated the K^+ dependence of Ba^{2+}

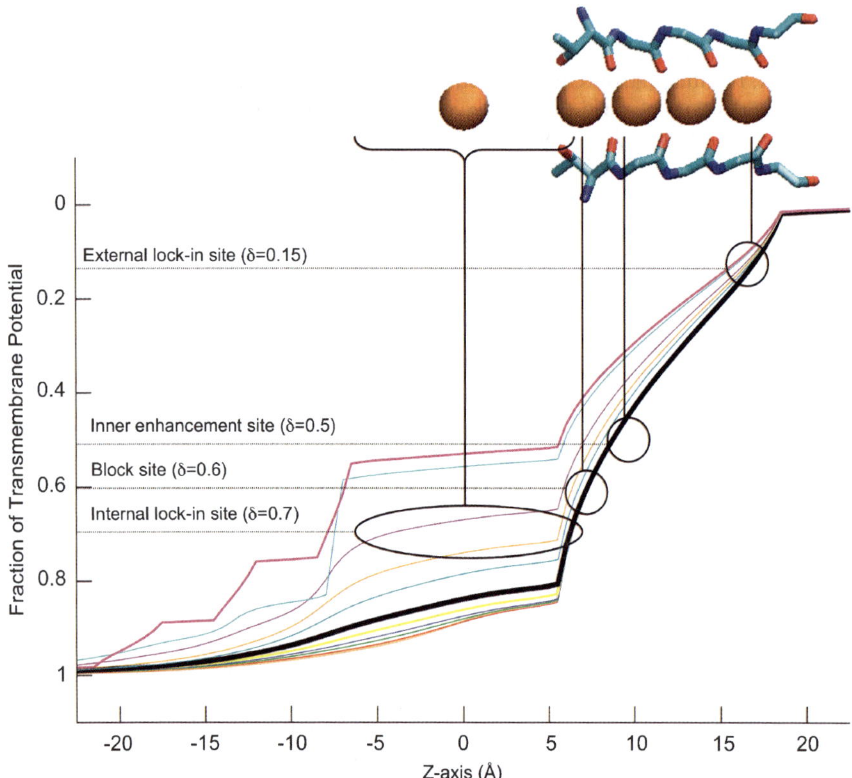

Figure 7.9   Calculated transmembrane potential along the pore of KcsA. The transmembrane potential is shown for various structures of the KcsA pore: closed (magenta), open (orange) and several intermediate models. The voltage profile of the hypothesized optimal open structure of that provides a cavity energy of $\sim 0\,kcal\,mol^{-1}$ is shown by the thick black line. The fractional distances, δ, of K^+ and Ba^{2+} binding sites determined from Neyton and Miller[20,21] and the positions of K^+ binding sites corresponding to the crystallographic structures are plotted alongside the calculated voltage profiles for a comparison of the structural, electrophysiological and computational results. The circled regions give estimates of the positions of the different K^+/Ba^{2+} binding sites.

blockade in BK channels and through this were able to determine the fractional electrical distance (δ) of four ion-binding positions along the pore: (i) an external lock-in K^+ selective site at $\delta = 0.15$, (ii) an inner enhancement K^+ selective site at $\delta = 0.5$, (iii) the Ba^{2+} block site located at $\delta = 0.6$ and (iv) an internal lock-in K^+/Na^+ sites that is $\delta = 0.7$ across the membrane field. With the crystal structures of closed-state KcsA in the presence and absence of Ba^{2+} determined over a decade later,[3,43,44] the identities of these sites were suggested as (i) S1, (ii) S3, (iii) S4 and (iv) central cavity. Since the functional results are from the open conducting state and also the electric field in the pore is dependent on the state of the channel, the predictions based on the crystal structure need to be evaluated in the light of the open-state conformation. The internal lock-in site is shown to be weakly selective and modulated by voltage in BK channel. These channels can open to have a wide vestibule to produce a large flux, resulting in an electric field concentrated mostly across the selectivity filter. Under these conditions, the cavity site is non-selective and does not show a strong voltage dependence. Hence the internal lock-in site should be situated closer to the S4 binding site. The results of Ba^{2+} blockade studies are plotted alongside the calculated transmembrane potential from continuum electrostatics, demonstrating the remarkable agreement between the open-state KcsA model and experimental data (Figure 7.9).

The voltage-gated subfamily of K^+ channels contains a central pore domain similar to KcsA, surrounded by a voltage-sensing domain formed by four transmembrane helices (S1–S4). The S4 helix, specifically designed to respond to changes in the transmembrane voltage, contains four positively charged arginine residues (R1–R4) distributed every three amino acids. These four arginines have been identified as the main charges that are responsible for the gating current that accompanies closed to open conformational changes.[6] Recently, a high-resolution X-ray crystallographic structure of a mammalian voltage-dependent K^+ channel, Kv1.2, was determined at zero potential, reflecting the open-activated or open-inactivated conformation.[5] MD studies of Kv1.2 in explicit solvent and membrane show that the structure is stable and forms interactions with lipid headgroups, which adds to the stability of this conformation.[45] To investigate the nature of the voltage field in such a system, continuum electrostatics calculations of the transmembrane potential were made for several snapshots taken from the MD simulation with explicit solvent and membrane (Figure 7.10). In this case, the dielectric boundary created by the membrane and protein is explicitly defined by the lipids. This allows for a more detailed and presumably better representation of the membrane structure around the voltage sensor.

The calculations show that the transmembrane field is 'focused' (Figure 7.10B, colored points) rather than being roughly distributed over the entire thickness of the membrane (Figure 7.10B, dotted line). The aqueous regions on the intracellular side, particularly the wide vestibular cavity at the entrance of the pore, and also the decreased membrane thickness near the channel induced by the multiple interfacial basic residues, are responsible for the extra-cellularly focused field. The field in S1–S4 and the field along the central pore

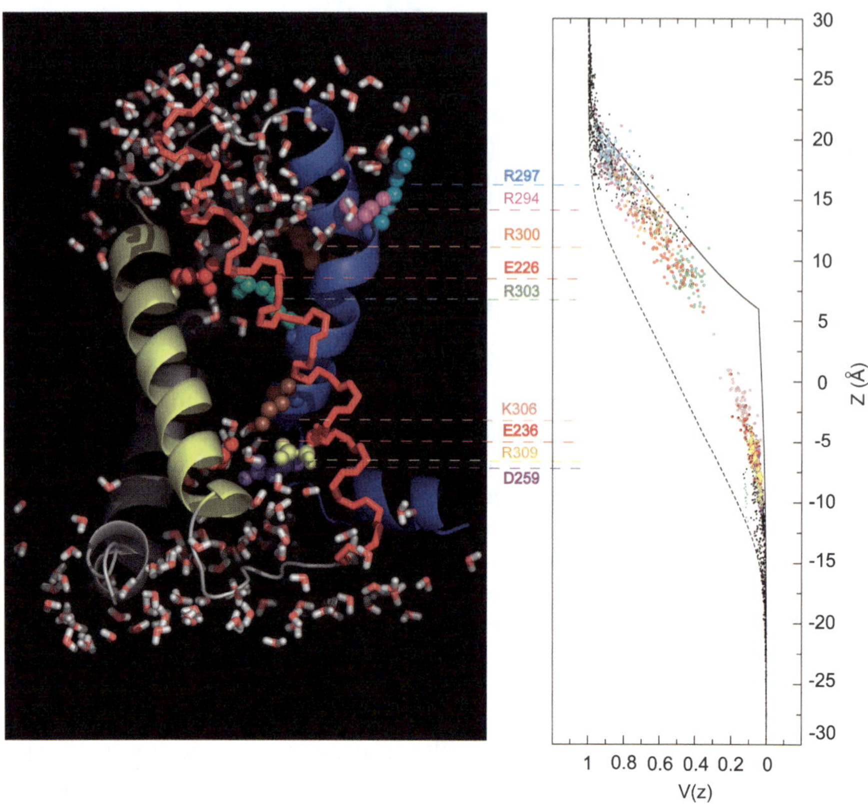

Figure 7.10 The voltage sensor of Kv1.2. Left: Snapshot of the voltage sensor of Kv1.2 from a molecular dynamics simulation in a DPPC bilayer. Charged residues in the S4 helix are labeled as shown. Right: The fractional transmembrane potential calculated from continuum electrostatics. The dotted line represents the potential in DPPC alone and the solid line represents the potential along the pore axis. The dots correspond to the fraction of potential sensed by a particular charged residue in S4 and are color coded according to (A). Adapted from Jogini and Roux, 2007.[50]

(Figure 7.10B, solid line) are qualitatively similar. The latter provides an important reference because it is well established that the membrane potential, acting as a driving force for ion permeation in K$^+$ channels, drops over the length of the narrow selectivity filter, and it demonstrates some coupling of the voltage sensor to this function.

The first four arginine residues in the S4 (R294, R297, R300 and R303) are located in the region that senses 40–80% of the transmembrane potential, the steepest part of the voltage gradient. Interestingly, these residue positions have been identified as contributing the majority of the gating charge during voltage activation.[6] As a consequence of the focused field, these four arginines along S4 are coupled strongly to the forces driving voltage gating; for example, if the charges were too far away on the extracellular side, they would not feel a driving force as soon as the membrane is repolarized. Although open–closed equilibrium might remain unaffected, this would certainly slow the kinetic rates. According to calculations, the field is pulling on each S4 with a force of $\sim 0.75 \, \text{kcal mol}^{-1} \text{Å}^{-1}$ at 100 mV. This translates into an incremental gating charge of $\sim 0.3e \, \text{Å}^{-1}$ for each S4. By a crude estimate, the experimentally observed gating charge of $13e$ would require a movement of $\sim 10 \, \text{Å}$ for all S4s in the Z-direction assuming that V_{mp} remains focused as in the open state. For comparison, a full translation of S4 over a distance of 25–30 Å would be required if the field was not focused.

No complete structures of the resting and activated state of voltage-gated channels are currently available from X-ray crystallography. Models of the closed (resting) and open (activated) states of the Kv1.2 channel were constructed using the X-ray structure of the Kv1.2 channel[5] together with all functional and spectroscopic data.[46] The PB-V equation offers a powerful route to assess the physical validity of putative 3D models of the open and closed channel conformations by allowing the calculation of the associated gating charge.[46,47] The gating charge for the models of the open and closed states is estimated to be ~ 14 elementary charges (Figure 7.11), close to the ~ 13 elementary charges estimated experimentally in Shaker.[48,49] The first four arginines in S4 contribute the most to the cumulative gating charge, with the per

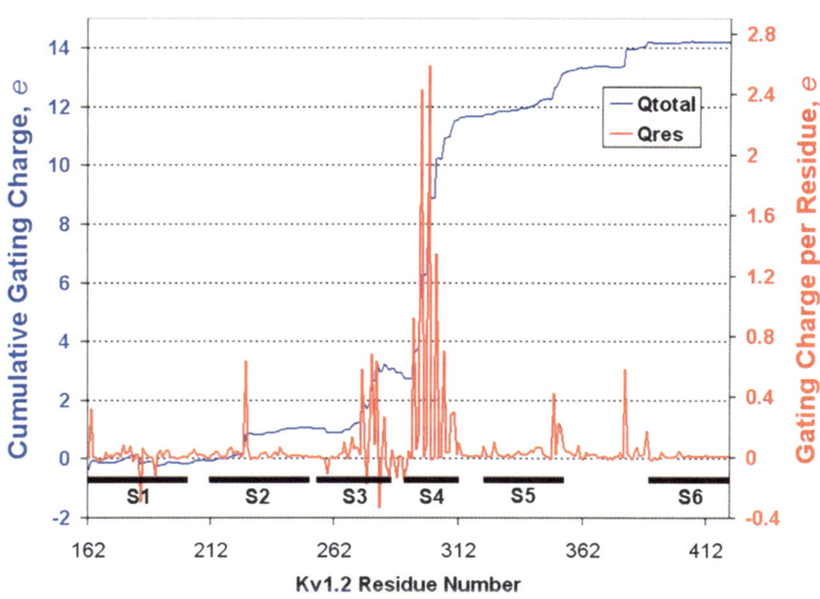

Figure 7.11 Cumulative gating charge in a tetramer (blue) and gating charge per residue summed over the 4 subunits (red) transferred across the membrane electric field between the resting to open state plotted as a function of Kv1.2 residue position. Adapted from Pathak *et al.*, 2007.[46]

subunit contributions being $0.23e$ from R1, $0.61e$ from R2, $0.65e$ from R3 and $0.33e$ from R4. This general conclusion agrees with earlier experimental work, although the values do not match the experimental values[48,49] precisely. These quantitative differences may arise from inaccuracies in side-chain packing by the structural modeling method or from the simplifying assumption of the PB-V calculations that the open and closed states do not change when the charge is neutralized. Analysis of the results shows that a much larger translocation of S4 would be required to account for the gating charge if the transmembrane field was not focused. This is achieved by high-dielectric aqueous regions in both the open and closed conformations. In the open state, the field is focused towards the extracellular half of the membrane by the large central vestibule at the intracellular pore entrance. In the closed state, the field is again focused towards the extracellular half of the membrane, this time by the wide aqueous crevice that opens up at the center of the VSD under the tilted S4 helix. The existence of aqueous crevices helping to focus the transmembrane field had been deduced indirectly from experiments.[47,50–54]

7.4 Conclusion

Continuum electrostatics applications of the PB equation and the modified PB-V equation to include the influence of a transmembrane potential have been reviewed with respect to three electrostatic features of K^+ channels. These studies enable us to highlight how structural features of K^+ channels are related to their function. Regarding permeation, the static field energy from the pore helices that surround the selectivity filter is key in providing a favorable stabilization for an ion inside the pore. The magnitude of the static field is more prominent for the closed state and reduces when the intracellular gate is opened. One of the most striking features is the large aqueous cavity located near the center of the bilayer, which helps to reduce the electrostatic penalty of transferring an ion from bulk solution into the membrane. When the intracellular gate is open, the shape of the cavity is altered into an opened vestibule, which then further reduces the magnitude of the unfavorable reaction field. In the case of Kir channels, an additional cytoplasmic domain lengthens the ion permeation pathway more than two-fold. The extension of the low dielectric into the cytoplasm potentiates the static field inside the entire pore, indicating that residues have long-range electrostatic effects on ion behavior inside the channel. The cytoplasmic domain creates a very favorable environment for cations that has both electrostatic and specific molecular roles in regulating conductance and rectification properties. Finally, the transmembrane potential calculated by solution of the PB-V equation along the K^+ channel pore is dependent on the channel state and corresponds well with experimental estimates of the voltage profile from permeation studies. Regarding voltage gating, the PB-V equation provides a powerful route to assess the physical validity of proposed structural models. Calculation of the transmembrane potential along the voltage sensor of Kv1.2 shows that the key gating charges are localized near the region where the voltage gradient is the steepest, as if optimized for sensitivity. Overall, the use of continuum electrostatics in the study of K^+ channel structure has provided useful and robust analysis in understanding how these channels facilitate the permeation of K^+ ions across the membrane and undergo voltage-dependent activation.

References

1. S. W. Cowan, T. Schirmer, G. Rummel, M. Steiert, R. Ghosh, R. A. Pauptit, J. N. Jansonius and J. P. Rosenbusch, *Nature*, 1992, **358**, 727.
2. G. Chang, R. H. Spencer, A. T. Lee, M. T. Barclay and D. C. Rees, *Science*, 1998, **282**, 2220.
3. D. A. Doyle, J. M. Cabral, R. A. Pfuetzner, A. Kuo, J. M. Gulbis, S. L. Cohen, B. T. Cahit and R. MacKinnon, *Science*, 1998, **280**, 69.
4. A. Kuo, J. M. Gulbis, J. F. Antcliff, T. Rahman, E. D. Lowe, J. Zimmer, J. Cuthbertson, F. M. Ashcroft, T. Ezaki and D. A. Doyle, *Science*, 2003, **300**, 1922.
5. S. B. Long, E. B. Campbell and R. Mackinnon, *Science*, 2005, **309**, 897.
6. F. Bezanilla, *Nat. Rev. Mol. Cell Biol.*, 2008, **9**, 323.
7. B. Honig and A. Nicholls, *Science*, 1995, **268**, 1144.
8. A. Warshel, P. K. Sharma, M. Kato and W. W. Parson, *Biochim. Biophys. Acta*, 2006, **1764**, 1647.
9. M. Born, *Z. Phys.*, 1920, **1**, 45.
10. J. G. Kirkwood, *J. Chem. Phys.*, 1939, **7**, 911.
11. L. Onsager, *J. Am. Chem. Soc.*, 1936, **58**, 1468.
12. J. Warwicker and H. C. Watson, *J. Mol. Biol.*, 1982, **157**, 671.
13. W. Im, D. Beglov and B. Roux, *Comput. Phys. Commun.*, 1998, **111**, 59.
14. B. Roux, *Biophys. J.*, 1997, **73**, 2980.
15. B. Roux, S. Berneche and W. Im, *Biochemistry*, 2000, **39**, 13295.
16. B. Roux and R. MacKinnon, *Science*, 1999, **285**, 100.
17. J. D. Faraldo-Gomez and B. Roux, *J. Mol. Biol.*, 2004, **339**, 981.
18. V. Jogini and B. Roux, *J. Mol. Biol.*, 2005, **354**, 272.
19. F. J. Sigworth, *Q. Rev. Biophys.*, 1994, **27**, 1.
20. J. Neyton and C. Miller, *J. Gen. Physiol.*, 1988, **92**, 569.
21. J. Neyton and C. Miller, *J. Gen. Physiol.*, 1988, **92**, 549.
22. E. Kutluay, B. Roux and L. Heginbotham, *Biophys. J.*, 2005, **88**, 1018.
23. H. G. Shin and Z. Lu, *J. Gen. Physiol.*, 2005, **125**, 413.
24. Y. Zhou, J. H. Morais-Cabral, A. Kaufman and R. MacKinnon, *Nature*, 2001, **414**, 43.
25. M. Nishida, M. Cadene, B. T. Chait and R. MacKinnon, *EMBO J.*, 2007, **26**, 4005.
26. J. L. Robertson and B. Roux, *Structure*, 2005, **13**, 1398.
27. M. Nishida and R. MacKinnon, *Cell*, 2002, **111**, 957.
28. S. Bernèche and B. Roux, *Biophys. J.*, 2002, **82**, 772.

29. Y. Jiang, A. Lee, J. Chen, M. Cadene, B. T. Chait and R. MacKinnon, *Nature*, 2002, **417**, 523.
30. Y. Jiang, A. Lee, J. Chen, V. Ruta, M. Cadene, B. T. Chait and R. MacKinnon, *Nature*, 2003, **423**, 33.
31. A. Kuo, C. Domene, L. N. Johnson, D. A. Doyle and C. Venien-Bryan, *Structure*, 2005, **13**, 1463.
32. C. Domene, D. A. Doyle and C. Venien-Bryan, *Biophys. J.*, 2005, **89**, L01.
33. J. L. Robertson, L. G. Palmer and B. Roux, *J. Gen. Physiol.*, 2008, **132**, 613.
34. T. I. Brelidze, X. Niu and K. L. Magleby, *Proc. Natl. Acad. Sci. USA*, 2003, **100**, 9017.
35. I. Carvacho, W. Gonzalez, Y. P. Torres, S. Brauchi, O. Alvarez, F. D. Gonzalez-Nilo and R. Latorre, *J. Gen. Physiol.*, 2008, **131**, 147.
36. F. C. Chatelain, N. Alagem, Q. Xu, R. Pancaroglu, E. Reuveny and D. L. Minor, *Neuron*, 2005, **47**, 833.
37. E. von Kitzing, *J. Gen. Physiol.*, 1999, **114**, 593.
38. D. Bichet, M. Grabe, Y. N. Jan and L. Y. Jan, *Proc. Natl. Acad. Sci. USA*, 2006, **103**, 14355.
39. J. Yang, Y. N. Jan and L. Y. Jan, *Neuron*, 1995, **14**, 1047.
40. Y. Fujiwara and Y. Kubo, *J. Gen. Physiol.*, 2006, **127**, 401.
41. H. T. Kurata, W. W. Cheng, C. Arrabit, P. A. Slesinger and C. G. Nichols, *J. Gen. Physiol.*, 2007, **130**, 145.
42. M. Taglialatela, B. A. Wible, R. Caporaso and A. M. Brown, *Science*, 1994, **264**, 844.
43. Y. X. Jiang and R. MacKinnon, *J. Gen. Physiol.*, 2000, **115**, 269.
44. S. W. Lockless, M. Zhou and R. MacKinnon, *PLoS Biol.*, 2007, **5**, e121.
45. V. Jogini and B. Roux, *Biophys. J.*, 2007, **93**, 3070.
46. M. M. Pathak, V. Yarov-Yarovoy, G. Agarwal, B. Roux, P. Barth, S. Kohout, F. Tombola and E. Y. Isacoff, *Neuron*, 2007, **56**, 124.
47. B. Chanda, O. K. Asamoah, R. Blunck, B. Roux and F. Bezanilla, *Nature*, 2005, **436**, 852.
48. S. K. Aggarwal and R. MacKinnon, *Neuron*, 1996, **16**, 1169.
49. S. A. Seoh, D. Sigg, D. M. Papazian and F. Bezanilla, *Neuron*, 1996, **16**, 1159.
50. L. D. Islas and F. J. Sigworth, *J. Gen. Physiol.*, 2001, **117**, 69.
51. C. A. Ahern and R. Horn, *J. Gen. Physiol.*, 2004, **123**, 205.
52. D. M. Starace and F. Bezanilla, *Nature*, 2004, **427**, 548.
53. C. A. Ahern and R. Horn, *Neuron*, 2005, **48**, 25.
54. F. Tombola, M. M. Pathak, P. Gorostiza and E. Y. Isacoff, *Nature*, 2007, **445**, 546.

Computational Approaches to Ionotropic Glutamate Receptors

RANJIT VIJAYAN,[a] BOGDAN IORGA[b] AND PHILIP C. BIGGIN[a]

[a] Department of Biochemistry, University of Oxford, South Parks Road, Oxford, OX1 3QU, UK; [b] Institut de Chimie des Substances Naturelles, CNRS UPR 2301, avenue de la Terrasse, F-91198, Gif-sur-Yvette, France

8.1 Introduction

The ionotropic glutamate receptors (iGluRs) are ligand-gated ion channels that mediate the responses at a large proportion of excitatory synapses. They are thought to underpin memory and learning and are likely drug targets for various neuropathological conditions such as ischaemic stroke, neuropathic pain, Parkinson's disease and epilepsy.[1] Each receptor is comprised of four large subunits (over 800 residues). These subunits can be broadly classified (Figure 8.1A and B) according to sequence similarity and by their response to the pharmacological agents N-methyl-D-aspartate (NMDA), α-amino-3-hydroxy-5-methyl-4-isoxazolepropionic acid (AMPA) and the naturally occurring compound kainate. The fourth class of subunit is termed orphan, as these proteins are not capable of forming functioning receptors on their own and their precise role remains to be determined. Despite the convenient classification, it should be remembered that *in vivo,* glutamate is the agonist at all iGluRs. In fast neurotransmission, glutamate is released from the presynaptic neuron and diffuses across the synaptic cleft, where it binds to glutamate receptors situated in the membranes of post-synaptic neurons. This binding

RSC Biomolecular Sciences No. 20
Molecular Simulations and Biomembranes: From Biophysics to Function
Edited by Mark S.P. Sansom and Philip C. Biggin
© Royal Society of Chemistry 2010
Published by the Royal Society of Chemistry, www.rsc.org

A

Ionotropic Glutamate Receptor Subunits

AMPA

GluA1 (GluR1)
GluA2 (GluR2)
GluA3 (GluR3)
GluA4 (GluR4)

Kainate

GluK1 (GluR5)
GluK2 (GluR6)
GluK3 (GluR7)
GluK4 (KA1)
GluK5 (KA2)

NMDA

GluN1 (NR1)
GluN2A (NR2A)
GluN2B (NR2B)
GluN2C (NR2C)
GluN2D (NR2D)
GluN3A (NR3A)
GluN3B (NR3B)

Orphan

GluD1 (GluRδ1)
GluD2 (GluRδ2)

B

GluA2 GluA1
GluA3   GluA4
GluN1
GluK3 GluK2
GluK1
GluK4
GluK5
GluN2C
GluN2D
GluN2B
GluN2A
GluD1
GluD2
GluN3A   GluN3B

C

NTD
D1
LBD
D2
OUT
TMD   M1   M2   M3
IN
P
CTD

D

NTD
LBD
TMD

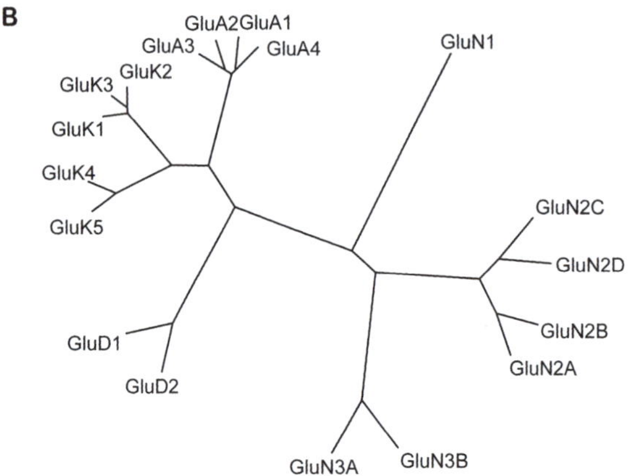

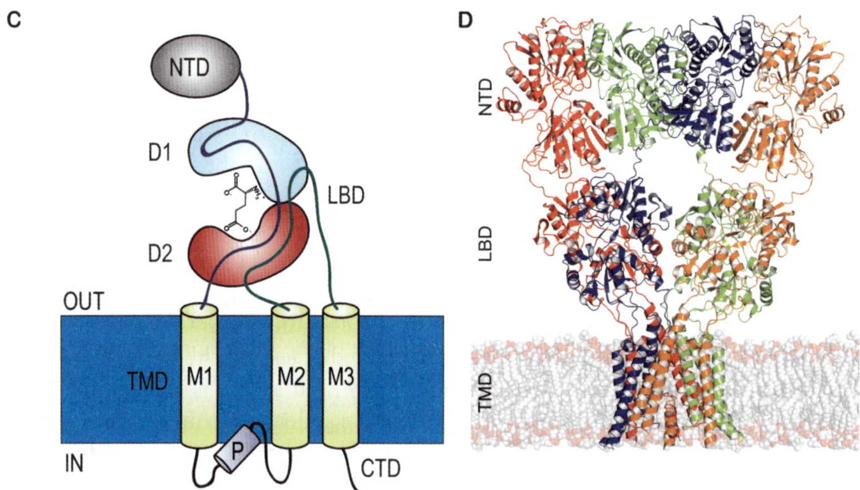

events causes the otherwise closed transmembrane domain to open and become permeable to cations. It is this flow of ions which causes depolarization of the membrane and is the physical basis for signalling throughout the central nervous system (CNS).

The four glutamate receptor subunits are arranged, in some part at least, as a dimer of dimers.[2–4] A single subunit is comprised (Figure 8.1C) of two extracellular domains (an N-terminal domain and the ligand-binding domain), a transmembrane domain and C-terminal intracellular domain which is thought to be predominantly unstructured. The development of a soluble construct of the ligand-binding domain[5] has led to extensive crystallographic[5–15] and NMR structural studies[16–20] on a large number of complexes from different subunits. It is on the back of these structural data that a large proportion of, but not all, computational studies have been performed. More recently, Gouaux and colleagues have managed to solve the structure of a complete tetrameric GluA2 receptor (Figure 8.1D),[21] a feat which has significantly improved our understanding of these receptors. Although the crystal structures have provided a large step forward in our understanding of ionotropic glutamate receptors, computational work can provide extensive complementary information, especially concerning the dynamics of these systems.

The nomenclature for iGluR subunits has recently been reviewed by the Nomenclature Committee of the International Union of Basic and Clinical Pharmacology (NC-IUPHAR).[22] They have proposed a standardized nomenclature (Figure 8.1A) which we will adopt throughout this chapter.

8.2 The Amino-terminal Domain

Until recently,[23] there was no structure available for the N-terminal domain (NTD) and therefore much effort has been spent constructing homology models. The NTD shows homology to the periplasmic binding protein (PBP), leucine isoleucine valine binding protein (LIVBP) and related folds, for which there are many candidate template structures that have been used in the past. The homology to the PBPs also implies that the NTD has the venus fly-trap fold[24]

Figure 8.1 (A) Classification of ionotropic glutamate receptors. The nomenclature is as described in reference 21 with one of the older nomenclature systems commonly used previously in parentheses. (B) Phylogenetic tree illustrating the 18 iGluR genes found in rat. (C) Topology of ionotropic glutamate receptors. The first domain is the N-terminal domain (NTD). The peptide chain (S1, shown as a blue line) then makes part of the D1 and D2 lobes of the ligand-binding domain (LBD) before part of the transmembrane domain (TMD). After exit from M3, the peptide chain (S2, shown as a green line) completes the formation of D2 and D1 before entering the membrane for the final time in the form of the M3 transmembrane helix. The final section of the protein is the C-terminal domain (CTD), which is thought to be dynamic in structure and contains the binding sites for many partner proteins involved in regulation of iGluRs. (D) Crystal structure of the tetrameric GluA2 receptor.

(which has been confirmed by the crystal structures) with the cleft between the two lobes of the fly-trap potentially forming an additional binding site. Intriguingly however, the crystal structures suggest that these domains, at least for AMPA and kainate receptors, may have lost the ability to bind any additional ligands and have evolved to become specialized association domains in their own right. The situation is probably different in NMDA receptors, where several modulatory compounds are known to act and probably do so *via* the binding cleft in the NTD domain. Several studies in which the NTD has been either partly or completely removed have shown that these truncated subunits are capable of forming functional receptors and in some cases are functionally indistinguishable from the full-length subunits.[25,26] However, this domain is involved in subunit-specific assembly and the regulation of the receptor and for these reasons represents a viable drug target. Most of the work has concentrated on the NTD of the NMDA receptor, in particular the NR2 subtype.[27] The main reason is that this is the site of many different allosteric modulators including Zn^{2+} and $pH^{28,29}$ and neuroprotectants such as ifenprodil.[30–33]

Paoletti *et al.*[34] built models of the NTD from NR2A based upon LIVBP to help rationalize the binding of Zn^{2+} by six critical residues. Huggins and Grant[35] built models of GluN1/GluN2A and GluN1/GluN2B NTD dimer combinations on the basis of the ligand-binding domain from rat metabotropic glutamate receptor 1 (mGluR1) for which the NTD of iGluRs is homologous. Using mGluR1 as a template has the advantages that there are both open- and closed-cleft structures available[36] and that they are dimeric in nature, thus providing a testable hypothesis for the interface between the monomers.

Ifenprodil is selective for the NTD from the GluN2B subunit and experiments have mapped the binding site using homology models again based on the LIVBP protein.[32] Other researchers[37] employed a previously derived homology model of GluN2B based on LIVBP[38] and pharmacophore modelling to identify new hit compounds that showed significant activity in functional assays and also *in vivo* anticonvulsant efficacy in DBA/2 mice.

Marinelli *et al.*[39] built two sets of ensemble of GluN2B homology models using the metabotropic glutamate receptor 1 (mGluR1) structures as the templates. The two sets of ensembles reflect models of the open- and closed-cleft forms of the NTD. Ifenprodil was docked to these ensembles and subsequent short molecular dynamics (MD) and molecular mechanics–Poisson–Boltzmann surface area (MM-PBSA) calculations were performed. The authors concluded that ifenprodil stabilizes the closed-cleft form of the NTD and this should therefore be taken into account for future virtual screening studies.

More recently, Mony *et al.*[40] used a variety of both computational and experimental methods to explore the binding of ifenprodil to the NTD. Their docking results suggested two possible orientations of ifenprodil within the closed-cleft NTD of GluN2B. Through site-directed mutagenesis and cysteine affinity labelling, they were able to narrow this down to one orientation that could account for nearly all of the reported experimental data. Further, they were able to suggest five new residues that are key for a high-affinity interaction of ifenprodil. It was also noted that this model does not agree with the model proposed by Marinelli *et al.*,[39] a difference perhaps attributable to the two groups using different sequence alignments. An intriguing observation from Mony *et al.*[40] was that despite near perfect conservation of residues involved in ifenprodil binding in GluN2A and GluN2B, ifenprodil is specific to GluN2B. The authors proposed that other residues scattered through the NTD are crucial in determining the correct positioning of ifenprodil.

8.3 The Ligand-binding Domain (LBD)

Immediately after the NTD is the ligand-binding domain (LBD), sometimes also called the agonist-binding domain (ABD). This region is by far the best characterized region of the receptor, not only structurally, but also functionally. The domain is formed by two discontinuous sections of polypeptide chain (termed S1 and S2) that form the clam-shell structure whereby the agonist binds in the middle (Figure 8.1C).

Prior to the resolution of crystal structures of iGluR LBD, owing to remarkable sequence homology, structures of periplasmic polar amino acid binding proteins such as glutamine binding protein (QBP), histidine binding protein (HBP) and lysine arginine ornithine binding protein (LAOBP) were used to generate models of the LBD of iGluRs with varying degrees of success.[41–43] A review of these earlier modelling efforts has been presented by Paas *et al.*[42,44] Following the publication of the structures of the rat GluA2 LBD,[2,5] numerous modelling studies have been undertaken to generate models of the iGluRs based on these structures. Such models, often combined with MD, have been used to explain how variations in hydrogen bond formation and hydrophobic packing produced by amino acid substitutions in the binding site can result in altered selectivity in iGluRs.[45–54]

8.3.1 Selectivity and Modulation

The design of subtype selective ligands offers therapeutic promise, but the design of selective compounds poses difficult challenges as there are very few differences within the agonist binding site between subtypes within AMPA and kainate receptors or between them. Crystal structures that are representative of all the different classes of subtypes are now available and in some cases several different subtypes within class could be crystallized. These structures enable aspects of subtype selectivity to be considered at the atomic level. In the absence of X-ray structures, one can use homology modelling and indeed several groups have used this technique in the past to good effect.[55–58] For example, ligand docking and site-directed mutagenesis informed by homology models of human GluA4 based on the GluA2 structure suggested that conformational changes in helix F could play a critical role in receptor activation.[58] The cumulative dipole moment of the helix F produces a positively charged N-terminus that interacts with the polar moieties of agonists glutamate, kainate and AMPA. Mutations in this region did not appear to affect antagonist affinity and perhaps explains its inability to activate the channel due to a lack of direct interaction. Homology models of all non-NMDA iGluRs (GluA1–4 and GluK1–5) using rat GluA2 as the template showed that selectivity is largely determined by seven residues in the binding site: P478, T480, L650, S654, T686, Y702 and M708.[50] MD has also been used by the same group to help rationalize the selectivity between GluA2 and GluK1 observed for these compounds.[51]

Water is an important consideration when performing any docking study. As we expand upon later, the waters in the closed-cleft form of the ligand-binding core are particularly important and probably must be considered for accurate descriptions of docking poses. For example, Jacobsen has shown that the antagonist DNQX can be docked in the correct pose with no consideration of water molecules, with a root mean square deviation (RMSD) of 0.29 Å compared with the crystal structure.[59,60] However, when the same authors examined the role of water in docking AMPA, they found that the docking program Glide[61] could not predict the correct pose unless crystallographic water molecules were included. The role of certain water molecules was also highlighted in a study by Lampinen *et al.*[58] The importance of the water molecules in selectivity has also been addressed in studies by Banke *et al.*[55] Through work examining the selectivity of Br-HIBO for GluA1 over GluA3, Banke *et al.* demonstrated that position 702, which is a tyrosine in GluA1/GluA2 and a phenylalanine in GluA3/GluA4, is the critical residue in conferring selectivity between these subtypes. This is particularly interesting as this residue does not make direct contact with the agonist. It appears to mediate an effect through a water molecule, an effect which was supported by density functional theory (DFT) calculations. More recently, the GluA4 structure was obtained by X-ray crystallography,[62] confirming the validity of that model.

Several groups have developed models of NMDA ligand-binding cores,[63–68] primarily in order to rationalize binding data and develop more subtype specific drugs. As more X-ray data become available, the accuracy of the models can be improved, particularly if the structures are of high enough resolution. For example, Clausen *et al.*[63] were able to exploit the structure of the GluN2A subunit[9] by including the position of two water molecules to improve their previous model of the GluN2B subunit.[65]

Prior to the crystal structures of LBDs from GluN3A and GluN3B,[15] homology modelling and docking were used to examine the binding site of the GluN3 subunit.[69–71] The study by Nilsson *et al.*[71] was performed in combination with experimental measurements of binding affinity of a series of compounds to the GluN3A subunit. The experimental measurements along with those reported by Yao and Mayer[72] confirm that glycine is preferred over glutamate at this subunit binding site. These results, in the absence of crystallographic information, were rationalized through the homology model where Nilsson *et al.* were not able to dock glutamate satisfactorily to GluN3A due to steric clashes created by Ala802 and Met844. Unlike GluN1, however, neither the D-cycloserine nor the antagonist 7-chlorokynurenic acid could displace [^{3}H]glycine from the binding site. Similarly, Yao and Mayer[72] also found low affinity for these compounds at GluN3A.

8.3.2 Dynamics

As they are supposed to convert ligand binding into a channel-opening event through allosteric movements, the ionotropic glutamate receptors are, by necessity, very dynamic entities. Computational approaches provide one of the most appropriate methods that one can use to study the dynamics of proteins. The dynamics of iGluRs have been investigated on a range of time scales (from picoseconds to nanoseconds) with a variety of different techniques (from quantum mechanics to umbrella sampling with molecular mechanics).

Ab initio quantum mechanical calculations have been used to investigate the electronic and structural states of glutamate (and other neurotransmitters) and also the nature of the interaction between glutamate and the receptors. For example, Odai and co-workers have shown that glutamate has two global energy minima in solution.[73,74] However, glutamate adopts an intermediate conformation in crystal structures of GluA2.[2] Hence the binding and reorientation of the ligand in the binding site appear to be a dynamic and possibly multi-stage process. The precise pathway from ligand binding to receptor activation is yet to be elucidated, but various plausible hypotheses have been postulated. Through vibrational coupling, a ligand could transfer energy to the receptor inducing conformational change. The symmetric stretching vibrational mode of the carboxyl group of an agonist has been shown to interact with the bending vibrational mode of the guanidinium group of the conserved arginine (Arg485 in GluA2) that forms hydrogen bonds with the α-carboxyl group of amino acid ligands such as kainate and glutamate.[75] However, an antagonist (CNQX) was shown not to produce a similar effect, which may partly explain its inability to activate the receptor. Picosecond time scale MD simulations have also been used to show that the vibrational energy from the ligand to Arg485 is transferred to Lys730 through electrostatic interactions and then further on to helix J through van der Waals interaction.[75] This perturbation of helix J, located in the subunit interface, may play a role in receptor activation.

8.3.2.1 Nanosecond Time Scale Movements of the Ligand-binding Domain

The dynamics of the LBD on the nanosecond time scale have received considerable attention in recent years. Molecular dynamics in particular has provided a lot of insight into various aspects, including what the determinants of cleft closure and opening are and to what extent the water molecules found in the binding pocket play a role in stabilizing bound ligands. Water is often found to act as a bridging molecule in binding pockets, mediating many protein–ligand indirect interactions. The iGluRs are no exception and the majority of the structures contain water molecules within the binding pocket that adopt similar positions within the pocket (Figure 8.2). The water molecules can be considered part of any pharmacophore that might be generated and it has been shown that the ability to replace some of these water molecules facilitates the

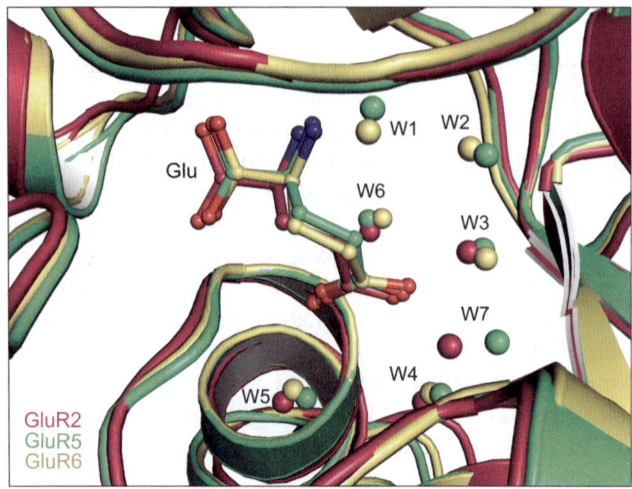

Figure 8.2 Conserved water molecules (oxygen atoms only) observed across a range of glutamate receptors structures: GluA2 (pink), GluK1 (green) and GluK2 (yellow).

binding of different ligands.[76] The best example of this is provided by the positioning of AMPA compared with the orientation of glutamate (Figure 8.3), where water molecules can be considered as part of the pharmacophore. Whereas, in crystal structures, the binding site is isolated from bulk solvent, MD simulations showed that intra-domain motion permits many water exchanges in the binding site.[77–79] Furthermore, the extent of water exchange appears to be related to the nature of the agonist bound.[77,78] It seems likely that the consideration of water not only within the agonist-binding pockets but also in the modulatory sites of these receptors will continue to be important for future drug-design problems.

The presence of ligand in the binding pocket stabilizes the closed cleft of the LBD. In this closed-cleft form it is becoming increasingly clear that both inter-lobe (D1 and D2) interactions and protein–ligand interactions are essential in stabilizing the closed state of the receptor.[15,80] The role of D1–D2 cross-cleft interactions is supported by the observation that removing glycine from the closed-state structure of GluN1[78] and glutamate from the GluA2 structure[77] did not bring about any significant opening of the cleft in 20 ns simulations. However, this does not preclude such an event over longer time scales.

The role of the residue E705 in the binding site of GluA2 has received much attention. Formation of the salt bridge E705–K730 is thought to stabilize the cleft in the absence of glutamate (Figure 8.4).[2,7] Using continuum electrostatic calculations, it was argued that the repulsive effect of a net positive charge on domains D1 and D2 may be partially compensated by the negatively charged glutamate in the binding site.[81] Furthermore, although unphysical, a

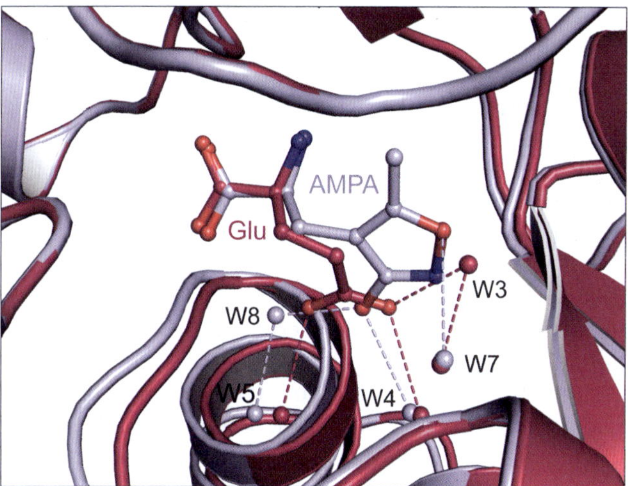

Figure 8.3 Comparison of the orientation of glutamate (pink) with AMPA (slate blue) in the GluA2 binding pocket. It can be seen how the waters labelled W3 (observed in the glutamate-bound structure) and W8 (found in the AMPA-bound structure) act in a mutually exclusive manner to alter the binding mode for these two agonists.

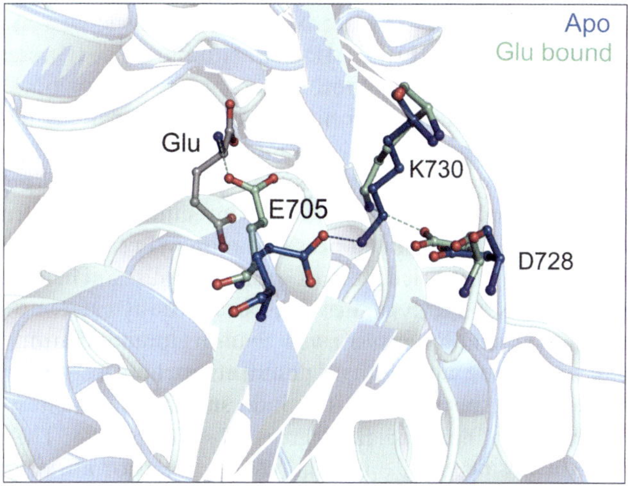

Figure 8.4 Salt bridge conformation in two different states of the ligand-binding core. In the open cleft (Apo) state, E705 forms a salt bridge with K730. When the cleft is closed and glutamate is bound, K730 forms a salt bridge with D728, allowing E705 to form part of the binding site and interact with the nitrogen of the bound glutamate.

cleft opening (which would normally be difficult to observe in an unconstrained simulation) was induced by artificially lowering the negative charge on the E705. Although this residue plays an important role in binding sites, it stabilizes the closed conformation by the formation of a salt bridge. Homology modelling and docking aided by radioligand assays and mutational studies in GluA4 suggest that the interaction of this residue with a bound ligand ranges from weak to no interactions.[48] Depending on the ligand, the binding site remained effective when the residue was mutated to an aspartate or glutamine.

In another study, where the LBD cleft was forced open using a constraint, the protein–ligand interactions were observed to break in a different sequence in kainate-bound compared with glutamate-bound simulations.[82] In the kainate-bound simulation, interactions of the γ-carboxylate group of the ligand were observed to break ahead of the E705–ligand interactions, whereas in the glutamate-bound simulations they occurred simultaneously. Hence E705 appears to have a very complex role in the binding site, both controlling protein–ligand interactions and stabilizing the protein in the closed conformation.

Although it has often been thought that the D1–D2 domain closure resembles a hinge-like motion, it has also been proposed, using rigidity analysis, that the structure may in fact not possess a hinge in strict terms.[83] A hinge generally separates two rigid regions. However, whereas D1 remains rigid, domain D2 was shown to exhibit a degree of flexibility (in agreement with other studies) and the authors proposed a refined term to describe the closure event: the 'load-and-lock' mechanism. Principal component analysis has also shown that the binding cleft undergoes two significant motions, *viz.*, movement of subdomains D1 and D2 about a 'hinge' (Figure 8.5A) and a twisting motion of D2 (Figure 8.5B) with respect to D1.[77] MD simulations generate trajectories that are multi-dimensional datasets. Often, reducing the dimensionality by projecting it on to dimensions that contribute most to the variability in the data set can provide insights into the predominant motions of the protein. Such analysis employed in simulations of GluA2, bacterial GluR0 and LAOBP indicated that major motions are conserved across proteins of similar fold.[84] Furthermore, distinct states of iGluRs and related proteins in an MD simulation can also be discerned by analysing the eigenvectors from a Gaussian network model.[85] Finally, rigid body essential dynamics analysis of X-ray crystal structure and snapshots from MD simulations showed that the predominant motions are hinge bending and lobe twisting,[86] as described earlier. The twist was particularly prevalent in the partial-agonist bound structures, which in the case of GluA2 complexes have intermediate degrees of cleft opening. The results suggest that the pathway between open and closed may not always be a simple hinge-bending closure as first thought.

Cleft movement has been less well studied in the NMDA receptors. They have the added complication of being susceptible to redox modulation via formation/destruction of disulfide bridges. MD simulations have been employed to study the effect of breaking a series of disulfide bridges in the

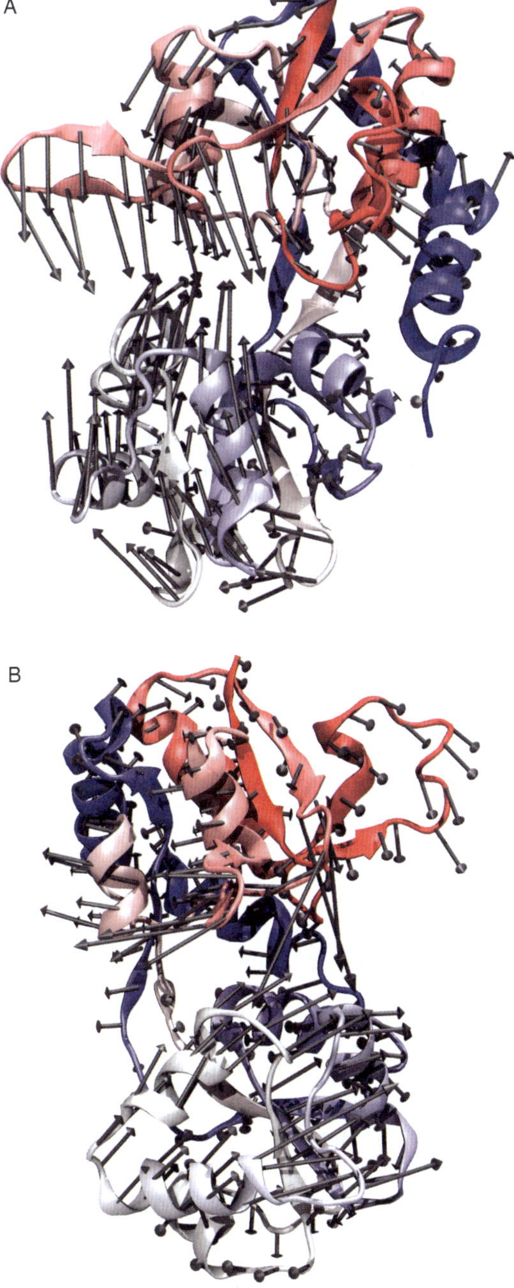

Figure 8.5 Predominant motions of ligand-binding domain from ionotropic gluta-mate receptors. (A) A hinge-bending motion; (B) twisting of D1 with respect to D2.

NMDA receptor subunit NR1.[87] Mutating the cysteines in a conserved disulfide bridge C744–C798 produced increases in both inter-lobe mobility and hinge flexibility. However, removal of the disulfide bridge C454–C420 in loop 1 did not produce any significant difference compared with wild-type simulations. The impact on the dimer-of-dimers assembly remains an open question.

One of the major limitations of conventional MD simulations is limited sampling due to available computer time. Various enhanced sampling techniques such as thermodynamic integration[88] and umbrella sampling[89] have been employed to probe the thermodynamics of ligand binding. Using umbrella sampling and two order parameters that denote cross-cleft distances, the 2D free energy landscape of cleft opening and closing was computed for apo-6,7-dinitro-2,3-quinoxalinedione (DNQX)-bound and glutamate-bound structures.[90] Whereas the crystal structure resided in a distinct minimum in the DNQX-and glutamate-bound structures, the free energy landscape of the apo structure showed that the structure could exist in a more open state without a significant energetic penalty. Förster resonance energy transfer (FRET)-based studies have also shown that *in vivo*, the cleft closure is lower than is observed in X-ray crystal structures.[91] Furthermore, starting from the apo X-ray crystal structure, ~ 9 kcal mol^{-1} is made available upon ligand binding and cleft closure. However, this increases to ~ 12 kcal mol^{-1} if the more open state observed in the 2D free energy landscape is used instead. On the other hand, ~ 9 kcal mol^{-1} would be required moving the DNQX-bound structure to the closed state, partly due to steric clashes inline with DNQX acting as an antagonist through a 'foot-in-the-door' mechanism. Independently, through a combination of thermodynamic integration, umbrella sampling and MD/PBSA methods, the binding free energy of a glutamate was computed to be $\simeq 9$ kcal mol^{-1},[81] which is also comparable to experimental findings.[92]

8.3.2.2 *Partial Agonists and Dynamics*

In the case of AMPA receptors, partial agonists have been shown to bind with a lobe closure between that of apo (or antagonist-bound) structures and structures with full agonists.[93] The degree of cleft closure (Figure 8.6) and consequently the distance between the artificial linkers (where the TM helices would connect in the full length structure) within the dimers have been shown to correlate with the efficacy of the agonist bound. Single-channel studies of homomeric GluA2 with various ligands have shown that the subconductance states are similar, but with a key difference in frequency of opening. The partial agonist iodowillardiine was shown to induce the same subconductance levels and open periods as glutamate, but with a different distribution in their frequency, leading to a submaximal response on average for multiple receptors.[93] Therefore, whether a ligand will behave as a full or partial agonist appears to be related to how efficiently it can open the transmembrane channel. Interestingly, MD simulations of the LBD of GluA2 with full and partial agonists showed

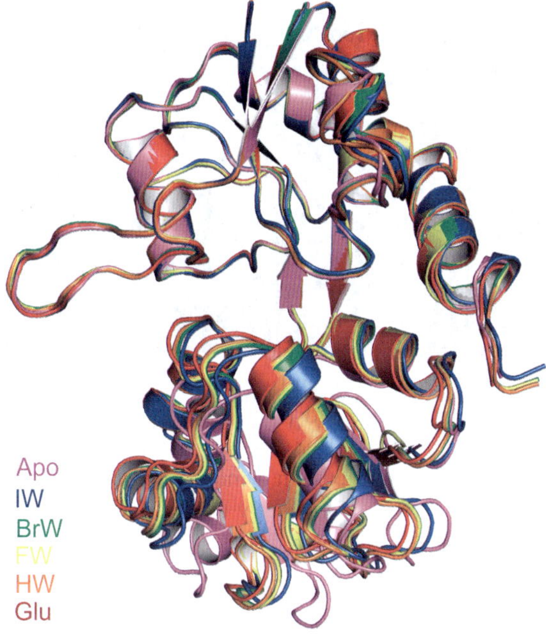

Figure 8.6 An overlay of GluA2 X-ray structures with different agonists. Apo (pink) shows the largest degree of cleft opening compared with the glutamate-bound structure (red). The degree of cleft opening is proportional to the efficacy of a series of partial agonists called willardiines.[93] The efficacy and degree of cleft opening depend on the size of the substituent at the 5-position, ranging from hydrogen (HW, orange), through fluorine (FW, yellow) and bromine (BrW, green) to iodine (IW, blue).

that the partial agonist kainate was more flexible in the binding pocket.[77,94] This flexibility may play a role in determining how efficient the signal is transmitted from the LBD to the transmembrane domain. The RMSD of the protein Cα from the crystal structure over the course of the simulations was also found to be higher in the partial agonist bound simulations.

On the other hand, such a correlation between agonism and cleft closure does not apply to NMDA receptors.[95] Although the hinge region of the partial agonist D-cycloserine-bound crystal structure of the NR1 receptor[8] resembles a full-agonist-like state, simulations have shown that the hinge could adopt a conformation resembling other partial agonist-bound crystals.[78] The authors suggested that the torsion angles could perhaps be used as an indicator for predicting partial agonism in this receptor subtype. Partial agonists were found to adopt different binding modes of which only some influence the conformation of the hinge. However, extensive replica exchange simulations of glycine and ACPC in simulations of NR3A did not appear to exhibit such motion.[15] Hence there remains much work to be done to establish the exact nature of how the signal is transmitted from the ligand-binding domain to the transmembrane domain.

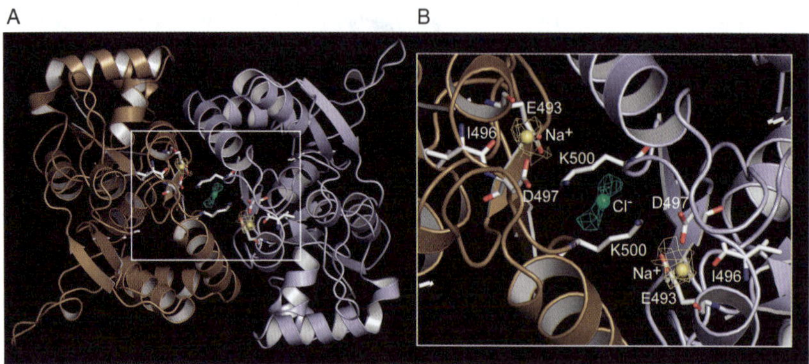

Figure 8.7 The program GRID was used to identify the cation and anion modulatory sites in kainate receptors. (A) The view from the synapse looking at the top of the ligand-binding domain dimers. (B) A zoomed-in section where the GRID predictions for anion and cation binding sites are shown in green and yellow, respectively. The positions of the ions observed in the crystal structures are shown as spheres (not to scale).

8.3.2.3 *Consideration of the Dimer Assembly*

The majority of simulation studies reported above have been performed on monomers, mostly for reasons of computational cost. However, as the receptor is a tetramer, it seems likely that dynamics beyond the monomer must be considered. Certainly for the ligand-binding domain, the dimer of dimers model accounts for many observations.[96] As an example, kainate receptors have been shown to be modulated by both anions and cations.[97,98] Computational approaches, including the use of the adaptive Poisson–Boltzmann solver (APBS[99]) and the GRID program,[100] have been used to identify possible cation-binding sites (Figure 8.7), which were subsequently confirmed by high-resolution X-ray crystallography.[101] These sites are located close to a modulatory anion-binding site previously identified in the dimer interface.[102] Furthermore, to elucidate the rank order of binding, rigorous free energy calculations were employed to calculate relative binding free energies of monovalent cations from the alkali metal family.[103] The selectivity of Na^+ over K^+ was shown to be due to both a rigid binding cavity and high negative charge density. Additionally, the cations were also shown to make anion binding more favourable in the modulatory binding site in the subunit interface.

8.4 The Transmembrane Domain

The TM domain of NMDA receptors has attracted particular attention in recent years as it contains the site of block for compounds such as memantine which have found use in the treatment of dementia.[104] The recent crystal structure[21] confirms that this region has a similar architecture to the pore region of potassium channels. It was recognized during the early 1990s that the P-segment from K-channels and the P-segment of iGluRs had sequence similarity,[105,106] suggesting that, despite the fact that mammalian iGluRs are not potassium selective, these two protein families may be evolutionarily related. The functional characterization of a potassium-selective prokaryotic glutamate receptor, GluR0, reported in 1999[107] strengthened the argument. Further evidence was subsequently provided by searching microbial sequence databases where several putative bacterial glutamate receptors were identified.[108] Taken together, this information has been used to derive reasonable homology models of the transmembrane region of iGluRs.

Work in our laboratory[109] examined the behaviour of homology models of the transmembrane domain of GluR0 based upon the structure of the bacterial potassium channel, KcsA. We found that the conformational stability of the model in a membrane was similar to that of the TM domain of KcsA in comparable simulations. Furthermore, concerted translocation of K^+ ions and water molecules along the filter region took place in a manner similar to that observed in simulations of KcsA itself. We also reported that the tetramer was 'stabilized' by the presence of an inter-subunit salt bridge, something that was not immediately apparent from the sequence alignment alone.

Models of the TM domain of eukaryotic iGluRs pose significantly harder challenges due to the lack of the GYG motif (and indeed they are not K^+ selective). The problem is amplified by the increasing evidence that the extracellular domains exhibit two-fold symmetry (dimer of dimers model), whereas the TM domain exhibits four-fold symmetry.[4,21] Hence there appears to be a symmetry mismatch. Studies by Sobolevsky *et al.*[110] have indicated that the two-fold symmetry may extend some way into the TM domain, but it is unclear how symmetry mismatch changes with the state of the receptor. In a different study, Sobolevsky *et al.*[111] also suggested that the actual mode of gating in eukaryotic iGluRs may be different to that in bacterial potassium channels. In bacterial potassium channels, a glycine residue midway along the M2 helix (lining the pore) is thought to act as a pivot point that allows the helix to swing open during gating. The equivalent glycine is not present in eukaryotic iGluRs, hence the mechanism of gating is likely to differ.

Despite these problems, the overall topology of the transmembrane region is similar to potassium channels and in the absence of data at the time this has formed the basis of many homology modelling studies in an effort to generate models of iGluRs.[112–116] Tikhonov and colleagues have developed models of GluA1 using the potassium channel MthK as a template. In order to improve the model in the region of the selectivity filter, a dicationic adamantane derivative known to block the open state of the channel was included to provide additional restraints. The model was also used to corroborate experimental data on philanthotoxin block.[117]

Tikhonov also reported models of NMDA receptors.[114] This is an even more difficult challenge as the NMDA receptors are comprised of at least two different subunits. Models were built from a combination of templates; the KvAP structure was used for GluN1 whereas the MthK structure was used for the

GluN2A subunit. The resultant model gives good agreement with the experimental data reported by Sobolevsky *et al.*[118] and provides a good framework for which to think about other blocking compounds which in the absence of receptor models have previously been considered in the context of 3D-QSAR studies.[119] It is satisfying to note that the pharmacophoric model from those earlier studies is consistent with the more recent models and the experimental data, indicating that models provide a useful device for which to build and test hypotheses on.

More recently, Kaczor *et al.*[113] built models of GluK1 with a view to incorporating recent experimental data[110] on the extent of the symmetry mismatch to generate models where the TM region exhibited two-fold symmetry to near the Q/N/R site.

8.5 Conclusion

Computational methods have made a significant contribution to our understanding of ionotropic glutamate receptor selectivity and dynamics. It is likely that these and related methods will continue to enhance our knowledge of this receptor family. There are a number of key questions that still remain unanswered, some of which may ultimately only be addressable by structural studies and others that may allow for progress in the absence of a structure. For example, what is the exact nature of the conformational change that converts the binding of an agonist to an open channel? How does this differ across the three different subtypes? Even in the case of AMPA receptors where there is an intuitive correlation between LBD cleft closure and agonist efficacy, it has been shown that there are mechanisms other than domain closure that can influence agonist efficacy.[120] Related to these issues is the question of how the NTD may regulate the behaviour of the agonist binding domain.

With the advent of a full tetrameric structure, computational methods may be used to develop models that can provide testable hypotheses, particularly with respect to dynamic properties. Such models will help our understanding of how these receptors can be modulated, by molecules in the synapse, the membrane and in the cytosol. Finally, the full model of the tetramer[21] may also offer potential routes for the design of more specific drugs.

Acknowledgements

We thank Michelle Sahai for useful discussions and the Wellcome Trust for support. P.C.B. is an RCUK Fellow.

References

1. D. Bowie, *CNS Neurol. Disord. Drug Targets*, 2008, **7**, 129–143.
2. N. Armstrong and E. Gouaux, *Neuron*, 2000, **28**, 165–181.
3. M. L. Mayer, *Curr. Opin. Neurobiol.*, 2005, **15**, 282–288.

4. W. Tichelaar, M. Safferling, K. Keinanen, H. Stark and D. R. Madden, *J. Mol. Biol.*, 2004, **344**, 435–442.

5. N. Armstrong, Y. Sun, G.-Q. Chen and E. Gouaux, *Nature*, 1998, **395**, 913–917.

6. A. H. Ahmed, Q. Wang, H. Sondermann and R. E. Oswald, *Proteins Struct. Funct. Bioinf.*, 2008, **75**, 628–637.

7. N. Armstrong, M. L. Mayer and E. Gouaux, *Proc. Natl. Acad. Sci. USA*, 2003, **100**, 5736–5741.

8. H. Furukawa and E. Gouaux, *EMBO J.*, 2003, **22**, 2873–2885.

9. H. Furukawa, S. K. Singh, R. Mancusso and E. Gouaux, *Nature*, 2005, **438**, 185–192.

10. C. Kasper, K. Frydenvang, P. Naur, M. Gajhede, D. S. Pickering and J. S. Kastrup, *FEBS Lett.*, 2008, **582**, 4089–4094.

11. M. L. Mayer, R. Olson and E. Gouaux, *J. Mol. Biol.*, 2001, **311**, 815–836.

12. M. H. Nanao, T. Green, Y. Stern-Bach, S. F. Heinemann and S. Choe, *Proc. Natl. Acad. Sci. USA*, 2005, **102**, 1708–1713.

13. P. Naur, K. B. Hansen, A. S. Kristensen, S. M. Dravid, D. S. Pickering, L. Olsen, B. Vestergaard, J. Egebjerg, M. Gajhede, S. F. Traynelis and J. S. Kastrup, *Proc. Natl. Acad. Sci. USA*, 2007, **104**, 14116–14121.

14. P. Naur, B. Vestergaard, L. K. Skov, J. Egebjerg, M. Gajhede and J. S. Kastrup, *FEBS Lett.*, 2005, **579**, 1154–1160.

15. Y. Yao, C. B. Harrison, P. L. Freddolino, K. Schulten and M. L. Mayer, *EMBO J.*, 2008, **27**, 2158–2170.

16. M. K. Fenwick and R. E. Oswald, *J. Mol. Biol.*, 2008, **378**, 673–685.

17. A. S. Maltsev, A. H. Ahmed, M. K. Fenwick, D. E. Jane and R. E. Oswald, *Biochemistry*, 2008, **47**, 10600–10610.

18. R. L. McFeeters and R. E. Oswald, *Biochemistry*, 2002, **41**, 10472–10481.

19. R. L. McFeeters and R. E. Oswald, *FASEB J.*, 2004, **18**, 428–438.

20. E. R. Valentine and A. G. Palmer III, *Biochemistry*, 2005, **44**, 3410–3417.

21. A. I. Sobolevsky, M. P. Rosconi and E Gouaux, *Nature*, 2009, 745–756.

22. G. L. Collingridge, R. W. Olsen, J. Peters and M. Spedding, *Neuropharmacology*, 2009, **56**, 2–5.

23. (a) R. Jin, S. K. Singh, S. Gu, H. Furukawa, A. I. Sobolevsky, J. Zhou, Y. Jin and E. Gouaux, *EMBO J.*, 2009, advanced online publication; (b) J. Kumar, P. Schuck, R. Jin and M. L. Mayer, *Nat. Struct. Mol. Biol.*, 2009, **16**, 631–638; (c) A. Clayton, C. Siebold, R. J. Gilbert, G. C. Sutton, K. Marlos, R. A. McIlhinney, E. A. Jones and A. R. Aricescn, *J. Mol. Biol.*, 2009, **392**, 1125–1132; (d) E. Karakas, N. Simorowski and H. Furukawa, *EMBO J.*, 2009, **28**, 3910–3920.

24. C. B. Felder, R. C. Graul, A. Y. Lee, H.-P. Merkle and W. Sadee, *PharmSci*, 1999, **1**, E2.

25. G. Ayalon and Y. Stern-Bach, *Neuron*, 2001, **31**, 103–113.

26. A. Pasternack, S. K. Coleman, A. Jouppila, D. G. Mottershead, M. Lindfors, M. Pasternack and K. Keinanen, *J. Biol. Chem.*, 2002, **277**, 49662–49667.

27. G. A. Herin and E. Aizenman, *Eur. J. Pharmacol.*, 2004, **500**, 101–111.
28. C. J. Hatton and P. Paoletti, *Neuron*, 2005, **46**, 261–274.
29. C. Madry, I. Mesic, H. Betz and B. Laube, *Mol. Pharmacol.*, 2007, **72**, 1535–1544.
30. I. Borza and D. G., *Curr. Top. Med. Chem.*, 2006, **6**, 687–695.
31. M. J. Gallagher, H. Huang, D. B. Pritchett and D. R. Lynch, *J. Biol. Chem.*, 1996, **271**, 9603–9611.
32. F. Perin-Dureau, J. Rachline, J. Neyton and P. Paoletti, *J. Neurosci.*, 2002, **22**, 5955–5965.
33. K. Williams, *Mol. Pharmacol.*, 1993, **44**, 851–859.
34. P. Paoletti, F. Perin-Dureau, A. Fayyazuddin, A. Le Goff, I. Callebaut and J. Neyton, *Neuron*, 2000, **25**, 683–694.
35. D. J. Huggins and G. H. Grant, *J. Mol. Graphics*, 2005, **23**, 381–388.
36. N. Kunishima, Y. Shimada, Y. Tsuji, T. Sato, M. Yamamoto, T. Kumasaka, S. Nakanishi, H. Jingami and K. Morikawa, *Nature*, 2000, **407**, 971–977.
37. R. Gitto, L. DeLuca, S. Ferro, F. Occhiuto, S. Samperi, G. DeSarro, E. Russo, L. Ciranna, L. Costa and A. Chimirri, *ChemMedChem*, 2008, **3**, 1539–1548.
38. P. Malherbe, V. Mutel, C. Broger, F. Perin-Dureau, J. A. Kemp, J. Neyton, P. Paoletti and J. N. C. Kew, *J. Pharmacol. Exp. Ther.*, 2003, **307**, 897–905.
39. L. Marinelli, S. Cosconati, T. Steinbrecher, V. Limongelli, A. Bertamino, E. Novellino and D. A. Case, *ChemMedChem*, 2007, **2**, 1498–1510.
40. L. Mony, L. Krzaczkowski, M. Leonetti, A. Le Goff, K. Alarcon, J. Neyton, H.-O. Bertrand, F. Acher and P. Paoletti, *Mol. Pharmacol.*, 2009, **75**, 60–74.
41. M. Lampinen, O. Pentikäinen, M. S. Johnson and K. Keinänen, *EMBO J.*, 1998, **17**, 4704–4711.
42. Y. Paas, *Trends Neurosci.*, 1998, **21**, 117–125.
43. M. J. Sutcliffe, Z. G. Wo and R. E. Oswald, *Biophys. J.*, 1996, **70**, 1575–1589.
44. Y. Paas, A. Devillers-Thièry, V. I. Teichberg, J.-P. Changeux and M. Eisenstein, *Trends Pharmacol. Sci.*, 2000, **21**, 87–92.
45. P. E. Chen, M. T. Geballe, P. J. Stansfeld, A. R. Johnston, H. Yuan, A. L. Jacob, J. P. Snyder, S. F. Traynelis and D. J. A. Wyllie, *Mol. Pharmacol.*, 2005, **67**, 1471–1484.
46. K. Erreger, M. T. Geballe, S. M. Dravid, J. P. Snyder, D. J. A. Wyllie and S. F. Traynelis, *J. Neurosci.*, 2005, **25**, 7858–7866.
47. K. Erreger, M. T. Geballe, A. Kristensen, P. E. Chen, K. B. Hansen, C. J. Lee, H. Yuan, P. Le, P. N. Lyuboslavsky, N. Micale, L. Jorgensen, R. P. Clausen, D. J. A. Wyllie, J. P. Snyder and S. F. Traynelis, *Mol. Pharmacol.*, 2007, **72**, 907–920.
48. A. Jouppila, O. T. Pentikainen, L. Settimo, T. Nyrunen, J.-P. Haapalahti, M. Lampinen, D. G. Mottershead, M. S. Johnson and K. Keinanen, *Eur. J. Biochem.*, 2002, **269**, 6261–6270.

49. L. L. Lash, J. M. Sanders, N. Akiyama, M. Shoji, P. Postila, O. T. Pentikainen, M. Sasaki, R. Sakai and G. T. Swanson, *J. Pharmacol. Exp. Ther.*, 2008, **324**, 484–496.

50. O. T. Pentikainen, L. Settimo, K. Keinanen and M. S. Johnson, *Biochem. Pharmacol.*, 2003, **66**, 2413–2425.

51. U. Pentikainen, L. Settimo, M. S. Johnson and O. T. Pentikainen, *Org. Biomol. Chem.*, 2006, **4**, 1058–1070.

52. J. M. Sanders, K. Ito, L. Settimo, O. T. Pentikainen, M. Shoji, M. Sasaki, M. S. Johnson, R. Sakai and G. T. Swanson, *J. Pharmacol. Exp. Ther.*, 2005, **314**, 1068–1078.

53. J. M. Sanders, O. T. Pentikainen, L. Settimo, U. Pentikainen, M. Shoji, M. Sasaki, R. Sakai, M. S. Johnson and G. T. Swanson, *Mol. Pharmacol.*, 2006, **69**, 1849–1860.

54. Z. G. Wo, K. K. Chohan, H. Chen, M. J. Sutcliffe and R. E. Oswald, *J. Biol. Chem.*, 1999, **274**, 37210–37218.

55. T. G. Banke, J. R. Greenwood, J. K. Christensen, T. Liljefors, A. Schousboe and D. S. Pickering, *J. Neurosci.*, 2001, **21**, 3052–3062.

56. G. Campiani, E. Morelli, V. Nacci, C. Fattorusso, A. Ramunno, E. Novellino, J. R. Greenwood, T. Liljefors, R. Griffith, C. Sinclair, H. Reavy, A. S. Kristensen, D. S. Pickering, A. Schousboe, A. Cagnotto, E. Fumagalli and T. Mennini, *J. Med. Chem.*, 2001, **44**, 4501–4504.

57. P. Kizelsztein, M. Eisenstein, N. Strutz, M. Hollmann and V. I. Teichberg, *Biochemistry*, 2000, **39**, 12819–12827.

58. M. Lampinen, L. Settimo, O. T. Pentikainen, A. Jouppila, D. G. Mottershead and M. S. Johnson, *J. Biol. Chem.*, 2002, **277**, 41940–41947.

59. J. R. Greenwood and T. Liljefors, in: *Molecular Neuropharmacology: Strategies and Methods,* ed. A. Schousboe and H. Bräuner-Osborne, Humana Press, Totowa, NJ, 2004, pp. 3–25.

60. L. Jacobsen, Masters thesis, University of Copenhagen, 2002.

61. R. A. Friesner, J. L. Banks, R. B. Murphy, T. A. Halgren, J. J. Klicic, D. T. Mainz, M. P. Repasky, E. H. Knoll, M. Shelley, J. K. Perry, D. E. Shaw, P. Francis and P. S. Shenkin, *J. Med. Chem.*, 2004, **47**, 1739–1749.

62. A. Gill, A. Birdsey-Benson, B. L. Jones, L. P. Henderson and D. R. Madden, *Biochemistry*, 2008, **47**, 13831–13841.

63. R. P. Clausen, C. Christensen, K. B. Hansen, J. R. Greenwood, L. Jørgensen, N. Micale, J. C. Madsen, B. Nielsen, J. Egebjerg, H. Bräuner-Osborne, S. F. Traynelis and J. L. Kristensen, *J. Med. Chem.*, 2008, **51**, 4179–4187.

64. G. Grazioso, L. Moretti, L. Scapozza, M. De Amici and C. De Micheli, *J. Med. Chem.*, 2005, **48**, 5489–5494.

65. K. B. Hansen, R. P. Clausen, E. J. Bjerrum, C. Bechmann, J. R. Greenwood, C. Christensen, J. L. Kristensen, J. Egebjerg and H. Brauner-Osborne, *Mol. Pharmacol.*, 2005, **68**, 1510–1523.

66. M. Sivaprakasam, K. B. Hansen, O. David, B. Nielsen, S. F. Traynelis, R. P. Clausen, F. Couty and L. Bunch, *ChemMedChem*, 2009, **4**, 110–117.
67. I. G. Tikhonova, I. I. Baskin, V. A. Palyulin and N. S. Zefirov, *Dokl. Biochem. Biophys.*, 2002, **382**, 67–70.
68. I. G. Tikhonova, I. I. Baskin, V. A. Palyulin and N. S. Zefirov, *J. Med. Chem.*, 2003, **46**, 1609–1616.
69. M. S. Belenkin, G. Constantino, V. A. Palyulin, R. Pellicciari and N. S. Zefirov, *Dokl. Biochem. Biophys.*, 2003, **389**, 83–89.
70. M. C. Blaise, R. Sowdhamini, M. R. Rao and N. Pradhan, *J. Mol. Model.*, 2004, **10**, 305–316.
71. A. Nilsson, J. Duan, L. L. Mo-Boquist, E. Benedikz and E. Sundström, *Neuropharmacology*, 2007, **52**, 1151–1159.
72. Y. Yao and M. L. Mayer, *J. Neurosci.*, 2006, **26**, 4559–4566.
73. K. Odai, T. Sugimoto, D. Hatakeyama, M. Kubo and E. Ito, *J. Biochem.*, 2001, **129**, 909–915.
74. K. Odai, T. Sugimoto, M. Kubo and E. Ito, *J. Biochem.*, 2003, **133**, 335–342.
75. M. Kubo, E. Shiomitsu, K. Odai, T. Sugimoto, H. Suzuki and E. Ito, *Proteins Struct. Funct. Genet.*, 2004, **54**, 231–236.
76. A. Hogner, J. S. Kastrup, R. Jin, T. Liljefors, M. L. Mayer, J. Egebjerg, I. K. Larsen and E. Gouaux, *J. Mol. Biol.*, 2002, **322**, 93–109.
77. Y. Arinaminpathy, M. S. P. Sansom and P. C. Biggin, *Mol. Pharmacol.*, 2006, **69**, 11–18.
78. L. S. Kaye, M. S. P. Sansom and P. C. Biggin, *J. Biol. Chem*, 2006, **281**, 12736–12742.
79. K. Speranskiy and M. Kurnikova, *Biochemistry*, 2005, **44**, 11508–11517.
80. W. Maier, R. Schemm, C. Grewer and B. Laube, *J. Biol. Chem.*, 2007, **282**, 1863–1872.
81. T. Mamonova, M. J. Yonkunas and M. G. Kurnikova, *Biochemistry*, 2008, **47**, 11077–11085.
82. J. Mendieta, G. Ramirez and F. Gago, *Proteins Struct. Funct. Genet.*, 2001, **44**, 460–469.
83. T. Mamonova, K. Speranskiy and M. Kurnikova, *Proteins Struct. Funct. Bioinf.*, 2008, **73**, 656–671.
84. A. Pang, Y. Arinaminpathy, M. S. P. Sansom and P. C. Biggin, *Proteins Struct. Funct. Bioinf.*, 2005, **61**, 809–822.
85. B. A. Hall, S. L. Kaye, A. Pang, R. Perera and P. C. Biggin, *J. Am. Chem. Soc.*, 2007, **129**, 11394–11401.
86. E. J. Bjerrum and P. C. Biggin, *Proteins Struct. Funct. Bioinf.*, 2008, **72**, 434–446.
87. S. L. Kaye, M. S. P. Sansom and P. C. Biggin, *Biochemistry*, 2007, **46**, 2136–2145.
88. P. A. Kollman, *Chem. Rev.*, 1993, **93**, 2395–2417.
89. G. M. Torrie and J. P. Valleau, *J. Comput. Phys.*, 1977, **23**, 187–199.
90. A. Y. Lau and B. Roux, *Structure*, 2007, **15**, 1203–1214.
91. G. Ramanoudjame, M. Du, K. A. Mankiewicz and V. Jayaraman, *Proc. Natl. Acad. Sci. USA*, 2006, **103**, 10473–10478.

92. D. Deming, Q. Cheng and V. Jayaraman, *J. Biol. Chem.*, 2003, **278**, 17589–17592.
93. R. Jin, T. G. Banke, M. L. Mayer, S. F. Traynelis and E. Gouaux, *Nat. Neurosci.*, 2003, **6**, 803–810.
94. Y. Arinaminpathy, M. S. P. Sansom and P. C. Biggin, *Biophys. J.*, 2002, **82**, 676–683.
95. A. Inanobe, H. Furukawa and E. Gouaux, *Neuron*, 2005, **47**, 71–84.
96. M. L. Mayer, *Nature*, 2006, **440**, 456–462.
97. D. Bowie, *J. Physiol.*, 2002, **539**, 725–733.
98. A. Y. C. Wong, D. M. MacLean and D. Bowie, *J. Neurosci.*, 2007, **27**, 6800–6809.
99. N. A. Baker, D. Sept, S. Joseph, M. J. Holst and J. A. McCammon, *Proc. Natl. Acad. Sci. USA*, 2001, **98**, 10037–10041.
100. P. J. Goodford, *J. Med. Chem.*, 1985, **28**, 849–857.
101. A. J. R. Plested, R. Vijayan, P. C. Biggin and M. L. Mayer, *Neuron*, 2008, **58**, 720–735.
102. A. Plested and M. L. Mayer, *Neuron*, 2007, **53**, 829–841.
103. R. Vijayan, A. J. R. Plested, M. L. Mayer and P. C. Biggin, *Biophys. J.*, 2009, **96**, 1751–1760.
104. S. A. Lipton, *Nat. Rev. Drug Discov.*, 2006, **5**, 160–170.
105. Z. G. Wo and R. E. Oswald, *Trends Neurosci.*, 1995, **18**, 161–168.
106. M. H. Wood, H. M. A. VanDongen and A. M. J. VanDongen, *Proc. Natl. Acad. Sci. USA*, 1995, **92**, 4882–4886.
107. G.-Q. Chen, C. Cui, M. L. Mayer and E. Gouaux, *Nature*, 1999, **402**, 817–821.
108. T. Kuner, P. H. Seeburg and H. R. Guy, *Trends Neurosci.*, 2003, **26**, 27–32.
109. Y. Arinaminpathy, P. C. Biggin, I. H. Shrivastava and M. S. P. Sansom, *FEBS Lett.*, 2003, **553**, 321–327.
110. A. I. Sobolevsky, M. V. Yelshansky and L. P. Wollmuth, *Neuron*, 2004, **41**, 367–378.
111. A. I. Sobolevsky, M. V. Yelshansky and L. P. Wollmuth, *J. Neurosci.*, 2003, **23**, 7559–7568.
112. S. Bachurin, S. Tkachenko, I. Baskin, N. Lermontova, T. Mukhina, L. Petrova, A. Ustinov, A. Proshin, V. Grigoriev, N. Lukoyanov, V. Palyulin and N. S. Zefirov, *Ann. N. Y. Acad. Sci*, 2001, **939**, 219–236.
113. A. A. Kaczor, U. A. Kijkowska-Murak and D. Matosiuk, *J. Med. Chem.*, 2008, **51**, 3765–3776.
114. D. B. Tikhonov, *Mol. Membr. Biol.*, 2007, **24**, 135–147.
115. D. B. Tikhonov, J. R. Mellor, P. N. Usherwood and L. G. Magazanik, *Biophys. J.*, 2002, **82**, 1884–1893.
116. I. G. Tikhonova, I. I. Baskin, V. A. Palyulin and N. S. Zefirov, *Dokl. Biochem. Biophys.*, 2004, **396**, 181–186.
117. T. F. Andersen, D. B. Tikhonov, U. Bolcho, K. Bolshakov, J. K. Nelson, F. Pluteanu, I. R. Mellor, J. Egebjerg and K. Stromgaard, *J. Med. Chem.*, 2006, **49**, 5414–5423.

118. A. I. Sobolevsky, L. Rooney and L. P. Wollmuth, *Biophys. J.*, 2002, **83**, 3304–3314.
119. R. T. Kroemer, E. Koutsilieri, P. Hecht, K. R. Liedl, P. Riederer and J. Kornhuber, *J. Med. Chem.*, 1998, **41**, 393–400.
120. M. M. Holm, M. L. Lunn, S. F. Traynelis, J. S. Kastrup and J. Egeberg, *Proc. Natl. Acad. Sci. USA*, 2005, **102**, 12053–12058.

Molecular Dynamics Studies of Outer Membrane Proteins: a Story of Barrels

SYMA KHALID[a] AND MARC BAADEN[b]

[a] School of Chemistry, University of Southampton, Highfield, Southampton, SO17 1BJ, UK; [b] Institut de Biologie Physico-Chimique, Laboratoire de Biochimie Théorique, CNRS UPR 9080, 13 rue Pierre et Marie Curie, F-75005, Paris, France

9.1 Introduction

The outer membrane proteins (OMPs) of Gram-negative bacteria play a key role in the function and structural integrity of the outer membrane. The OMPs cover a number of different functions, including membrane pores, passive and active transporters, recognition proteins and membrane-bound enzymes. From a biomedical perspective, OMPs are of some interest as potential targets for novel antimicrobial drugs and vaccines. Currently only ∼30 high-resolution structures of OMPs are known, revealing all but two of them to be based on a transmembrane (TM) barrel architecture, with sizes ranging from 8 to 22 strands in the barrel. Recently, two helical OMPs from *Escherichia coli*[1] and *Corynebacterium glutamicum*[2] have been reported. The availability of structural data for these proteins is benefiting from advances in NMR techniques and high-throughput crystallographic methods, which are beginning to provide structural data at a greater rate. However, currently, whereas these proteins are ubiquitous in Gram-negative bacteria and represent an important class of

RSC Biomolecular Sciences No. 20
Molecular Simulations and Biomembranes: From Biophysics to Function
Edited by Mark S.P. Sansom and Philip C. Biggin
© Royal Society of Chemistry 2010
Published by the Royal Society of Chemistry, www.rsc.org

future drug targets, there are relatively few structural data available to help identify their structure–function relationships.

Molecular dynamics (MD) simulations enable us to extend the available structural data by exploring the conformational dynamics of membrane proteins in environments that mimic either experimental (*in vitro*) or *in vivo* conditions. In this chapter, we illustrate such simulations of β-barrel proteins from the outer membranes of Gram-negative bacteria. To facilitate this discussion, we focus on examples from our own work and related studies.

Our intention is not to provide a comprehensive list of OMP MD simulations but rather to focus on simulations that (i) aid our understanding of the structure–function relationships of OMPs and (ii) exploit this understanding for applications in bionanotechnology. We restrict our attention to atomistic and coarse-grain simulations, in which protein, lipid and water atoms are all treated explicitly. Recent years have seen advances in the use of continuum solvent models to study membrane protein insertion into lipid bilayers.[3]

9.2 Outer Membrane Proteins

The outer membrane (OM) of Gram-negative bacteria controls the influx and efflux of solutes while also serving as a protective barrier against the external environment. Outer membrane proteins (OMPs) fulfil a variety of roles, including passive and active transport, host/pathogen recognition, signal transduction and enzymatic catalysis. It has been predicted that 2–3% of the genes in Gram negative bacteria encode integral OMPs.[4,5] Whereas inner membrane proteins usually have an α-helical fold,[6] most OMPs have a β-barrel architecture.[7–9] The barrels are composed of anti-parallel β-strands that are connected by short turns on the periplasmic side of the membrane and by long loops on the extracellular side. The bacterial OM is asymmetric in nature. The inner leaflet, which faces the periplasmic space, is similar in phospholipid composition to the inner (cytosolic) membrane. In contrast, the outer leaflet is rather more complicated. It is composed of complex lipopolysaccharides (LPS),[10] which are anionic oligosaccharides cross-linked by divalent cations with multiple saturated fatty acid tails. The structure of LPS varies substantially from species to species and can be modified within a single cell in response to changes in the local environment. The combination of highly charged sugars and tightly ordered hydrocarbon side chains contributes to the low permeability of the OM, which enables it to protect the cell against the influx of toxic agents.

To permit the influx and efflux of solutes across the OM, it is rendered selectively permeable to molecules smaller than ∼600 Da by the presence of proteins known as porins.[11,12] These porins can be either non-specific or solute-specific for, *e.g.*, sugars[13,14] or phosphate.[15] A number of other OMPs with varying transport functions are also found in bacterial OMs. The non-porin OMPs range in function from the transport of specific solutes (*e.g.* transport of siderophores by TonB-coupled OMPs[16]), through to peptide autotransport.[17]

There are also a number of OMPs that are not involved in transport; instead, they play a role in, *e.g.*, target cell recognition by pathogenic bacteria or are enzymes. A summary of earlier simulation studies is available.[18] In recent years, the potential of OMPs as components of biosensors has been recognized. Hence in this chapter we will focus on simulations of OMPs from a biophysics and bionanotechnology perspective.

9.3 Simple Barrels

9.3.1 OmpA and Its Homologues

The small and relatively simple protein OmpA from *E. coli* has been extensively studied by MD simulations; indeed, it has provided something of a test bed for simulations of more complex OMPs. One reason for this is the availability of a range of structural data with which to validate simulation results. The structure of the N-terminal TM domain of OmpA has been solved by X-ray diffraction[19,20] and by NMR studies (the latter in detergent micelles).[21,22] OmpA is a small, monomeric protein. The N-terminal TM domain is an eight-stranded β-barrel whereas the periplasm-residing, C-terminal domain is globular. The β-barrel is connected by large loops on the extracellular side and by short turns on the periplasmic side. To the best of our knowledge, the structure of the C-terminal domain has not been determined.

Several functional studies have shown that OmpA forms low-conductance pores when reconstituted into planar lipid bilayers.[23–25] However, the X-ray structure of OmpA revealed several cavities separated by polar and charged side chains pointing into the interior of the β-barrel, rather than a single continuous pore. Hence there appeared to be a discrepancy between the functional studies and the X-ray structure. A combination of modelling and MD simulations of the OmpA N-terminal domain in a dimyristoylphosphatidylcholine (DMPC) bilayer[26] suggested that the charged side chains within the interior of the β-barrel may form a gate which controls the transition between the closed OmpA pore seen in the crystal structure and an open-pore state that corresponds to the 60 pS conductance pores observed in functional studies. Small conformational changes in the side chains would be sufficient to switch between the two states. MD simulations revealed a degree of flexibility in the side chains lining the aqueous cavities within the β-barrel. The R138–E52 salt bridge was proposed as the gate controlling the transition between the open and closed states of the pore. This salt bridge was stable throughout the 5 ns simulation and constricted the pore to prevent the passage of, *e.g.*, water molecules. Molecular modelling of the proposed gate in an alternative conformation yielded an open state of the pore in which the R138 side chain was oriented towards the side chain of the nearby E128 rather than E52. Water molecules were able to diffuse impeded from one mouth of the β-barrel to the other in simulations of this putative open state. Estimates of the conductance based on dimensions of the pore were in good agreement with experimental conductance data. Hence it was proposed that the switching of OmpA between the open and closed states may be effected by changes in the rotameric state of the R138 side chain.

In a related study, the dynamics of OmpA in a DMPC lipid bilayer and in a detergent [dodecylphosphocholine (DPC)] micelle were compared for 10 ns MD simulations.[27] A greater flexibility of the protein ($\sim$1.5$\times$ from Cα atom fluctuations) was observed in the micelle environment compared with the lipid bilayer. The increased mobility is most likely a consequence of the reduced packing constraints in the micellar environment, with a small contribution from the slight differences in the hydrophobic chain properties of the detergents and lipids. A functional consequence of embedding OmpA in a detergent micelle was that in this environment, conformational rearrangements of side chains located within the β-barrel lead to the formation of a continuous pore through the centre of the protein barrel. This reinforced the proposal of a gating mechanism for OmpA pores involving disruption and reformation of salt bridges within the interior of the β-barrel. This model has received support from recent mutagenesis and functional studies of OmpA.[28]

Recent studies suggest that this mechanism may be extended to OmpA homologues from other bacterial species. For example, two homology models of OprF,[29,30] the main OMP of *Pseudomonas aeruginosa*, revealed aqueous cavities formed by pore-occluding residues within the β-barrel. These cavities do not form a single channel spanning the length of the β-barrel, yet experimental data suggest that OprF has pore-forming capabilities.[31] MD simulations of both OprF models in a DMPC bilayer (see Figure 9.1) displayed remarkably similar dynamics to OmpA.[30] Water molecules were unable to pass from one mouth of the barrel to the other in any of the 6$\times$10 ns simulations reported. In particular, a persistent salt bridge near the periplasmic mouth of the β-barrel formed by E8–K121 prevented the passage of water through the β-barrel in all of the simulations and was identified as a likely gate, analogous to the R138–E52 salt bridge in OmpA.

In contrast to the simulations described above, which were designed to mimic *in vitro* experimental studies of OprF, recently the conformational dynamics of the Hancock homology model of OprF embedded within a membrane, relevant to the *in vivo* environment, have been reported.[32] The OprF model was embedded within an asymmetric membrane composed of lipopolysaccharide (LPS) in the outer leaflet and phosphatidylethanolamine (PE) in the inner leaflet. Compared with the OprF/DMPC simulations, reduced lateral diffusion of the protein in the LPS membrane was observed. The β-barrel exhibited lower flexibility relative to the simulations in DMPC, whereas the loop flexibility was similar in the two membrane environments. Encouragingly, the pore-constricting side-chain interactions within the barrel were similar to those found in the OprF/DMPC simulations. The diameter of the barrel was observed to widen slightly at the extracellular mouth. This widening is likely to be a consequence of the interaction of the highly charged LPS with the long, flexible protein loops. However, despite the overall diameter of the barrel increasing in this region, the pore remained occluded throughout the simulation. The effect

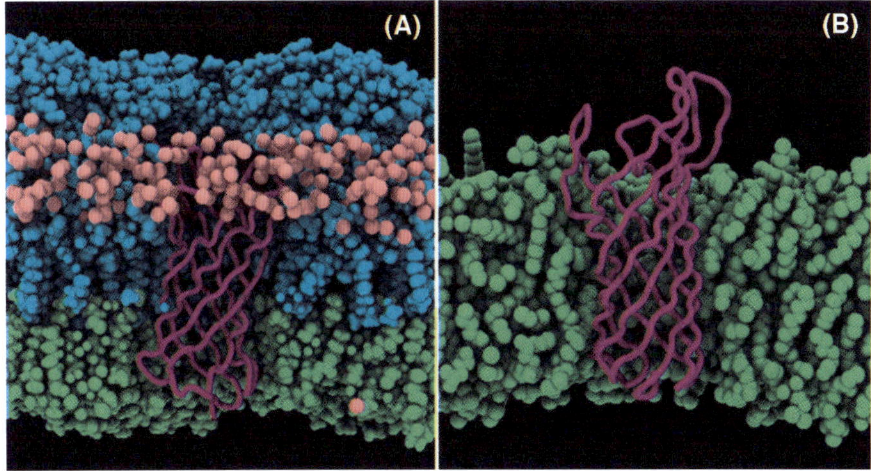

Figure 9.1   Snapshots of MD simulations of a homology model of OprF (protein shown in purple). (A) Oprf embedded in a detailed model of the bacterial outer membrane.[32] The outer leaflet is composed of LPS, coloured cyan, and phospholipids in the inner leaflet, coloured lime. Ca^{2+} ions are coloured pink. In contrast, in (B) the homology model is embedded in a bilayer with phospholipids in both leaflets.[30] Water molecules have been omitted for clarity.

of the protein on its local environment was investigated. The asymmetric nature of LPS membranes results in a potential gradient across the membrane. A negative surface charge density builds on the LPS side and a positive charge density on the PE side of the membrane. This is altered by the presence of OprF, which induces an increase in charge polarization on the membrane surface and leads to a small electrically positive patch in the vicinity of the protein.

Modelling and simulation studies of an OmpA homologue have been used to demonstrate that computational approaches may be applied to more complex, multi-domain OMPs, rather than just to TM β-barrels.[33] PmOmpA is the major OMP found in the outer membrane of *Pasteurella multocida*. It is a two-domain outer membrane protein whose N-terminal domain is a homologue of the transmembrane β-barrel domain of OmpA from *E. coli*, whereas the C-terminal domain of PmOmpA is a homologue of the extra-membrane *Neisseria meningitidis* RmpM C-terminal domain. As X-ray structures of both the N- and C-terminal domain homologues of PmOmpA have been determined, it was possible to construct a model of a complete two-domain PmOmpA and to explore its conformational dynamics *via* MD simulations. A degree of water penetration into the interior of the β-barrel suggested the formation of a TM pore. The PmOmpA model was conformationally stable over a 20 ns simulation, whereas substantial flexibility was observed in the short (four-residue) linker region between the N- and C-terminal domains. Simulations of the PmOmpA model in

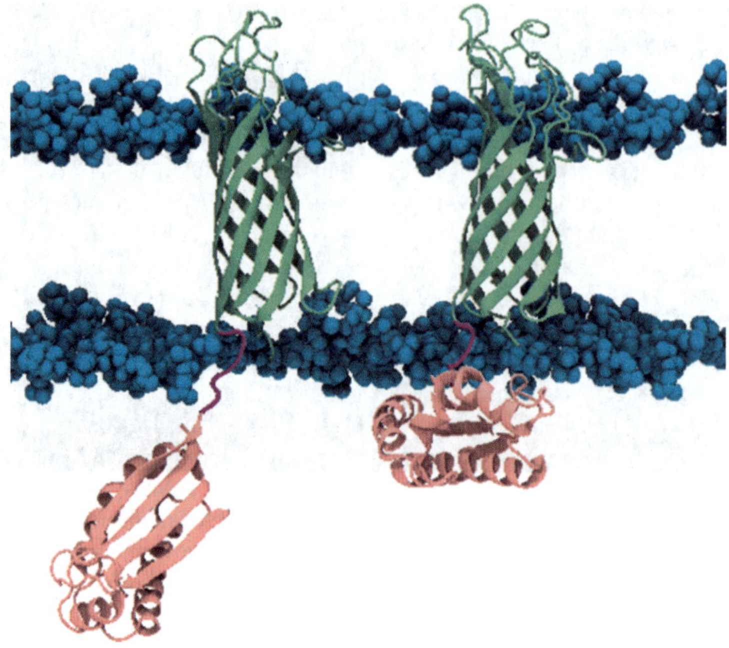

Figure 9.2   Snapshots of the intact PmOmpA protein embedded in DMPC after 20 ns
of molecular dynamics (left) in just neutralizing counterions and (right) in
1 M NaCl. The barrels are coloured lime, the linkers are purple and the
periplasmic domains are pink. The lipid headgroups are coloured cyan.
Water molecules have been omitted for clarity.

1 M NaCl, and also in just neutralizing ions, exhibited different behaviours of the
linker and the C-terminal domain in the two environments. The C-terminal
domain was observed to interact strongly with the lipid bilayer headgroups under
low-salt conditions, whereas in 1 M NaCl, the long-range electrostatic forces of
attraction were shielded and therefore interactions of the C-terminal domain
with the polar lipid headgroups were reduced (see Figure 9.2). This suggests that
interactions of extra-membrane domains of OMPs with lipid bilayers may be
rather complex and modulated by local environmental conditions.

9.3.2   Simple OMPs in Diverse Environments

As we have seen in the case of OmpA and OprF, MD simulations of OMPs are
not restricted to phospholipid bilayer environments. Indeed, simulations of
OMPs in diverse environments have permitted exploration of the conformational dynamics of OMPs in a range of experimental environments. In particular, OmpA has been studied experimentally in lipid bilayers, in detergent
micelles and in crystals containing the protein plus a small number of bound
detergent molecules.[20] Comparisons of MD simulations in these three

environments have revealed differences in the conformational dynamics of the protein; this study was reviewed by Khalid *et al.*[34]

A number of OMP structures (*e.g.* OmpA,[21] OmpX[35] and PagP[36]) have been determined by NMR and by X-ray crystallography. Often these two experimental techniques yield structures with conformational differences. The ability of MD to explore the dynamics of proteins makes it an invaluable tool for helping to identify the origins of such conformational differences in experimentally determined structures. A comparison of the X-ray and NMR structures of the TM domains of the small OMPs OmpA, OmpX and PagP revealed the mobility of the proteins to be largely dependent upon the 'quality' of the initial structure. A set of 15 ns simulations of proteins embedded in DMPC showed that while all three proteins were generally more mobile in the simulations based on NMR structures, the overall mobilities of the residues were qualitatively similar for the corresponding X-ray and NMR structures. These comparative studies represent an important step in relating the quality of a given structure to the conformational dynamics of OMPs; however, there remains a need to extend such studies to a wider range of membrane proteins in order to understand more fully the influence of both environment and structure quality on the dynamics of OMPs and membrane proteins in general.

As we have discussed, MD simulations can play a key role in helping to relate the structure quality to dynamics of OMPs; however, it is equally important to explore the relationship between the structure, conformational dynamics and the biological function of the protein. In this context, changes in conformational dynamics revealed by MD simulations have been used to explore the relationship between the X-ray structure and the dynamic function of OpcA from *N. meningitides*. OpcA is an adhesion protein which forms a 10-stranded β-barrel.[37] The X-ray structure revealed a wide, water-filled β-barrel that is occluded by the extracellular loop L2. This prevents the formation of a continuous pore. Zn^{2+} ions were required for the formation of OpcA crystals; three of these are identified in the X-ray structure, one in the interior of the β-barrel and two near the loops on the extracellular surface. MD simulations of OpcA in a DMPC bilayer in the *absence* of bound Zn^{2+} revealed the loop regions to be extremely flexible. In particular, large conformational changes of loop L2 resulted in opening of the putative pore. Hence it was proposed that the Zn^{2+} ions binding loops L2 and L4 together in the crystal structure may force the protein to adopt a non-physiological conformation. The conformational changes observed in the MD studies also suggest that the proteoglycan binding site formed by the extracellular loops may be rather dynamic. This may suggest an induced-fit mechanism of ligand binding. Thus, these simulations have extended understanding of the mechanism of ligand binding. In a related study, simulations of OpcA in palmitoyloleoylphosphatidylcholine (POPC) and a crystal lattice were compared. Increased mobility of the loops in the lipid bilayer compared with the crystal lattice was observed. Simulations of the crystal lattice in the presence and absence of Zn^{2+} ions revealed the role of Zn^{2+} ions in stabilizing the loops, thus providing further evidence to suggest a non-physiological conformation of the OpcA protein in the crystal structure.

9.4   Leaking Barrels

Porins can be divided into two classes, the general porins (also known as diffusion porins) and selective porins. The general porins provide a simple diffusion pathway across the membrane for molecules smaller than $\sim 1\,kDa$, with little substrate selectivity. OmpF is a general porin from *E. coli*, which provides a translocation pathway for small molecules, water and ions across the outer membrane. One of the first MD simulations of an OMP[38] was of OmpF embedded in a phosphatidylethanolamine (POPE) bilayer. Water molecules within the protein pore were observed to have reduced mobility in comparison with the bulk solvent. Although this simulation was short (1 ns) by current standards, it was encouraging that the reduced rate of water diffusion within pores agreed with earlier empirical approaches.[39] This has since been observed in a number of simulations of OM channels. The origin of cationic selectivity of OmpF has been explored *via* MD simulations.[40,41] It was observed that chloride and potassium ions followed two well-separated pathways along the axis of the pore with potassium having a greater propensity to occupy the pore. Ion–ion interactions played a key role in the permeations of these ions. Whereas single potassium ions were free to permeate the pore, chloride ions were only able to pass the constriction zone when paired with potassium ions. The translocation of a range of penicillins through OmpF was studied by a combined MD/single channel electrical recording approach. The simulations were able to suggest a possible pathway of these drugs through the OmpF channel and rationalize the experimental findings. In particular for specific penicillins, favourable interactions with the protein suggested possible binding sites near the constriction zone. This suggests that interaction with a set of high-affinity binding sites located near the constriction zone of the OmpF channel improve a drug's ability to cross the membrane *via* the pore. MD[42] and steered MD simulations[43] have been used to simulate the transport of a range of other molecules through OmpF. More recently, MD studies of OmpF have revealed the influence of side-chain protonation states on the conformational dynamics of the protein.[44] In particular, varying the protonation state of a key side chain (D127) substantially influenced the cross-sectional area profile of the pore. These observations have received support from single-molecule experimental studies, which have demonstrated that the conductance and selectivity of OmpF are rather sensitive to the charge state of the residue at position 127. This further emphasizes the importance of electrostatic interactions within the central core of the β-barrel domains of OMPs.

9.5   Transporting Barrels

Although most OMPs are β-barrels, there are considerable variations in the size and geometry of these β-barrels. Many of the larger and more complex OMPs contain additional domains located within the interior of the β-barrel. For example, all known structures of the TonB-dependent transporter family reveal a globular 'plug' or 'cork' domain located within the β-barrel.[45] Similarly, a common feature of the two structures of bacterial autotransporters is the presence of α-helices within their β-barrels. The dynamic interplay between the β-barrel and the 'additional' domain is a key aspect of several proposed transport mechanisms.[17] In the following section, we describe a few key MD simulation studies in which the nature of the dynamic interactions between the two domains has been investigated.

9.5.1   TonB-dependent Transporters

Large solutes (> 600 Da) are unable to cross the outer membrane *via* simple passive diffusion; instead, they must utilize some other means to enter the periplasmic space. A number of specialized transport pathways have evolved to allow the transport of these large, scarce nutrients across the OM. For example, the TonB system actively transports a number of ions and their chelates *via* a complex mechanism. This mechanism involves outer membrane transporters coupled to a periplasmic protein (TonB), which in turn is coupled to an energy-transducing protein complex anchored in the inner membrane.[16] Thus, TonB-dependent transporters embedded in the outer membrane mediate the active transport of metal siderophores into the periplasm. The crystal structures of a number of TonB-dependent transporters, with and without bound ligand, are known (a summary listing of all membrane protein structures is provided at http://blanco.biomol.uci.edu/Membrane_Proteins_xtal.html). All known structures of TonB-dependent transporters are 22-stranded β-barrels with a globular 'plug' domain located within the barrel (see Figure 9.3). The X-ray structure of the soluble TonB protein in complex with FhuA and BtuB has revealed a four-stranded β-sheet. The TonB protein contributes three strands and the outer membrane transporter the fourth.[46,47] Although the availability of crystal structures of both the outer membrane and periplasmic components of the TonB transport machinery has furthered our understanding of these complexes, the molecular-level details of the mechanism of the siderophore transport pathway remain unclear.

Some insights into various aspects of this mechanism have been achieved through MD simulations. FhuA embedded in a DMPC bilayer was studied in the ligand-free and iron/siderophore-bound states. These simulations revealed two functionally important aspects of the conformational dynamics of FhuA. First, some ligand-induced changes in the mobilities of the extracellular loops were observed. In particular, when the ligand is bound, the most flexible of the loops, L8, was observed to block access to the binding site, effectively closing it. The crystal structure of the ferric citrate receptor, FecA, exhibited a similar loop arrangement in the ligand-bound state.[48,49] The second key finding concerns the movement of water molecules within the core of FhuA. The presence of the plug domain within the β-barrel reduces the permeability of FhuA by water. These results support the view that translocation of the siderophore, either actively or passively, requires a substantial conformational change in the plug domain. This hypothesis is further supported by steered MD simulations

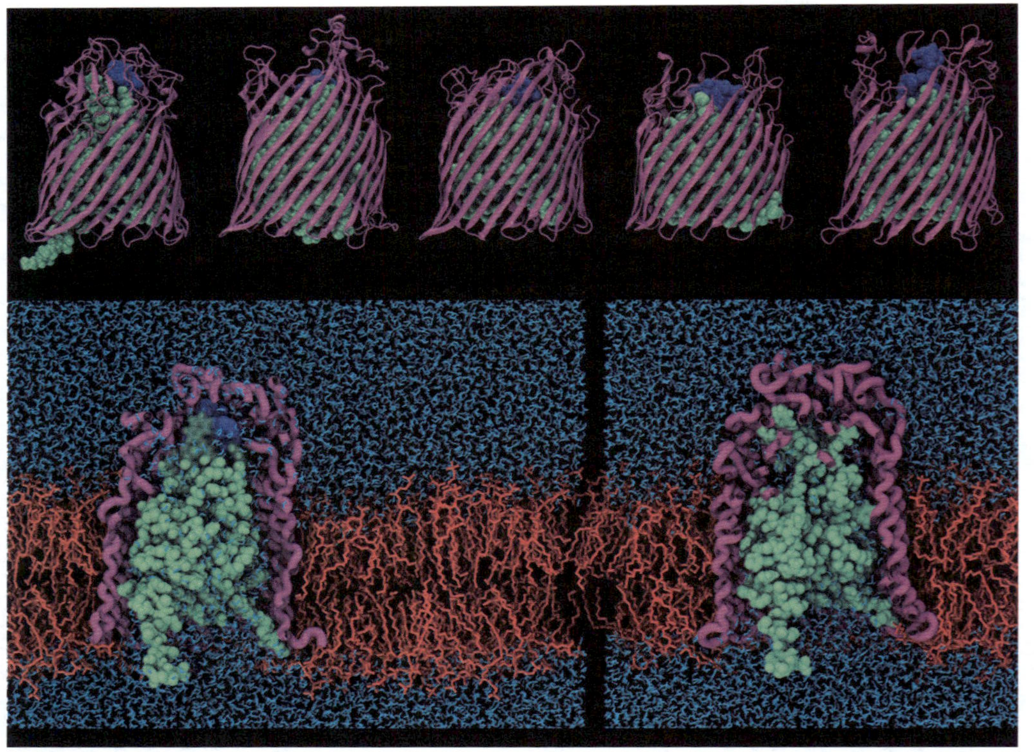

Figure 9.3 The top row illustrates five TonB-dependent transporter structures determined by crystallography: FepA, FhuA, FecA, BtuB and FpvA. The 22-stranded β-barrel is shown in purple, the plug domain in lime and the transported molecule in blue. The FepA holo structure corresponds to a model derived from the crystallized apo structure. The bottom row shows snapshots from molecular dynamics simulations of the FepA holo (left) and apo system (right), inserted in a fully hydrated lipid bilayer (red). Water molecules are shown in cyan. In the visual representation, the barrel is cut open and selected lipid and water molecules have been omitted for clarity.

of the TonB–BtuB complex based on X-ray structures, which suggested that force transduction initiates partial unfolding of the plug domain. These simulations suggested that partial unfolding of the plug domain is more likely than translocation of the intact, folded domain, which is held within the barrel by an extended network of hydrogen bonds and salt bridges. Steered MD simulations were also employed to explore the transmission of force to the outer membrane transporter. In the X-ray structure, two hydrogen-bonded β-strands hold TonB and BtuB together, one from each protein. Steered MD simulations in which TonB is pulled away from BtuB (towards the cytoplasmic membrane) revealed that force applied to TonB can be transmitted to BtuB without disrupting the interaction between the two proteins. Thus a mechanical mode of coupling was proposed.

9.5.2 Autotransporters

Autotransporter proteins are a family of bacterial proteins that are secreted across the bacterial membrane system to the cell surface *via* a translocator domain located in the outer membrane. This is known as the type V or auto-transport pathway. Autotransporters may be divided into two subfamilies: conventional (or classical) autotransporters and trimeric autotransporters. All autotransporters are expressed as precursor proteins with three basic functional domains: an N-terminal signal peptide, an internal passenger domain and a C-terminal translocator domain (a β-barrel TM domain). The passenger domain is often cleaved at the cell surface following translocation. The crystal structures of two outer membrane autotransporters have been determined.[50,51] A common structural feature of both proteins is the presence of α-helices within the β-barrel. NalP from *N. meningitidis* is a conventional autotransporter. The translocation domain of NalP is a 10-stranded β-barrel, with an attached single α-helix located centrally within the hydrophilic interior of the β-barrel. The α-helix is connected to the β-barrel by a short linker. In contrast, HiA from *Haemophilus influenzae* is a trimeric autotransporter. Three monomers each contribute one α-helix and four β-strands to form three α-helices located within a 12-stranded β-barrel (translocator domain). In comparison with NalP, the α-helices of HiA are connected to the β-barrel by long linkers, which extend into the barrel (see Figure 9.4).

The conformational dynamics of the β-barrel pore and the role of the central helix of NalP have been studied *via* MD simulations of the protein in DMPC.[52a] Comparative simulations of the intact TM translocator domain and this domain without the central α-helix revealed that both domains were conformationally stable on a 10 ns time scale. Although the pore of the intact TM domain was flexible and exhibited a degree of 'breathing', there were no systematic drifts in its dimensions throughout the simulation. In contrast, a degree of narrowing of the pore at both mouths of the β-barrel was observed upon removal of the α-helix. This narrowing of the pore was largely a consequence of the enhanced mobility of the loops and turns in the absence of the α-helix. This

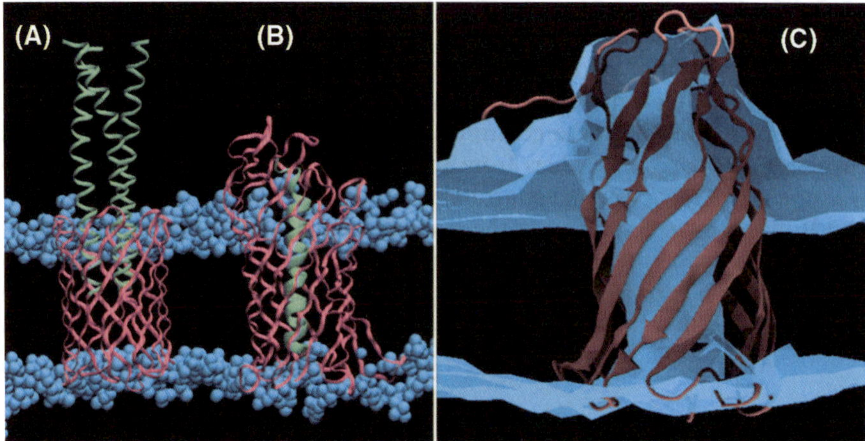

Figure 9.4   The left panel illustrates snapshots from simulations of the outer membrane autotransporter proteins (A) NalP[52a] and (B) HiA. Lipid headgroups are coloured cyan. The β-barrels and α-helices are coloured pink and lime, respectively. In the right panel, areas of high water density in and near to the protein NalP in which the central helix was removed prior to simulation are indicated in cyan (C). Water and lipid molecules have been omitted for clarity.

was most pronounced at the extracellular mouth, where loops L2 and L5 were partially folded into the pore. Once folded into the pore, these loops then formed H-bonds across the mouth of the β-barrel, reducing the dimensions of the pore. In particular, a persistent H-bond was detected between residues Y1017 and E873. Similarly, H-bonds between Asn815 and Glu1052 and Arg820 led to narrowing of the pore on the periplasmic side. In the presence of the α-helix, the reduced mobility of the loops and turns prevented formation of these H-bonds, resulting in a slightly wider pore at both mouths. The reduced dimensions of the pore at the mouths of the β-barrel did not prevent entry of water into the pore region. An increased number of water molecules were seen to enter the pore when the helix was removed, an observation that suggests a plug-like role for the helix (see Figure 9.4C). The pore radius profile revealed a pore just wide enough to allow translocation of the passenger domain in an extended or unfolded conformation. Overall, these simulations support an (auto)transport role for the *monomeric* form of NalP.

As noted above, HiA is a trimeric autotransporter. Although sharing a similar general topology to NalP (α-helices surrounded by a β-barrel), the translocator domain of HiA differs from NalP in that the loops connecting the helices to the β-strands are large, extending over one-third of the way into the interior of the β-barrel. Furthermore, the α-helices connected to these loops protrude from the barrel into the extracellular medium. MD simulations of HiA[52b] revealed a degree of β-barrel distortion when the helices were removed from the barrel interior. Further distortion was observed when the loops connecting the helices to the barrel were also removed. Interestingly, the greatest distortion was observed when the helices were removed one by one rather than all three at once. Persistent H-bonds thought to play a role in stabilizing the β-barrel were identified. In particular, a glycine residue from each loop, located near the centre of the barrel, forms two H-bonds to the corresponding glycine residues from the two other loops. This pattern of H-bonding leads to a triangular arrangement of glycines, stabilized by six H-bonds, which help to maintain the stability of the rather flexible trimeric β-barrel. Thus, in agreement with experimental studies,[51] simulations of HiA suggest that the flexible loops connecting the α-helices to the β-barrel play a key role in maintaining the geometry of the HiA translocator domain.

9.5.3   TolC

The TolC family of proteins play a key role in the type I secretion of small peptides, toxins and drug molecules in Gram-negative bacteria. The type I secretion apparatus includes an efflux pump complex composed of three proteins: AcrA (periplasmic 'adapter' protein), AcrB (inner membrane efflux pump) and TolC (located in the outer membrane and extending into the periplasm). This tripartite arrangement of the complex enables the direct passage of the solute from the cytoplasm to the external medium. Indeed, as a complex, TolC, AcrB and AcrA are capable of forming a ~ 260 Å long molecular tunnel. This arrangement of the proteins is supported by the X-ray structures of TolC[53] and of its homologues OprM[54] and VceC,[55] AcrB,[56,57] AcrA[58] and its homologue MexA.[59,60]

The outer membrane protein TolC is a cylindrical trimer, 140 Å long and containing a TM and a periplasmic domain. The TM domain is a 12-stranded β-barrel about ~ 35 Å long, while the periplasmic domain contains 12 α-helices and is ~ 100 Å long and ~ 30 Å in diameter. In the export of solutes, the main obstacle appears to be at the periplasmic mouth. Whereas the extracellular mouth of the TM pore domain is open in the X-ray structure, the overall pore is closed at the periplasmic mouth, with a pore radius of only ~ 2 Å. The gating of TolC is a key aspect of the function of the efflux machinery, but the mechanism by which this occurs remains to be completely understood. A mechanism of opening has been proposed that involves a twisting motion of the helices accompanied by the disruption of H-bonds and electrostatic interactions.[53]

MD simulations have been performed in order to explore the conformational flexibility of TolC and OprM (a TolC homologue from *P. aeruginosa*) in DMPC bilayers[61] (Figure 9.5). These included simulations of the isolated TolC TM (*i.e.* β-barrel) domain, of the intact TolC protein and of the OprM β-barrel. Substantial flexibility was observed in the extracellular loop regions of TolC. During the simulations, collapse of the loops into the extracellular mouth resulted in closure of the pore by reducing its radius to <2 Å. A network of H-bonds between residues located on the loops stabilized the closure of the pore. A comparable role of loops in gating has been suggested based on recent X-ray structures of OmpG at different pH values[62] and MD simulations.[63]

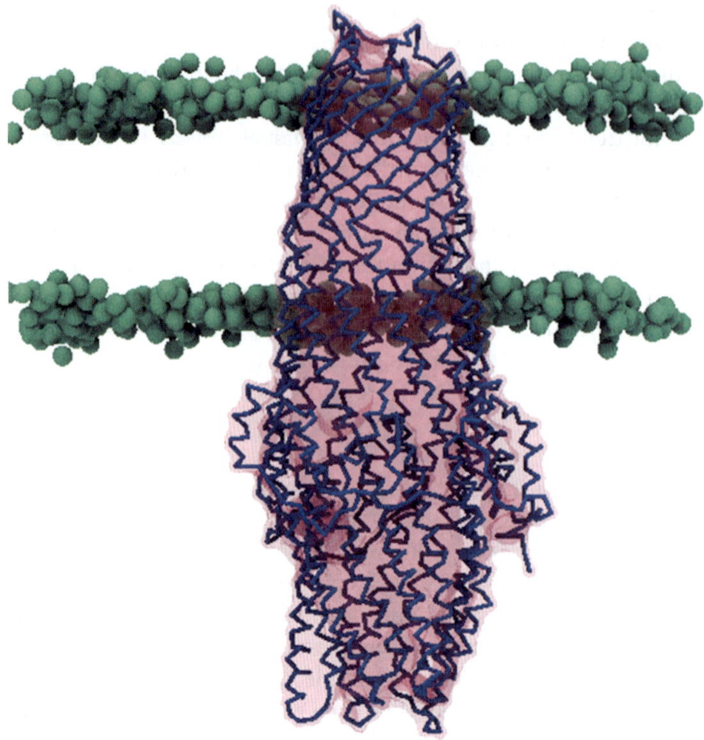

Figure 9.5   Snapshot from the end of a 200 ns coarse-grain simulation of the intact TolC protein in a self-assembled DPPC bilayer.[83] The lipid headgroups are coloured lime, the surface of the protein is shown in magenta and the protein backbone is coloured dark blue. The rest of the solvent and bilayer have been omitted for clarity.

Another key observation from the MD simulations was the change in the cross-sectional conformation of the β-barrel domain of TolC. Comparison of the X-ray structures of TolC and its homologue, OprM, reveals that the β-barrel domain of TolC is cylindrical in cross-section whereas that of OprM is closer to a triangular prism. Simulations of the intact TolC and to a lesser extent, the isolated TM domain revealed the β-barrel to switch from a cylindrical to a triangular prism conformation (similarly to OprM). The more marked conformational change in the intact TolC may indicate that the transition between the two conformations is coupled to the presence of the periplasmic domain.

To address the issue of incomplete convergence, coarse-grain (CG) simulations were employed to explore the conformational dynamics of TolC. In this method of reduced resolution, the internal dynamics of the protein are modelled *via* an elastic network. The CG approach permitted extended simulations of 200 ns to be performed. Encouragingly, the CG-MD simulations revealed both the 'collapse' of the extracellular loops and the transition to a triangular prismatic cross-section β-barrel observed in the atomistic simulations.

A major characteristic of both TolC and OprM is the presence of the α-helical domain, which protrudes ~ 100 Å into the periplasmic space where it interacts directly with the membrane transport protein, *e.g.* AcrB. Whereas the X-ray structure of TolC shows the tunnel to have an internal pore wide enough to allow unimpeded passage of solutes, MD simulations have shown that the dynamics of this region may be rather complex, involving both a 'breathing motion' of the pore and an iris-like motion of the whole protein.

MD simulations have revealed a complex range of conformational dynamics in TolC. Putative gates at both mouths of the barrel have been identified and complex 'breathing' motions in the intermediate domains have been observed. Overall, it appears that models in which the gating of TolC is restricted to an iris-like motion at the periplasmic mouth may be too simplistic and that the conformational dynamics of TolC are more complex than might have been expected.

9.6   Reacting Barrels

The study of enzymatic OMPs adds complexity to the already challenging simulation approach for membrane proteins and is a particularly delicate task. In addition to the previously mentioned issues of sampling and fine-tuning of membrane interactions, the investigation of an enzymatic reaction by classical simulation methods is intrinsically limited. Strictly, it is beyond the capacities of classical MD simulations because bond breaking and the creation of new chemical links are not possible within this framework. This means that one has to learn about the reaction by investigating specific states, prior or subsequent to the reaction, at a transition state or blocked by an inhibitor. Three OMP enzymes have so far been investigated by MD simulations: outer-membrane phospholipase A, a bacterial outer-membrane enzyme which degrades phospholipids,[64] outer-membrane protease T, a peptide hydrolase,[65,66] and PagP, an outer-membrane enzyme which transfers a palmitate chain from a phospholipid to lipid A and is strictly an acyl transferase.[67] These enzymes do not have related folds and differ in various aspects, but they do have similar active sites consisting of a catalytic triad of amino acid residues. The active sites of OMPLA, OmpT and PagP furthermore bear similarity to acetylcholinesterase, which is not a membrane protein. A structural comparison of the active site dynamics of these systems has been carried out.[68] By comparing multiple MD simulations of these enzymes, it was possible to explore potentially functionally significant patterns of catalytic side-chain dynamics and attempt to relate their mobility to the catalytic mechanism. In this way, 'functional' and 'non-functional' catalytic triads could indeed be distinguished.

Given the above-mentioned limitations in the study of enzymatic systems, it is important to establish a dialogue between simulation and experiment in order to corroborate the implications of the theoretical studies. OMPLA is a nice example where such a dialogue seems well engaged (Figure 9.6). Initial MD simulations established the conformational dynamics of three functionally

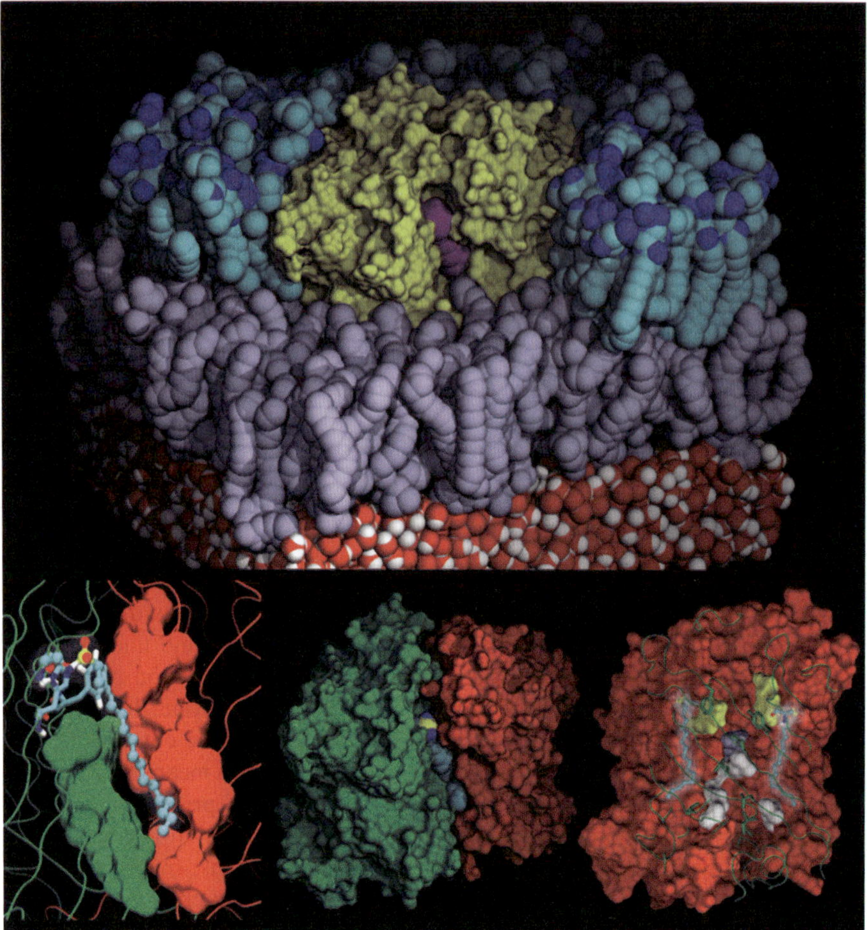

Figure 9.6   Snapshots from MD simulations of the OMPLA enzyme. The top image shows the position of the HDS substrate (purple) in the binding pocket of the OMPLA dimer (yellow). Outer and inner leaflets of the bilayer are coloured differently and part of the upper layer and also all extracellular water molecules have been omitted for clarity. The bottom row highlights several specific features with the position of the HDS substrate at the OMPLA monomer–monomer interface shown in a close-up view on the left. The overall monomer–substrate–monomer system is illustrated in the central panel. Key interactions at the dimer interface are highlighted on the right-hand side with stacked aromatic rings in yellow, four leucine bulges in white and a central polar glutamine in pale blue.

interesting OMPLA states in a lipid bilayer environment: the inactive monomer, the inactive dimer and the inhibited, potentially active dimer.[64] The stability of the active site and in particular of the substrate-binding cleft varied between these states. Furthermore, the stabilizing interactions such as hydrogen bonds were characterized. Subsequent experimental studies have investigated the energetics of OMPLA dimerization,[69] lipid chain selectivity[70] and the role of a hydrogen-bonding network.[71] Structural implications of these experimental studies were in part based on comparison to the simulation results. For instance, using sedimentation equilibrium ultracentrifugation, it was shown that the highly conserved glutamine side chain (Q94) participates in an intermolecular hydrogen bond that plays only a very modest role in stabilizing the OMPLA dimer. In the original MD simulations, the Q94 hydrogen bond appeared as a weak interaction, already hinting at its humble role in stabilizing the OMPLA dimer.

Simulation time scale is another important factor in modelling enzymatic OMPs, as has been illustrated in the case of OmpT. In an initial all-atom model, the properties of the apo-enzyme inserted in a lipid bilayer were investigated.[65] A model for the complex of OmpT with a short ARRA tetrapeptide substrate was also suggested. This model was then refined in subsequent all-atom simulations and finally simulated with a hybrid molecular mechanics/coarse-grained approach on the microsecond time scale.[66] The results indicated that large-scale motions and fluctuations of the electric field on this extended microsecond time scale may impact on the biological function of OmpT. Interestingly, such a conclusion cannot be drawn within the shorter 10–100 ns time scale typical of current all-atom molecular dynamics simulations of OMPs. Furthermore, a structural explanation for the drop in the catalytic activity of two known OmpT mutants, S99A and H212A, could be provided.

As a conclusion on reactive barrels, it should be pointed out that conformational fluctuations of OMP enzymes may play an important role for substrate recognition and/or catalysis. Simulations are a fast and reliable tool for providing structure–function relationships for both wild-type OMP enzymes and their mutants.

9.7 Technological Barrels

In recent years, the pore-like properties of membrane proteins have been exploited for applications in bionanotechnology. Much of the research in this area has been focused on the staphylococcal toxin α-haemolysin (HL).[72] The pore formed by wild-type HL consists of seven identical subunits arranged around a central axis (Figure 9.7A).[73] The transmembrane domain is a 14-stranded β-barrel with two anti-parallel strands contributed by each subunit. The extramembraneous domain contains a large internal cavity. Engineered versions of the HL pore have been used as stochastic sensors for a range of organic molecules.[74,75] MD simulations of HL have been employed to investigate the osmotic permeability, the selectivity, the voltage dependence and the pH dependence of the channel.[76] In particular, it was found that changing the protonation states of the seven His144 residues from neutral to positively charged resulted in an increase in the ion conductance at both positive and negative biases and an increased selectivity for chloride ions. Thus it was

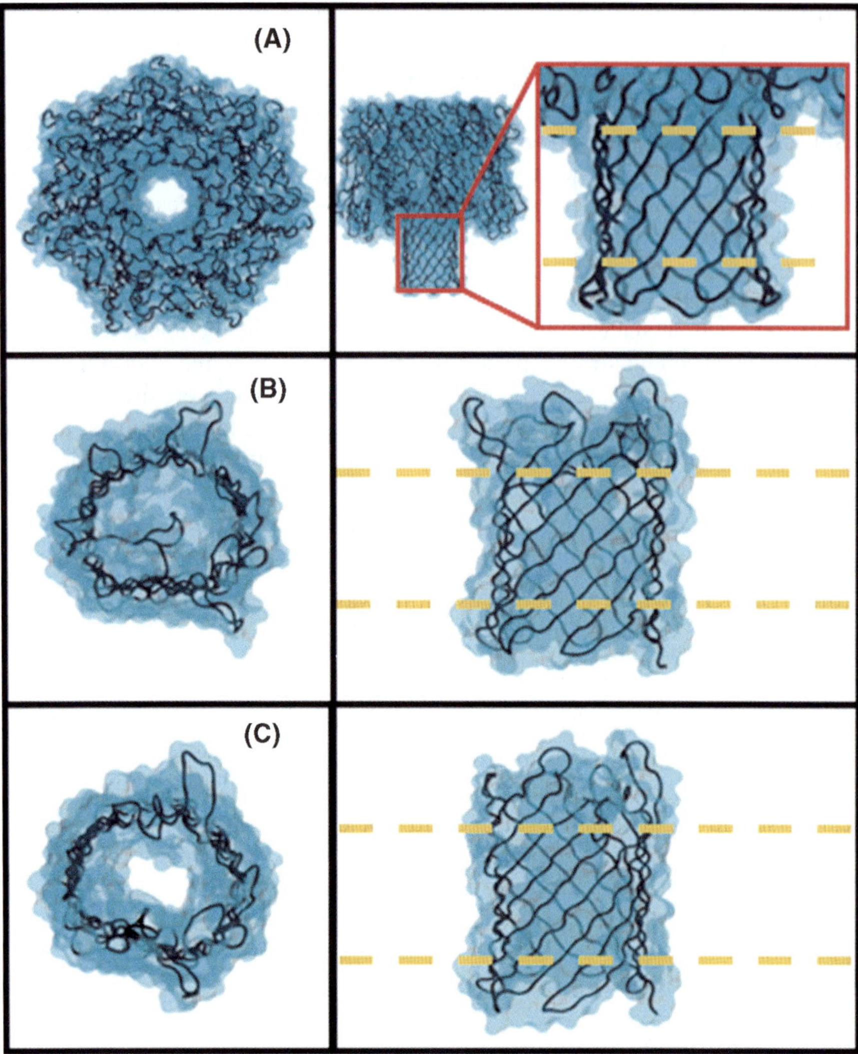

Figure 9.7   Top and side views of αHL and OmpG. (A) Top view (viewed from the extracellular side) of the bacterial toxin αHL. The side view is shown alongside this panel with the TM β-barrel indicated by the red box. In the close-up view of the barrel, the dashed yellow lines indicate the putative location of the membrane. (B) and (C) show the closed (pdb code 2IWW) and open (pdb code 2IWV) X-ray structures of OmpG, respectively.

proposed that the seven His144 residues comprise the pH sensor that gates the conductance and selectivity of the α-haemolysin channel. Steered MD simulations have been employed to study the translocation of nucleic acids through the HL pore.[77] In recent years, the practical difficulties associated with selectively modifying the HL pore have led to the search for a similarly sized

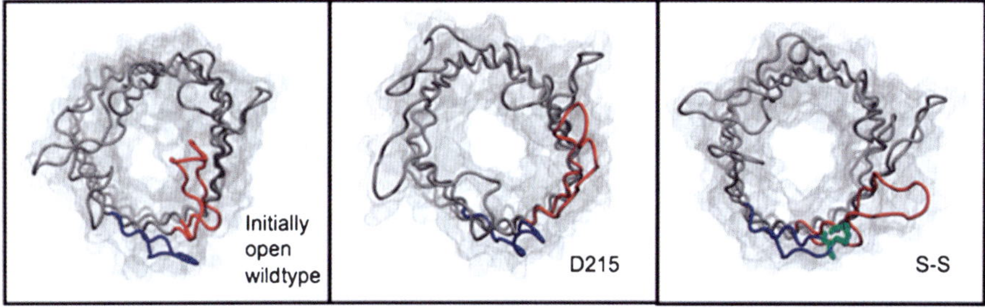

Figure 9.8   The three panels depict the top view (from the extracellular side) of the OmpG wild-type protein and two mutants after 10 ns of simulation. Loops L6 and L7 are shown in red and blue, respectively. The disulfide bond introduced in mutant S–S is shown in green. L6 is partially occluding the pore after 10 ns in the wild-type simulation, but not in the two mutant simulations.

alternative pore. A combined MD simulation and single-channel electrical recording approach was used to design and engineer a mutant of the *E. coli* protein OmpG in which spontaneous gating was sufficiently reduced to allow it to be used as a biosensor (see Figure 9.7 for a comparison of HL and OmpG structures). MD simulations starting from the X-ray structures of the open and closed states of wild-type OmpG suggested that strands 11 and 12 and loop 6 were implicated in the spontaneous gating of the protein. Simulation of mutants in which the inter-strand hydrogen bonding between strands 11 and 12 was optimized and the mobility of loop 6 was reduced by tethering to the nearest strand *via* a disulfide bridge did not demonstrate any propensity to switch to the closed state (Figure 9.8). Single-channel recording of this mutant revealed a 95% reduction in gating activity. Furthermore, detecting ADP at the single molecule level demonstrated the suitability of this mutant for biosensing applications.

9.8   Conclusion

These studies provide a snapshot of the current status of MD simulations of outer membrane proteins. It is evident that such simulations can now be performed with a reasonable degree of accuracy and can provide valuable molecular-level details of the relationship between protein structure and function. Indeed, simulation methodology and force field accuracy have progressed such that we are now at the stage where simulations can be considered as a standard tool for analysis of new outer membrane protein structures. However, despite these advances, limitations of sampling and time scale remain. As more structures and simulations emerge, sampling issues[78] may be addressed by *comparative* simulation studies. This approach has been explored in, *e.g.*, comparative studies of membrane protein–lipid interactions.[79] Furthermore, recently a database of outer membrane protein simulations has been developed in order to allow comparison of OMPs in terms of, *e.g.*, their specific lipid interactions and also their general effect on the local bilayer (sbcb.bioch.ox.-ac.uk/ompdb/). Accelerated simulation techniques such as steered MD[77] and metadynamics[80] also offer a route to enhanced sampling of the conformational landscape. In terms of time scales, coarse-grained simulations[81] and elastic network models[82] offer promise for addressing longer time scale events, albeit more approximately. By combining the coarse-grain and atomistic levels of detail, a more integrative computational biology of outer membrane proteins will emerge.

Acknowledgements

S.K. thanks the University of Southampton, Life Sciences Interface for funding and Professor Mark Sansom for helpful discussions. M.B. thanks ANR for funding research on the relation of structural fluctuations and biological function (ANR-06-PCVI-0025).

References

1. C. Dong, K. Beis, J. Nesper, A. L. Brunkan-Lamontagne, B. R. Clarke, C. Whitfield and J. H. Naismith, *Nature*, 2006, **444**, 226–229.
2. K. Ziegler, R. Benz and G. E. Schulz, *J. Mol. Biol.*, 2008, **379**, 482–491.
3. S. Tanizaki and M. Feig, *J. Phys. Chem. B*, 2006, **110**, 548–556.
4. M. P. Molloy, B. R. Herbert, M. B. Slade, T. Rabilloud, A. S. Nouwens, K. L. Williams and A. A. Gooley, *Eur. J. Biochem.*, 2000, **267**, 2871–2881.
5. W. C. Wimley, *Protein Sci.*, 2002, **11**, 301–312.
6. J. L. Popot and D. M. Engelman, *Annu. Rev. Biochem.*, 2000, **69**, 881–922.
7. S. K. Buchanan, *Curr. Opin. Struct. Biol.*, 1999, **9**, 455–461.
8. R. Koebnik, K. P. Locher and P. Van Gelder, *Mol. Microbiol.*, 2000, **37**, 239–253.
9. W. C. Wimley, *Curr. Opin. Struct. Biol.*, 2003, **13**, 404–411.
10. W. T. Doerrler, *Mol. Microbiol.*, 2006, **60**, 542–552.
11. T. Schirmer, *J. Struct. Biol.*, 1998, **121**, 101–109.
12. W. Achouak, T. Heulin and J. M. Pages, *FEMS Microbiol. Lett.*, 2001, **199**, 1–7.
13. T. Schirmer, T. A. Keller, Y. F. Wang and J. P. Rosenbusch, *Science*, 1995, **267**, 512–514.
14. D. Forst, W. Welte, T. Wacker and K. Diederichs, *Nat. Struct. Biol.*, 1998, **5**, 37–46.
15. T. F. Moraes, M. Bains, P. E. W. Hancock and N. C. J. Strynadka, *Nat. Struct. Mol. Biol.*, 2007, **14**, 85–87.
16. J. D. Faraldo-Gómez and M. S. P. Sansom, *Nat. Rev. Mol. Cell Biol.*, 2003, **4**, 105–115.
17. I. R. Henderson, F. Navarro-Garcia, M. Desvaux, R. C. Fernandez and D. Ala'Aldeen, *Microbiol. Mol. Biol. Rev.*, 2004, **68**, 692–744.

18. P. J. Bond and M. S. P. Sansom, *Mol. Membr. Biol.*, 2004, **21**, 151–162.
19. A. Pautsch and G. E. Schulz, *Nat. Struct. Biol.*, 1998, **5**, 1013–1017.
20. A. Pautsch and G. E. Schulz, *J. Mol. Biol.*, 2000, **298**, 273–282.
21. A. Arora, F. Abildgaard, J. H. Bushweller and L. K. Tamm, *Nat. Struct. Biol.*, 2001, **8**, 334–338.
22. C. Fernandez, C. Hilty, S. Bonjour, K. Adeishvili, K. Pervushin and K. Wüthrich, *FEBS Lett.*, 2001, **504**, 173–178.
23. N. Saint, E. De, S. Julien, N. Orange and G. Molle, *Biochim. Biophys. Acta*, 1993, **1145**, 119–123.
24. E. Sugawara and H. Nikaido, *J. Biol. Chem.*, 1992, **267**, 2507–2511.
25. A. Arora, D. Rinehart, G. Szabo and L. K. Tamm, *J. Biol. Chem.*, 2000, **275**, 1594–1600.
26. P. J. Bond, J. D. Faraldo-Gómez and M. S. P. Sansom, *Biophys. J.*, 2002, **83**, 763–775.
27. P. J. Bond and M. S. P. Sansom, *J. Mol. Biol.*, 2003, **329**, 1035–1053.
28. H. Hong, G. Szabo and L. K. Tamm, *Nat. Chem. Biol.*, 2006, **2**, 627–635.
29. F. S. L. Brinkman, M. Bains and R. E. W. Hancock, *J. Bacteriol.*, 2000, **182**, 5251–5255.
30. S. Khalid, P. J. Bond, S. S. Deol and M. S. P. Sansom, *Proteins Struct. Funct. Bioinf.*, 2005, **63**, 6–15.
31. E. M. Nestorovich, E. Sugawara, H. Nikaido and S. M. Bezrukov, *J. Biol. Chem.*, 2006, **281**, 16230–16237.
32. T. P. Straatsma and T. A. Soares, *Proteins*, 2008, **74**, 475–488.
33. T. Carpenter, Khalid, S. and Sansom, M.S.P., *Biochemistry*, 2007, **1768**, 2831–2840.
34. S. Khalid, J. Holyoake and M. S. P. Sansom, in *Biophysical Approaches of Structure and Function of Membrane Proteins*, ed. E. Pebay-Peyroula, Wiley, New York, 2007, 161–179.
35. C. Fernandez, C. Hilty, G. Wider, P. Güntert and K. Wüthrich, *J. Mol. Biol.*, 2004, **336**, 1211–1221.
36. P. M. Hwang, W. Y. Choy, E. I. Lo, L. Chen, J. D. Forman-Kay, C. R. H. Raetz, G. G. Privé, R. E. Bishop and L. E. Kay, *Proc. Natl. Acad. Sci. USA*, 2002, **99**, 13560–13565.
37. S. M. Prince, M. Achtman and J. P. Derrick, *Proc. Natl. Acad. Sci. USA*, 2002, **99**, 3417–3421.
38. D. P. Tieleman and H. J. C. Berendsen, *Biophys. J.*, 1998, **74**, 2786–2801.
39. O. S. Smart, J. Breed, G. R. Smith and M. S. P. Sansom, *Biophys. J.*, 1997, **72**, 1109–1126.
40. W. Im and B. Roux, *J. Mol. Biol.*, 2002, **319**, 1177–1197.
41. W. Im and B. Roux, *J. Mol. Biol.*, 2002, **322**, 851–869.
42. K. Malek and A. Maghari, *Biochem. Biophys. Res. Commun.*, 2007, **352**, 104–110.
43. K. M. Robertson and D. P. Tieleman, *FEBS Lett.*, 2002, **528**, 53–57.
44. S. Varma, S. W. Chiu and E. Jakobsson, *Biophys. J.*, 2006, **90**, 112–123.
45. M. C. Wiener, *Curr. Opin. Struct. Biol.*, 2005, **15**, 394–400.

46. D. D. Shultis, M. D. Purdy, C. N. Banchs and M. C. Wiener, *Science*, 2006, **312**, 1396–1399.
47. P. D. Pawelek, N. Croteau, C. Ng-Thow-Hing, C. M. Khursigara, N. Moiseeva, M. Allaire and J. W. Coulton, *Science*, 2006, **312**, 1399–1402.
48. A. D. Ferguson, R. Chakraborty, B. S. Smith, L. Esser, D. van der Helm and J. Deisenhofer, *Science*, 2002, **295**, 1715–1719.
49. W. W. Yue, S. Grizot and S. K. Buchanan, *J. Mol. Biol.*, 2003, **332**, 353–368.
50. C. J. Oomen, P. van Ulsen, P. van Gelder, M. Feijen, J. Tommassen and P. Gros, *EMBO J.*, 2004, **23**, 1257–1266.
51. G. Meng, N. K. Surana, J. W. St Geme and G. Waksman, *EMBO J.*, 2006, **25**, 2297–2304.
52. (a) S. Khalid and M. S. Sansom, *Mol. Membr. Biol.*, 2006, **23**, 499–508; (b) S. Khalid and M. S. Sansom, *Biophys. J.*, (in preparation).
53. V. Koronakis, A. Sharff, E. Koronakis, B. Luisi and C. Hughes, *Nature*, 2000, **405**, 914–919.
54. H. Akama, M. Kanemaki, T. Tsukihara, A. Nakagawa and T. Nakae, *J. Biol. Chem.*, 2004, **279**, 52816–52819.
55. L. Federici, D. Du, F. Walas, H. Matsumura, J. Fernandez-Recio, K. S. McKeegan, M. I. Borges-Walmsley, B. F. Luisi and A. R. Walmsley, *J. Biol. Chem.*, 2005, **280**, 15307–15314.
56. S. Murakami, R. Nakashima, E. Yamashita and A. Yamaguchi, *Nature*, 2002, **419**, 587–593.
57. S. Murakami, R. Nakashima, E. Yamashita, T. Matsumoto and A. Yamaguchi, *Nature*, 2006, **443**, 173–179.
58. J. Mikolosko, K. Bobyk, H. I. Zgurskaya and P. Ghosh, *Structure*, 2006, **14**, 577–587.
59. M. K. Higgins, E. Bokma, E. Koronakis, C. Hughes and V. Koronakis, *Proc. Natl. Acad. Sci. USA*, 2004, **101**, 9994–9999.
60. H. Akama, T. Matsuura, S. Kashiwagi, H. Yoneyama, T. Tsukihara, A. Nakagawa and T. Nakae, *J. Biol. Chem.*, 2004, **279**, 25939–25942.
61. L. Vaccaro and M. S. P. Sansom, *Biophys. J.*, 2008, **95**, 5681–5691.
62. O. Yildiz, K. R. Vinothkumar, P. Goswami and W. Kühlbrandt, *EMBO J.*, 2006, **25**, 3702–3713.
63. M. Chen, S. Khalid, M. S. Sansom and H. Bayley, *Proc. Natl. Acad. Sci. USA*, 2008, **105**, 6272–6277.
64. M. Baaden, C. Meier and M. S. Sansom, *J. Mol. Biol.*, 2003, **331**, 177–189.
65. M. Baaden and M. S. P. Sansom, *Biophys. J.*, 2004, **87**, 2942–2953.
66. M. Neri, M. Baaden, V. Carnevale, C. Anselmi, A. Maritan and P. Carloni, *Biophys. J.*, 2008, **94**, 71–78.
67. K. Cox, P. J. Bond, A. Grottesi, M. Baaden and M. S. Sansom, *Eur. Biophys. J.*, 2008, **37**, 131–141.
68. K. Tai, M. Baaden, S. Murdock, B. Wu, M. H. Ng, S. Johnston, R. Boardman, H. Fangohr, K. Cox, J. W. Essex and M. S. Sansom, *J. Mol. Graphics Modell.*, 2007, **25**, 896–902.

69. A. M. Stanley, P. Chuawong, T. L. Hendrickson and K. G. Fleming, *J. Mol. Biol.*, 2006, **358**, 120–131.
70. A. M. Stanley, A. M. Treubrodt, P. Chuawong, T. L. Hendrickson and K. G. Fleming, *J. Mol. Biol.*, 2007, **366**, 461–468.
71. A. M. Stanley and K. G. Fleming, *J. Mol. Biol.*, 2007, **370**, 912–924.
72. H. Bayley and P. S. Cremer, *Nature*, 2001, **413**, 226–230.
73. L. Song, M. R. Hobaugh, C. Shustak, S. Cheley, H. Bayley and J. E. Gouaux, *Science*, 1996, **274**, 1859–1866.
74. H. C. Wu and H. Bayley, *J. Am. Chem. Soc.*, 2008, **130**, 6813–6819.
75. X. F. Kang, S. Cheley, X. Guan and H. Bayley, *J. Am. Chem. Soc.*, 2006, **128**, 10684–10685.
76. A. Aksimentiev and K. Schulten, *Biophys. J.*, 2005, **88**, 3745–3761.
77. D. B. Wells, V. Abramkina and A. Aksimentiev, *J. Chem. Phys.*, 2007, **127**, 125101.
78. J. D. Faraldo-Gómez, L. R. Forrest, M. Baaden, P. J. Bond, C. Domene, G. Patargias, J. Cuthbertson and M. S. P. Sansom, *Proteins Struct. Func. Bioinf.*, 2004, **57**, 783–791.
79. S. S. Deol, P. J. Bond, C. Domene and M. S. P. Sansom, *Biophys. J.*, 2004, **87**, 3737–3749.
80. F. L. Gervasio, M. Parrinello, M. Ceccarelli and M. L. Klein, *J. Mol. Biol.*, 2006, **361**, 390–398.
81. P. J. Bond, J. Holyoake, A. Ivetac, S. Khalid and M. S. P. Sansom, *J. Struct. Biol.*, 2007, **157**, 593–605.
82. I. H. Shrivastava and I. Bahar, *Biophys. J.*, 2006, **90**, 3929–3940.
83. K. A. Scott, P. J. Bond, A. Ivetac, A. P. Chetwynd, S. Khalid and M. S. Sansom, *Structure*, 2008, **16**, 621–630.

Molecular Mechanisms of Active Transport Across the Cellular Membrane

PO-CHAO WEN, ZHIJIAN HUANG, GIRAY ENKAVI, YI WANG, JAMES GUMBART AND EMAD TAJKHORSHID

Department of Biochemistry, Center for Biophysics and Computational Biology, and Beckman Institute, University of Illinois at Urbana–Champaign, 405 N. Mathews, Urbana, IL 61801, USA

10.1 Introduction

Active transport across the cellular membrane constitutes one of the most fundamental, mechanistically diverse and highly regulated processes in all living organisms. Membrane transporters are the principal players in the active exchange of materials across the cellular membrane in an energy-dependent manner. These complex proteins constitute highly sophisticated, fine-tuned molecular pumps that efficiently couple various sources of energy in the cell to the transport of a wide range of molecules across the membrane, often against their chemical gradient. Due to their fundamental role, membrane transporters are directly involved in a wide range of cellular activities, thus playing important roles in maintaining normal physiology of the organism. As expected, any defect in their function will have devastating consequences in the organism. Although such defects are usually lethal, inadequate function of membrane transporters also constitutes the molecular basis of a wide range of serious human diseases, such as cystic fibrosis, familial intrahepatic cholestasis,

RSC Biomolecular Sciences No. 20
Molecular Simulations and Biomembranes: From Biophysics to Function
Edited by Mark S.P. Sansom and Philip C. Biggin
© Royal Society of Chemistry 2010
Published by the Royal Society of Chemistry, www.rsc.org

schizophrenia, Alzheimer's disease and epilepsy. Furthermore, membrane transporters provide not only a common mechanism of acquiring nutrients by microbial cells, but also a potent mechanism of active drug efflux and, thus, a common mechanism of drug resistance in both bacterial and cancer cells. Investigating the molecular basis of these nano-machines is therefore not only of the utmost importance in various disciplines of biological sciences, but also highly relevant to medical and pharmaceutical applications.

Investigation of the molecular details of the transport mechanism in membrane transporters poses a major challenge to biophysical studies, primarily due to the structural complexity of these protein pumps and the high dimensionality of the energy-dependent transport processes. This is in sharp contrast to, *e.g.*, membrane channels, which generally provide a passive diffusion pathway for their substrates. Substrate binding and translocation along the transport pathway in membrane transporters are closely coupled to numerous stepwise protein conformational changes of various magnitudes that are induced by and/or coordinated with the energy-providing mechanisms. For instance, substrate transport can be driven by co-transport of protons and/or other ions along essentially different pathways across the protein or by binding and hydrolysis of ATP in a remotely situated domain of the transporter machinery. A detailed, mechanistic description of the transport cycle, therefore, relies on high-resolution methodologies that can describe the dynamics of the process at an atomic level. Despite significant advances in experimental biophysical methodologies, *e.g.*, advanced spectroscopic and single-molecule techniques, we are still very limited with respect to providing simultaneous temporal and spatial resolutions required for a detailed description of such dynamic phenomena. Due to these technical limitations, often we do not even know how to attempt experimentally to characterize the sequence of events involved in a transport cycle.

In conjunction with and highly complementary to experimental methodologies, molecular dynamics (MD) simulation remains a highly relevant approach with sufficient temporal and spatial resolutions to investigate such processes in atomic detail. Application of the method to membrane transporters, however, is an exceedingly young and unexplored area of research, since sufficient structural data required for such simulations has become available only recently. Furthermore and more importantly, in order to describe molecular events involved in the function of transporters, atomistic simulations of these large biomolecules on the order of at least 0.1–1 µs (sufficient to describe individual molecular events (steps) comprising the transport cycle) are necessary, a computational requirement which has also been met only recently, due to both a significant increase in computational resources and algorithmic advances permitting the exploitation of such resources. Although still extremely challenging, owing to the timely convergence of discoveries of membrane transporter structures and biophysical measurements and advances in computer hardware and software, we can now expand the scope of simulation studies into the realm of membrane transporters and study the molecular mechanism of their function.

This chapter presents some of the examples taken from our ongoing research on a mechanistically diverse group of membrane transporters using a combination of various computational methodologies. These include: (1) ATP-binding cassette (ABC) transporters, a large family of transporters, in which ATP binding and hydrolysis drive substrate transport across the membrane; (2) glutamate transporter (GlT), an ion-driven neurotransmitter transporter that uses proton and ionic gradients as the source of energy for its function; (3) glycerol phosphate transporter (GlpT), a member of the major facilitator superfamily (MFS) mediating exchange of organic and inorganic phosphate across the membrane; (4) the mitochondrial ADP/ATP carrier (AAC), which depends on the membrane electric potential for exchange of nucleotides between the cytoplasm and the mitochondrial matrix; and, (5) vitamin B_{12} transporter (BtuB), representing outer membrane transporters that mysteriously rely on remote sources of energy in the inner membrane to import essential metal ions.

These systems represent a number of membrane transporters for which high-resolution structures are currently available and span a wide range of energy coupling mechanisms involved in traffic of various molecular species across the membrane. Below, after a brief description of major computational methodologies employed, we present the results of some of our recent simulation studies investigating the mechanism of energy coupling in a diverse group of membrane transporters.

10.2 Computational Methodology

In this section, we describe major computational methodologies employed for the simulation and analysis of molecular systems reported in this chapter. The simulation systems and general conditions used for MD simulations will be described only briefly. We refer the reader to the references cited for more details.

Classical MD is the main method used for the simulations presented in this chapter. In general, membrane-embedded models of channels and transporters were used, with lipid bilayer and water explicitly included in the model, thus resulting in system sizes of 100 000–300 000 atoms. Experimentally solved, atomic-resolution structures were used to build the model and other elements (*e.g.* water, lipids and ions) were added to the system by modeling. The CHARMM27 parameter set[1] was used, while missing parameters and topology files for certain ligands, *e.g.* inorganic phosphate, were developed by adopting similar parameters from CHARMM. All calculations employ particle mesh Ewald (PME)[2] for calculation of electrostatic forces without truncation. The resulting Newtonian equations of motion, with modifications to control temperature and pressure, are integrated by symplectic and reversible methods using a time step of 1 fs. All of the simulations used NAMD,[3–5] a highly parallel, publicly available MD program, developed mainly with simulations of large biomolecular systems in mind. NAMD uses efficient algorithms, reducing the complexity in atom count (N) of the long-range electrostatic force evaluation to $O(N\log N)$ *via* the PME[2] algorithm and then increasing the period of PME evaluation from 1 to 4 fs *via* a symplectic multiple time-stepping method.

In addition to classical MD simulations, NAMD was also used to perform steered MD (SMD) simulations[6] to simulate processes that are otherwise too slow to be modeled. In these simulations, an external force is applied to a certain part of the system in order to accelerate the process of interest. This force is applied by either directly adding a constant force to a group of atoms, or by introducing a harmonic potential, ΔU, that restrains the atoms of interest to an imaginary particle moving at a constant velocity, v:

$$\Delta U(\vec{r}_1, \vec{r}_2, ..., t) = \frac{1}{2}k\left\{ vt - \left[\vec{R}(t) - \vec{R}_0\right] \cdot \vec{n} \right\}^2 \tag{10.1}$$

where k is the force constant, $\vec{R}(t)$ and $\vec{R}_0$ are the current and initial center of mass of the atoms, respectively, and $\vec{n}$ is a unit vector representing the direction of the applied force.[5]

10.2.1 Electrostatic Potential Calculation

In order to characterize the electrostatic features of the simulation system, we calculated its instantaneous electrostatic potential for each snapshot, from which an average electrostatic potential is then obtained.[7] The calculations were performed using the PMEPOT plugin of VMD,[8] which reproduces the internal electrostatic potential computed by NAMD[5] during an MD simulation. In these calculations, each point charge is approximated by a spherical Gaussian with a width determined by the Ewald factor β. The Poisson equation is then solved numerically on a 3D grid to yield the electrostatic potential:[8]

$$\Delta^2 \phi(\vec{r}) = 4\pi \sum_i \rho_i(\vec{r}) \tag{10.2}$$

where $\phi(\vec{r})$ is the electrostatic potential and $\rho_i(\vec{r})$ is the charge density of the ith atom in the system. The grid spacing determines the accuracy of the result.[8] In the ADP/ATP carrier (AAC) simulations, the calculation is performed on the apo-AAC system with a grid spacing of less than 1 Å and an Ewald factor of 0.25.[7] As Cl^- ions become bound with the protein after ~ 2 ns simulation, to eliminate the effect of the bound Cl^-, we performed the calculation using the first 2 ns of simulation.

10.2.2 Net Charge Density Distribution Calculation

In the study of AAC, we performed the net charge density calculation on the mitochondrial carrier family (MCF) proteins, of which AAC is the only one with a known crystal structure. To compare AAC with the rest of MCF members, we calculated the net charges of all 34 yeast MCF members based on their sequences from the Saccharomyces Genome Database (http:// www.yeastgenome.org/), assuming a generic titration state for side chains, *i.e.* glutamates and aspartates negative and lysines and arginines positive. In addition, to examine whether mitochondrial carriers show distinct charge characteristics from other yeast membrane proteins, we repeated the calculation for every protein in the Yeast Membrane Protein Library (http:// www.cbs.umn.edu/yeast/). As MCF members are ~300 amino acids (aa) long, among the 1593 sequences in the library we selected the 1067 sequences with a minimum of 200 aa for comparison with MCF members. Normalizing the net charge of each protein by its amino acid count, a histogram with a bin width of 0.003 e/aa is then obtained, which represents the net charge density distribution for yeast membrane proteins.

10.3 ATP-driven Transport in ABC Transporters

ATP-binding cassette (ABC) transporters constitute one of the most abundant families of membrane transporters. Thousands of proteins in this family have been identified in all kingdoms of life, facilitating transport of a variety of substrates ranging from ions and small molecules to lipids and polypeptides.[9] In the human genome, more than 40 genes encoding ABC transporters have been identified, whose products play essential roles in diverse cellular and physiological processes,[10] such as secretion of bile salt,[11] homeostasis of lipid/ cholesterol[12–16] and iron,[17–19] and mediating adaptive immunity.[20,21] Malfunction of ABC transporters has been associated with several disease conditions, *e.g.* cystic fibrosis,[22,23] Alzheimer's disease,[24–27] Tangier disease,[13,28] and adrenoleukodystrophy.[29,30] They are also involved in development of multidrug resistance in cancer cells.[31–34]

Eukaryotic ABC transporters are exclusively 'exporters', that is, they transport their substrate from the cytoplasm, either into certain organelles or outside the cell. In the prokaryotic domain, however, ABC transporters have evolved differently, branching into either the 'exporter' or the 'importer' categories. ABC importers play a vital role in the uptake of essential nutrients into prokaryotic cells, including ions,[35,36] sugars,[37,38] amino acids[39] and vitamins.[40,41] Prokaryotic ABC exporters play crucial roles in survival and virulence of the cell. In addition to providing drug resistance, these exporters are involved in toxin secretion,[42] and are responsible for the formation of the outer membrane in Gram-negative bacteria.[43] The essential role of prokaryotic ABC transporters renders them potential targets for development of novel antibiotics.

Structurally, ABC transporters consist of at least four basic components related with a two-fold symmetry (Figure 10.1). These four building blocks can exist either as different polypeptide chains or as different domains of the same protein: two transmembrane domains (TMDs, also known as the permease domains), providing the permeation pathway for the substrate, and two nucleotide-binding domains (NBDs), which facilitate active transport through ATP binding and hydrolysis. ABC importers, in addition, include specialized substrate binding proteins (BP) located outside the cytoplasmic membrane to acquire specific substrates.

The amino acid sequence of ABC transporters is highly conserved in the NBDs, implying a common mechanism for energy coupling regardless of the transport direction. In contrast, the diverse TMD sequences might relate to the variety of substrate species. Many crystal structures of isolated NBDs are available from different ABC transporters and the relationship between their nucleotide binding states and dimerization states is fairly well established.[44,45] Specifically, the NBDs of the *Escherichia coli* maltose transporter, MalK, have been crystallized in several different states, varying with respect to the bound nucleotide.[46,47] In general, once the NBDs are bound with ATP, they form a closely associated dimeric structure (closed form), which provides two symmetrically related catalytic sites at the interface of the two monomers. NBDs in ADP-bound or nucleotide-free forms, on the other hand, show either monomeric structures or as open, loosely associated dimers (open form). The dimeric states between the ATP-bound and ADP-bound NBD structures, therefore, indicate that either ATP hydrolysis or release of P_i, or a combination of both, results in separation of the two monomers.

It was not until the solution of several crystal structures of complete ABC transporters within the past 2–3 years[48–56] that a consistent transport mechanism with the motion of NBDs described above could be put forward. The substrate transport is apparently achieved by a 'flip-flop' conformational change of the TMDs, alternating substrate accessibility between the extracellular and the cytoplasmic sides, a process which is controlled by ATP binding and hydrolysis in the NBDs (Figure 10.1B).[57–59] It is generally accepted that the nucleotide-free transporter is in a 'resting state', in which the TMDs are open towards the cytoplasm (inward facing) and the two NBDs are either loosely associated or separated (state 1 in Figure 10.1B). Upon ATP binding and NBD dimerization, the transporters reverse the TMD conformation to an outward-facing conformation, which is annotated as the 'intermediate state' (state 4 in Figure 10.1B).[53] Due to the reversal of the TMD conformation, the substrate can be incorporated into the TMD lumen of importers from the BP or be excluded from the TMDs in the case of exporters. ATP hydrolysis opens the NBD dimer, which in turn flips the TMD conformation back to the resting state for the next cycle. During the cycle, 'occluded states' might be formed between the resting and intermediate states, in which the central lumen of TMDs is sealed from both ends (states 3 and 5 in Figure 10.1B).

MD simulations have been employed in several studies to interpret the transport mechanism since the availability of earlier crystal structures.[60–65] However, some of these researches did not yield consensus conclusions, presumably due to insufficient conformational sampling. For example, asymmetric nucleotide binding site movement has been observed in several MD studies,[60,61,63,65,66] but the normal mode analysis of an importer crystal

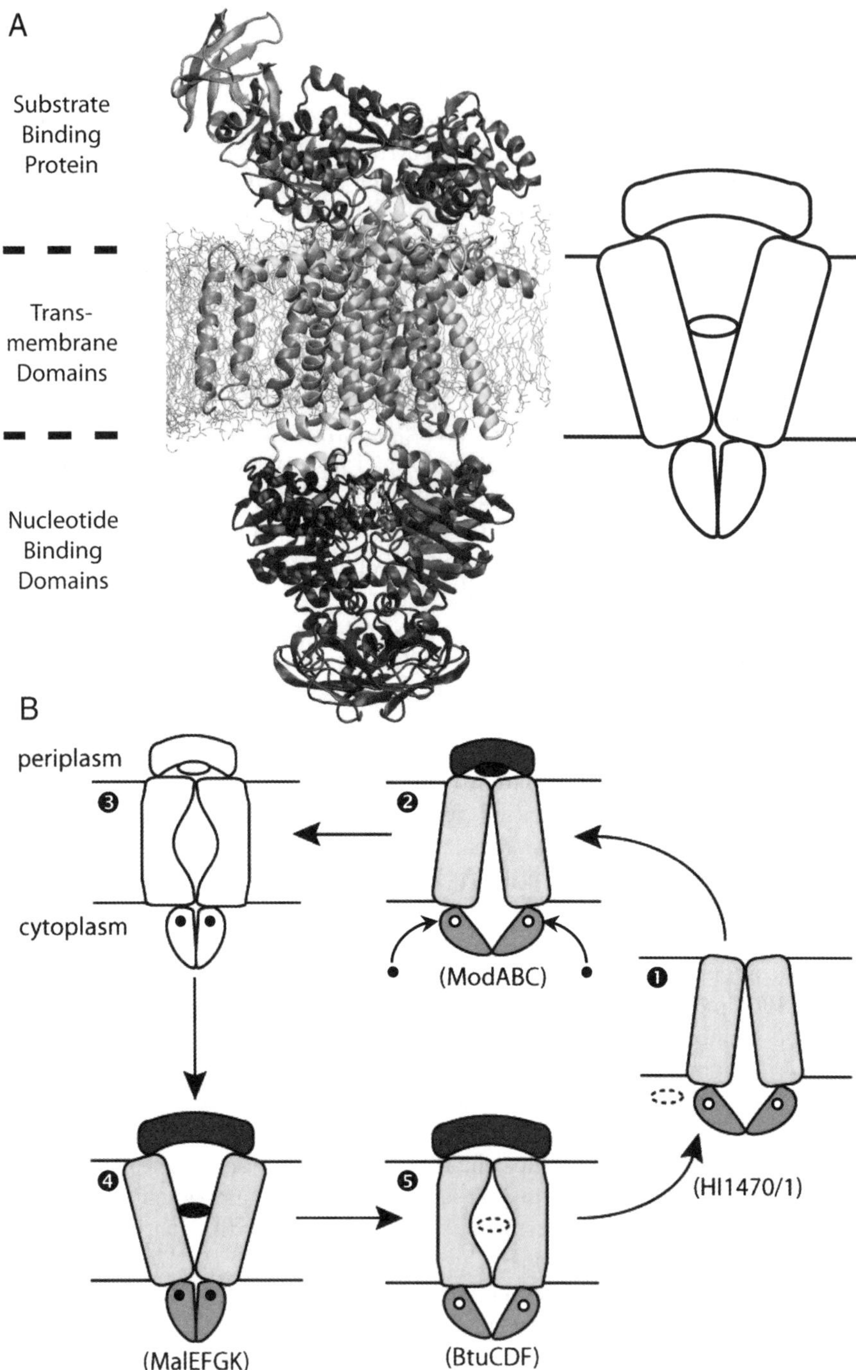
A
Substrate
Binding
Protein
Trans-
membrane
Domains
Nucleotide
Binding
Domains
B
periplasm
cytoplasm
3
2
(ModABC)
1
(HI1470/1)
4
(MalEFGK)
5
(BtuCDF)

structure indicates that the asymmetric motions are irrelevant to the transport mechanism.[67]

In order to investigate the mechanism of NBD opening upon ATP hydrolysis and to study specifically whether one or two ATP hydrolysis events are required for the opening, we performed a set of MD simulations to examine the conformation and dynamics of the NBD dimer of maltose transporter (MalK, Figure 10.1A) in different bound states.[68] Earlier simulation studies had investigated the NBD opening either by replacement of ATP with ADP or by complete removal of ATP from the closed NBD dimer.[61,62,65] In other words, these studies did not examine the immediate effect of ATP hydrolysis, which is studied in simulations reported here by replacing ATP with ADP and P_i in the binding site(s). Our investigation was composed of four independent equilibrium simulations, each for ~ 80 ns. NBD structures in the four simulation systems were constructed with different combinations of bound nucleotides (the two active sites are bound with ATP/ATP, ADP-P_i/ATP, ATP/ADP-P_i or ADP-P_i/ADP-P_i). The ATP/ATP system was based on the crystal structure of ATP-bound MalK dimer;[46] the other three ADP-P_i-containing systems were constructed starting with the ATP/ATP system at $t \approx 8$ ns and hydrolyzing either one or both ATP molecules.

The results show that despite the presence of two ATP binding sites, ATP hydrolysis in only one binding site is sufficient to trigger the opening of the NBD dimer (Figure 10.2).[68] The degree of NBD opening captured in these simulations resembles that observed in the crystal structure of the semi-open form,[46] *i.e.* ~ 3 Å, measured with respect to the ATP-bound, closed state. Since ATP hydrolysis in either active site can trigger the NBD opening, the two nucleotide binding sites do not need to hydrolyze ATP in a certain order or simultaneously, but likely in a stochastic manner. It does not rule out that the dimer opening can couple to two hydrolysis events, depending on the speed of NBD separation. In addition, the simulations provide evidence for the first time

---

Figure 10.1   Architecture and general mechanism of ABC transporters. (A) The maltose transporter embedded in a lipid bilayer, along with a schematic representation relating to the transport cycle depicted in (B). The transporter is composed of five polypeptide chains, namely a periplasmic binding protein, two transmembrane domains (TMDs) and two nucleotide-binding domains (NBDs). (B) Generalized transport mechanism for ABC importers based on crystal structures of full ABC transporters captured in different conformational states. States based on currently available crystal structures are shown in gray. The resting state (1) is suggested to be the cytoplasmic-open state. Substrate binding through the periplasmic binding protein prepares the NBDs for ATP-dependent dimerization (2); ATP binding and closure of the NBDs triggers the formation of the periplasmic-open state (4), most likely through an intermediate *occluded* state (3); ATP hydrolysis and opening of the NBDs results in the recovery of the resting state through a second occluded state (5). The nucleotide binding sites in NBDs are represented as empty (nucleotide-free) or filled (ATP-bound) circles. Substrate absent in crystal structures are represented as dashed ovals.

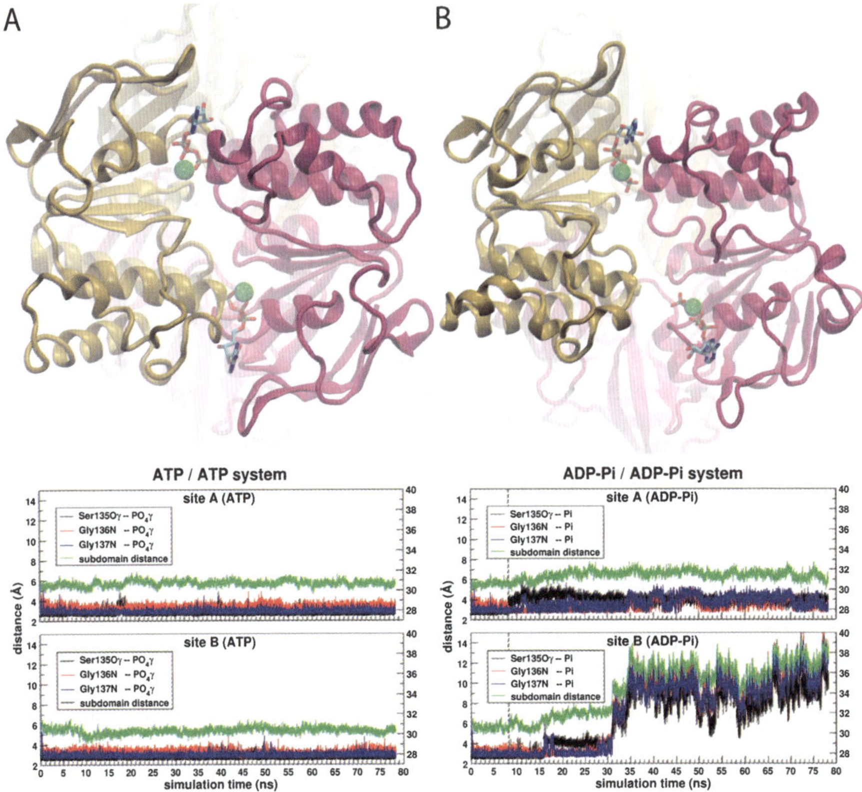

Figure 10.2   ATP hydrolysis induced NBD opening. Each panel represents a different nucleotide-bound configuration: (A) ATP/ATP, (B) ADP-P$_i$/ADP-P$_i$, (C) ADP-P$_i$/ATP and (D) ATP/ADP-P$_i$. The dimer structure in the end of each simulation is shown on top of each panel; the bottom panels show the dimer opening quantified by the time evolution of the distances between two subdomains across the active site, and also the distances between the terminal phosphate and the key nucleotide binding residues.

that the dimer opening is initiated immediately after the hydrolysis reaction, *i.e.* the NBD dimer opens before the P$_i$ dissociates from the active site.[68]

The simulations provide mechanistic details for hydrolysis-induced opening of the dimer. Once ATP has been hydrolyzed, the repulsion between the produced ADP and P$_i$ results in a fast rearrangement of the active site configuration, such that the bound Mg^{2+} moves in between the ADP and P$_i$, with the latter slightly moved away from its original position (the γ-phosphate in ATP). Due to rearrangement of the nascent P$_i$, the binding between the conserved ABC signature motif (LSGGQ) and P$_i$ is significantly weakened. In particular, the once tightly associated hydrogen bond between the γ-phosphate and the serine of the LSGGQ motif is completely ruptured (Figure 10.2, black traces in each panel). The impaired LSGGQ binding

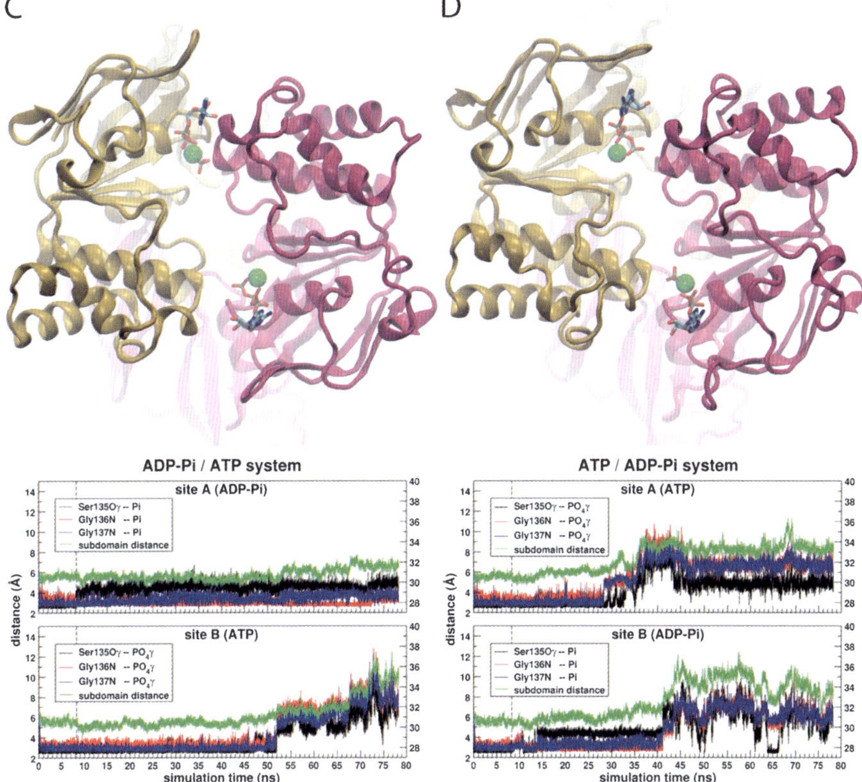

Figure 10.2 Continued.

destabilizes the dimer interface, so that the dimer becomes prone to opening, which can be caused by its thermal fluctuations. Remarkably, the simultaneous breaking of several weaker hydrogen bonds between the LSGGQ motif and the bound ATP or P_i is required for the two NBD monomer to dissociate completely from each other, which also explains the observed stochastic behavior of dimer opening. Meanwhile, the tight binding between Mg^{2+} and ADP-P_i, together with the strong interaction between the Walker-A motif (the major nucleotide binding motif in the NBD core) and the ADP, effectively prevent hydrolysis products from rapidly dissociating from the active sites. Consequently, the dimer opening is independent of the unbinding of either ADP or P_i.

These results provide a new perspective for a long-debated subject regarding the mechanism of ABC transporters, specifically regarding the stoichiometry of ATP hydrolysis in ABC transporters. Our one-ATP-hydrolysis model is consistent with the fact that NBDs in the cystic fibrosis transmembrane conductance regulator (CFTR, a human chloride channel in the ABC superfamily) can bind two ATP molecules, but only one active site is hydrolytic.[23] The model can also explain the measured residual activity when mutations are introduced into one of the two catalytic sites in murine multi-drug resistant protein (MDR3, P-glycoprotein)[69] and bacterial histidine transporter (HisP).[70]

It is noteworthy that the dimer opening cannot be captured in these equilibrium MD simulations until typically 30–50 ns (Figure 10.2). Therefore, to draw any conclusions about post-hydrolysis dynamics of NBDs, MD simulations on the order of 50 ns or longer need to be employed.

Despite the greatly advanced structural knowledge of ABC transporter in the recent years, many mechanistically important aspects of the transport cycle remain elusive. Some of the open questions regarding the mechanism of ABC transporters include the dynamics of the TMDs and their structural transitions during transport and, more importantly, conformational coupling of the NBDs and the TMDs such that the energy harvested from NBD can mediate the substrate transport. With more available computational resources, it is now possible to expand our scope to the complete ABC transporters in the membrane environment and extend the simulation time into the microsecond time scale in the near future. With better conformational sampling with better described systems, it is hoped that some of these open questions can be answered using MD or combined experimental–theoretical approaches.

10.4   Ion-driven Neurotransmitter Uptake by the Glutamate Transporter

Glutamate transporters (GlTs) are membrane transporters in neurons and glial cells that remove the neurotransmitters glutamate and aspartate from the synapse using pre-existing ionic gradients as a source of energy.[71,72] Glutamate is a major excitatory neurotransmitter that plays critical roles in fundamental processes such as learning and memory.[73] In order to maintain recurrent and selective signaling, the neurotransmitter must be rapidly removed after its release into the synaptic cleft.[74,75] The GlT family includes five human EAAT (excitatory amino acid transporter) subtypes (EAAT1–EAAT5), two neutral amino acid transporters and a large number of bacterial amino acid and dicarboxylic acid transporters.[76,77] Malfunction of these transporters has been implicated in several neurodegenerative diseases, such as schizophrenia,[73] Alzheimer's disease,[77] Huntington's disease,[78] and parkinsonism dementia complex.[79] EAATs also provide one of the main targets for antipsychotic drug development.[80]

GlT belongs to the family of secondary membrane transporters, which couple 'uphill' translocation of the substrate across the membrane to the energetically favorable flow of ions down their concentration gradient. By coupling to the co-transport of three Na^+ ions and one H^+ ion and the counter-transport of one K^+ ion, GlT transports one substrate across the membrane during each transport cycle.[81–84] According to this stoichiometry, glutamate transport by GlT is an electrogenic process, meaning that it is associated with net charge transport across the membrane.

The X-ray crystal structure of an archaeal homolog (Glt_{ph})[85] marks the beginning of a new chapter in the study of the structure and mechanism of GlT. Glt_{ph} shares about 37% amino acid identity with the EAATs, suggesting that it can serve as a structural model for understanding transport in the EAATs. The structure reveals a bowl-shaped trimer, with a solvent-accessible extracellular basin extending half way across the membrane. Each GlT monomer is composed of eight transmembrane helices and two highly conserved helical hairpins (HP1 and HP2) forming a lumen for binding and transport of substrate and Na^+ ions (Figure 10.3A). In each monomer, the substrate binding site is cradled by the two helical hairpins, HP1 and HP2, reaching from the opposite sides of the membrane. The functional importance of HP1 and HP2 is supported by biochemical experiments on bacterial[86] and mammalian transporters.[87–91] In a recent study,[92] HP2 was proposed to serve as the extracellular gate that adopts an open conformation in the apo state, thus exposing the substrate binding site to the extracellular solution.

A large number of experimental studies[72,76,85–116] have investigated various structural and functional properties of GlTs and shown that substrate binding induces conformational changes in GlT. However, the limited spatial resolution of these studies made it difficult to draw specific conclusions about the nature and magnitude of such conformational changes. Fluorimetric measurements of conformational changes[100,101] and the study of the pre-steady-state kinetics[113] in GlT suggest that binding of H^+ precedes the binding of the substrate. Based on the X-ray structure,[85] Grewer and co-workers proposed a model for EAAT3 in which one Na^+ ion binds to the empty transporter before the substrate.[105] They also proposed that conformational changes take place in two glutamate-dependent half-cycles: glutamate-induced closing of an extracellular gate and the subsequent opening of an unknown cytoplasmic gate that allows glutamate dissociation and diffusion into the cytoplasm.[103]

Although numerous experiments have provided insightful information about the putative transport cycle of GlT (Figure 10.3B), the details of the mechanism of extracellular gating, the sequence of binding of the substrate and cotransported Na^+ ions and the role of these ions in transport remain elusive. In order to investigate the coupling between binding and translocation of the substrate and the ions, we performed a set of MD simulations of membrane-embedded trimeric models of GlT.[117] Different combinations of the substrate and the two structurally resolved Na^+ ions (Na1 and Na2)[92] were used to investigate equilibrium dynamics of GlT at different bound states and the interdependence of binding of Na^+ ions and the substrate.[117] The system was simulated under nine different conditions, each simulation lasting 20–30 ns.[117] The results revealed two highly relevant mechanistic details regarding the transport cycle in GlT.

Comparison of the dynamics of the substrate-bound and the substrate-free (apo) states of GlT in our simulations suggests that the helical hairpin HP2 plays the role of the extracellular gate. Invariably, in all the simulations performed in the presence of the substrate, HP2 has a very stable conformation (Figure 10.4A). After removing the substrate, however, HP2 undergoes a large

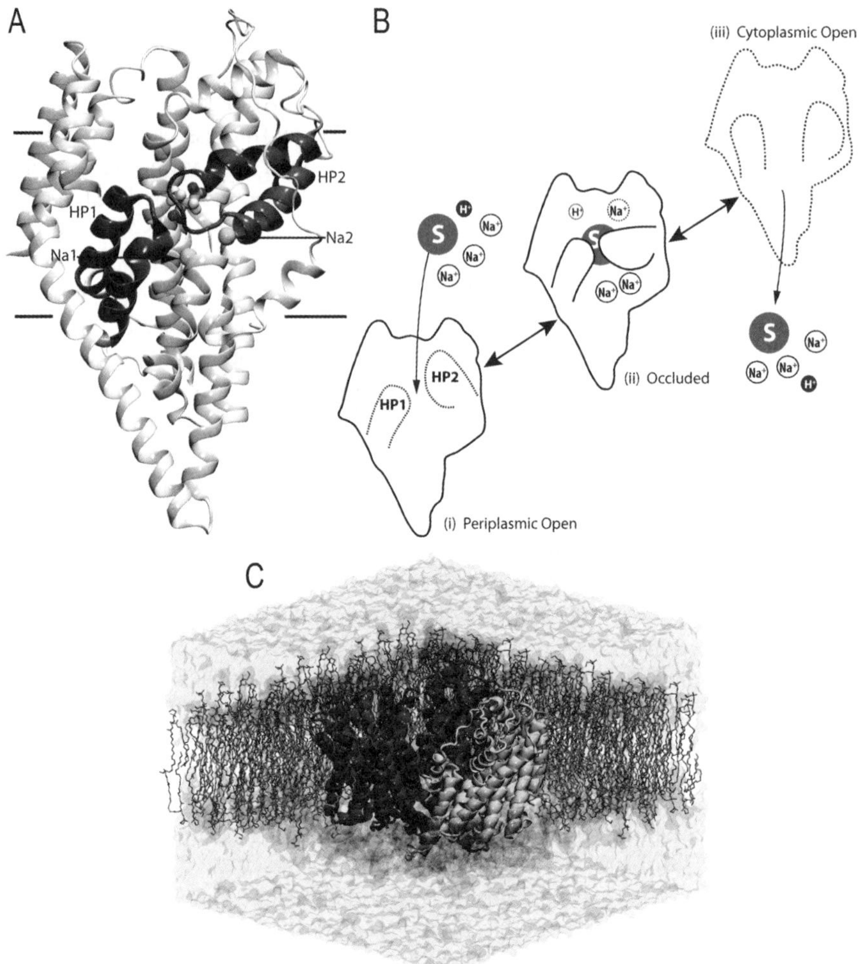

Figure 10.3   Structure and schematic transport cycle of GlT. (A) The structure of GlT monomer with bound substrate and two Na^+ ions. (B) Schematic transport cycle in GlT. (i) Apo state with HP2 in an open conformation. (ii) Three Na^+ ions, one H^+ ion and the substrate bind and induce closure of HP2, yielding the occluded state. (iii) Opening of a cytoplasmic gate (unknown) allows the release of three Na^+ ions, one H^+ ion and the substrate into the cytoplasm. The occluded state (ii) containing only substrate and two Na^+ ions is the only structurally known state of GlT. (C) The simulation system used in the study is composed of the bowl-shaped trimer of GlT embedded in a lipid bilayer and water.

opening motion, resulting in the complete exposure of the substrate-binding site to the extracellular solution (Figure 10.4A). Opening of the binding site is accompanied by its full hydration. These results suggest that HP2 plays the role of the extracellular gate and that, more importantly, its opening and closure of the gate is controlled by substrate binding.[117] A gating role for HP2 is supported by the structure of GlT in the presence of an inhibitor[92] and the results of rapid solution exchange and laser-pulse photolysis experiments.[105] Furthermore, recent inhibition studies in mutant EAAT2 using oxidative cross-linking of engineered cysteine pairs[118] suggest that HP2 serves as the extracellular gate of the transporter and that substrate induces distinct conformations of HP2. A recent MD simulation study[119] has also provided support for this idea.

Interestingly, despite its apparent structural symmetry to HP2, helical hairpin HP1 was found to exhibit a high level of conformational stability regardless of the presence of the substrate (Figure 10.4A).[117] This result, which we attribute to the shorter length of the loop of HP1 (compared with HP2), suggests that, at least during the extracellular half of the transport cycle, HP1 does not play a direct role and its involvement might be limited to stabilization of the structure of HP2 upon substrate binding.

Structural inspection of the trajectories shows that fluctuations of HP2 are confined mainly to its loop region (by loop region we refer to the polypeptide stretch of Gly351–Gly359 that connects the two helical regions of the hairpin structure), which contains four highly conserved glycines,[85] Gly351, Gly354, Gly357 and Gly359, and also five hydrophobic residues, Tyr352, Ala353, Val355, Pro356 and Ala358. The loop region has great flexibility, as described above. In the absence of the substrate, the loop region of HP2 has only one attractive interaction provided by the loop region of HP1 in the closed state. The attraction between the hydrophobic residues seems to be the main driving force that induces disordering the conformation of the flexible loop region of HP2. Therefore, after water molecules have disrupted the interaction between the tips of HP1 and HP2, the hydrophobic interactions disorder the conformation of the loop of HP2, resulting in the opening of the substrate binding site.

Substrate binding to GlT brings the HP2 and HP1 loops together, through establishing direct interactions between the charged groups of the substrate and the backbone groups of HP2. It should be noted that, upon substrate binding, only one half of HP2 (the Gly359 side) is completely sealed, a state that might be best characterized as a partially occluded state (Figure 10.4A). In this state, although the binding site is largely shielded from the extracellular region, water molecules can still move in and out of the binding pocket since the other half of HP2 (the Gly351 side) is not fully sealed (Figure 10.4B). Therefore, a complete occlusion of the binding site requires additional factors, likely binding of Na^+ ion(s) that will bring the extracellular gate to a completely closed state during the transport process (Figure 10.4B).

Another major consequence of substrate binding revealed by the simulations is the formation of a new Na^+ binding site.[117] In the crystal structure, one of the Na^+ ions (Na2) is bound to a binding site formed between two half-helical structures (HP2a and TM7a, see Figure 10.4A). In the apo state, the dipole moments of these half helices were found to be totally misaligned (Figure 10.4A). Upon substrate binding, the two opposing half-helices align such that

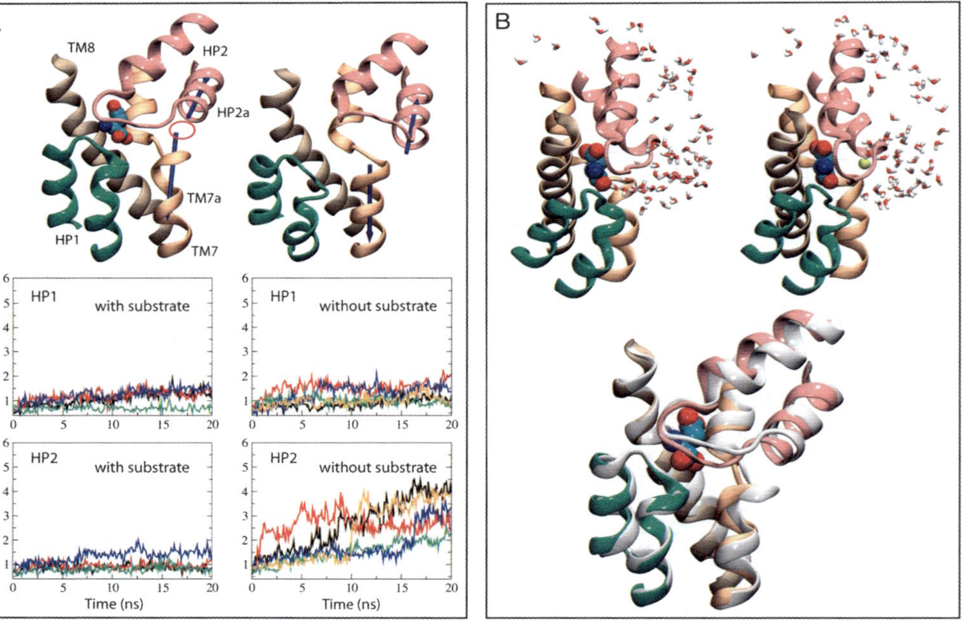

Figure 10.4 Dynamics of the extracellular gate in GlT. (A) Left and right panels show the results of the simulations performed in the presence and absence of the substrate, respectively. Left, top panel: substrate-bound state. Closing of HP2 (the extracellular gate) and formation of the Na2 binding site (marked with a circle; focusing of the helical dipole moments). Right, top panel: substrate-free state. Opening of HP2 and exposure of the binding site is evident. Note significant misalignment of the dipole moments of helices TM7a and HP2a (blue arrows). The middle and bottom panels show time evolution of the RMSDs of the helical hairpins HP1 and HP2. (B) Na2-induced formation of the occluded state. Binding of Na2 to GlT results in complete closure of the binding site to the extracellular solution. In the absence of Na2 (left), the binding pocket is accessible to water, whereas upon Na2 binding (right), the binding site is completely sealed. An overlay of the two states (white, before Na2 binding; colored, after Na2 binding) (bottom) highlights the small change in the conformation of HP2 upon Na2 binding.

their dipole moments converge on a single point resulting in the formation of the Na2 binding site (Figure 10.4A). These results have direct implications for the order and coupling of binding of Na^+ ions and the substrate; they strongly suggest that Na2 binding can only take place after binding of the substrate.

Na2 binding further stabilizes HP2, resulting in a completely occluded form of GlT, in which water molecules (and, therefore, H^+ and Na^+ ions) can no longer access the binding site from the extracellular side (Figure 10.4B). Thus, Na2 binding results in a complete closure of the extracellular gate. Our results are strongly supported by various experiments, including crystallographic and thermodynamic studies,[92] determination of the steady- and pre-steady kinetics[106] and measurements of transporter currents associated with stoichiometric and anion charge movements in GlT,[120] which have suggested that substrate binding permits the binding of one of the co-transported Na^+ ions.

Although we have investigated the extracellular gating mechanism and the coupling between substrate and one of Na^+ ions, the mechanism of transport in GlT consists of a large number of steps whose molecular details and sequence are largely unknown (Figure 10.3B), *e.g.* the sequence of binding and coupling of ions and the substrate and associated protein conformational changes and the cytoplasmic pathway for substrate release and the cytoplasmic gating mechanism. Further investigations are needed to elucidate the entire mechanism of transport in GlT.

10.5 Substrate Binding and Selectivity in Glycerol-3-Phosphate Transporter

Transporters of the major facilitator superfamily (MFS) constitute the largest group of secondary membrane transporters, ubiquitously found in all three kingdoms of life. More than 15 000 members of MFS have been sequenced and organized into various families, including a large number of medically and pharmaceutically highly relevant transporters.[121–129] Members of MFS, typically, consist of a single 400–600 aa polypeptide which folds into either 12, 14 or 24 transmembrane α-helices with both the N- and C-termini on the cytoplasmic side.[121,130,131] Structurally, they are characterized by similar transmembrane topologies, and also signature sequences in two of their cytosolic loops. Transmembrane helices are organized into two functionally relevant helix bundles, referred to as the N- and the C-terminal halves, respectively, which are connected to each other by a central loop of low sequence homology (Figure 10.5B).[129] Despite the importance of this family, X-ray structures have been resolved for only three members: glycerol-3-phosphate transporter (GlpT), lactose permease (LacY) and multidrug transporter EmrD.[121,132–134]

Transport of L-glycerol-3-phosphate (G3P) across the inner membrane in *E. coli* is mediated by G3P transporter (GlpT), which is the main focus of this section. G3P plays an essential role in the cell both structurally by incorporation into phospholipids and cardiolipin and as a nutrient entering

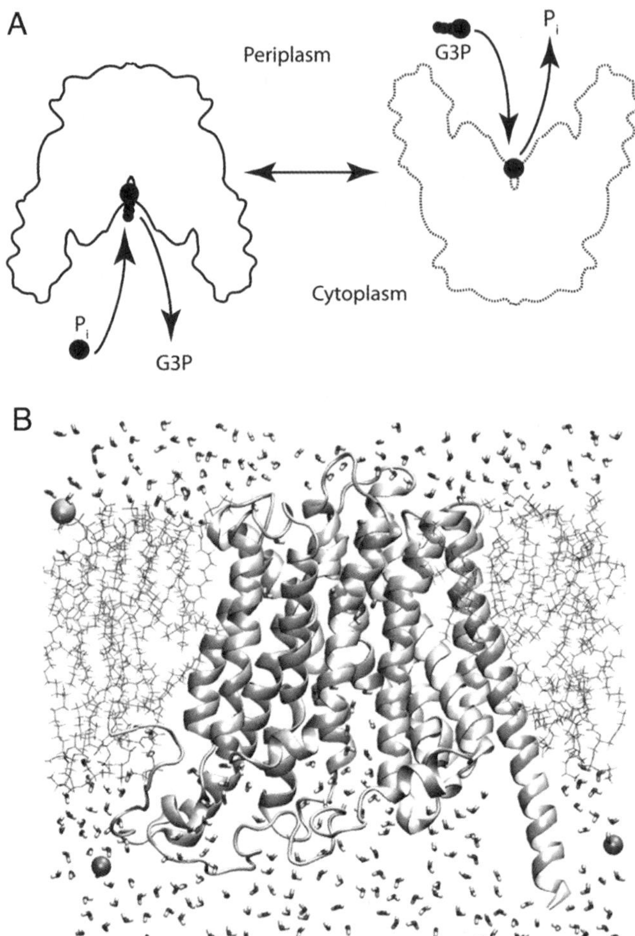

Figure 10.5   (A) Schematic mechanism of GlpT: binding of the substrate, P$_i$, from the cytoplasmic side, induces a series of conformational changes that lead to opening of the lumen to the periplasm. Release of P$_i$ and binding of G3P triggers the recovery of the initial state. (B) The simulation system after 5 ns equilibration: the protein is embedded into POPE membrane, solvated with water and ions. The system is simulated for 50 ns in the apo state and with P$_i^-$, P$_i^{2-}$, G3P$^-$, G3P^{2-}.

glycolysis.[123] GlpT belongs to the organophosphate/phosphate antiporters (OPA) subfamily of MFS.[125,135] A human homolog of GlpT, glycerol-3-phosphate permease (G3PP), mediates mitochondrial import of G3P and is involved in oxidative phosphorylation and lipid biosynthesis.[123,124] GlpT attracts attention not only due to its role in the transport of G3P and antibiotic resistance, but also as a structural model to study other MFS proteins; it has been used in homology modeling of a large number of other MFS members, *e.g.* human glucose transporter (GLUT1) and glucose-6-phosphate transporter (G6PT).[136,137]

GlpT is composed of two halves, each with six helices that are related by a pseudo-two-fold structural symmetry. The X-ray structure of GlpT shows a wide open cytoplasmic side gradually narrowing towards the periplasmic side (Figure 10.5B).[121,123,138] A long cytoplasmic loop connects the N- and C-terminal halves. The lumen formed between the two halves provides the substrate translocation pathway (Figure 10.5B). Two arginine residues (Arg45 and Arg269), one on each half, confer strong positive charges towards the closed end, representing the binding site, although the lumen is mostly composed of neutral residues. A conserved histidine (His165) is located between Arg45 and Arg269 and is suggested to play a role in substrate binding (Figure 10.6A).[129,138] Another charged residue lining the lumen, which is likely to be relevant to substrate binding, is Lys80. Mutation studies on Arg45, Arg269, Lys80 and His165 identified them as functionally indispensable residues for substrate binding and transport, whereas other arginines can be mutated to lysines in a close homolog of GlpT, glucose-6-phosphate transporter (UhpT), without loss of function.[121,123,125,129,138–141]

GlpT functions as a monomer under physiological conditions *in vitro*.[121,139] Based on structural and kinetic studies, an alternating-access/rocker-switch mechanism has been proposed (Figure 10.5A).[121,123,125] Phosphate (P_i) binding to the 'cytoplasmic-open' state results in closing of the cytoplasmic side and opening of the periplasmic side ('periplasmic-open' state). Replacement of P_i by G3P in the latter state restores the cytoplasmic-open state. Occluded states are likely to be involved during the transition between the cytoplasmic-open and periplasmic-open states.[121,123,125]

The rocker-switch model requires that the activation energy for the interconversion of the two states is overcome by thermal energy, a hypothesis that is supported by kinetic studies performed by Wang and co-workers, who showed that the interconversion of the states is rate limiting and temperature dependent (likely due to involvement of large protein conformational changes), whereas substrate binding is temperature independent and fast.[128] The alternating access mechanism has been proposed for many other MFS proteins such as UhpT, LacY and GLUT1.[128,139,142]

We performed several 50 ns MD simulations of membrane-embedded GlpT with two substrates (G3P and P_i) at all physiologically relevant protonation states (P_i^-, P_i^{2-}, $G3P^-$, $G3P^{2-}$), and also in the apo state, in order to characterize the unknown binding site and to probe substrate-induced protein conformational changes in detail. In all the simulations, the substrate was initially placed close to the mouth of the lumen in the beginning of the simulation (~ 15 Å away from the binding site). All these unbiased simulations revealed spontaneous binding of the substrate through a common pathway (Figure 10.6A). We also performed several shorter simulations (on the order of ~ 10 ns each) with the fully charged form of P_i, *i.e.* P_i^{3-}, which is known not to be transported by GlpT. These simulations all resulted in substrate diffusing out of the lumen. This indicates that although electrostatic interaction between

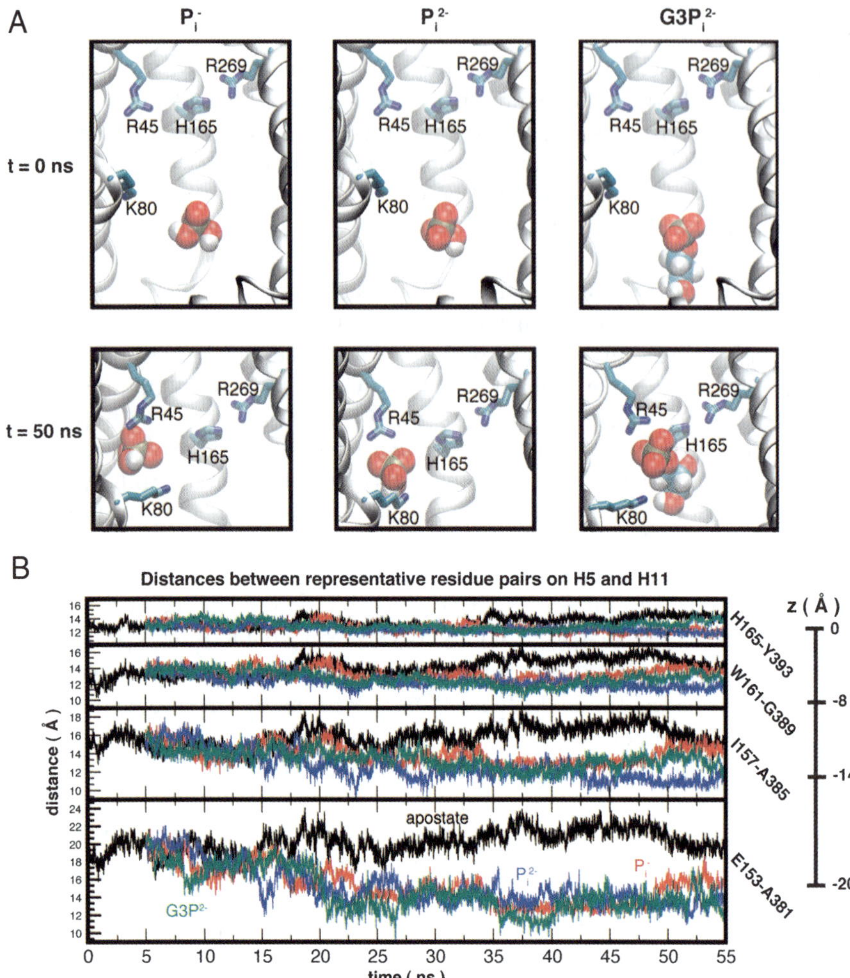

Figure 10.6 (A) Spontaneous binding of GlpT. Initial ($t = 0$ ns) and final ($t = 50$ ns) positions of the substrate in binding simulations for P_i^-, P_i^{2-} and $G3P^{2-}$ systems are shown. Arg45, Lys80, His165 and Arg269, which play direct roles in substrate binding are shown. (B) Substrate-induced closure of the cytoplasmic half of GlpT is shown using the distance change between the C_a of opposing residue pairs on helices H5 and H11. The scale on the right shows approximate positions of the residue pairs along the membrane normal with the mid-point of the membrane corresponding to $z = 0$.

P_i^{3-} and the protein is expected to be stronger than P_i^-, P_i^{2-}, it is not sufficient to break apart the solvation shell of P_i^{3-} and overcome its electrostatic interaction with the solvent.

In the following, we describe the results of simulations of P_i^-, P_i^{2-} and $G3P^{2-}$, which exhibited stable binding. Unlike these species, $G3P^-$ did not establish a stable bound state within the time scale of the simulation, an observation which is either related to the short time scale of the simulation (50 ns) or indicative of irrelevance of this species ($G3P^-$) to the transport cycle.

Our MD simulations capture the differences and similarities between the binding of different substrates and provide insight into how substrate specificity is achieved in GlpT. In all binding simulations, after initial interaction with Lys80, the substrate stably binds to Arg45 through establishing strong hydrogen bonds. Lys80 seems to be more important during initial binding as it steers the substrate deeper into the lumen towards its final binding site. Mutation of this residue is known to lead to a significant increase in K_d.[129] Similarly, simulations in which Lys80 was neutralized resulted in diffusion of the substrate (P_i^{2-}) out of the lumen, indicating that the electrostatic interaction of the substrate with Lys80 constitutes an important step in binding of the substrate to GlpT.

Arg45 and Arg269 also play important roles in substrate association and binding. In concordance with experimental mutation studies, simulations performed with neutralized Arg45 and Arg269 also impaired binding; in these simulations, the substrate either left the lumen or stayed there without proceeding toward the binding site.[129] Arg45 is identified to be the stable binding site and the final destination of the substrate. The guadinium group of this residue makes stable hydrogen bonds with the phosphate moiety of the substrate in all cases. The role of arginine has long been recognized as a paradigm in coordination and recognition of phosphate moieties by providing perfect geometry for hydrogen bond formation ('arginine fork').[143]

Coordination of the substrate in the bound state also involves His165. Close positioning of the charged substrate can change the protonation state of His165, which is most likely neutral in the absence of substrate. A substrate-induced change of the protonation of His165 might be an important step in the mechanism in GlpT.[129]

Arg269 also has been suspected to be a part of the substrate binding site.[129] It is symmetrically positioned to Arg45 in the N-terminal domain. However, in our simulations, none of the substrates established direct interaction with Arg269. This observation suggests that Arg45 and Arg269 play different roles in binding despite their near perfect structural symmetry.[121,123,138] Although we cannot rule out the role of Arg269 in later stages of the transport cycle, it is clear that it is not a part of the initial binding site. Arg269 might also contribute to the overall positive potential of the GlpT lumen, facilitating the initial recruitment of the substrate into the lumen.

Spontaneous binding of substrates to GlpT within a few nanoseconds, as observed in our simulations, is most likely due to the localized electrostatic potential. The potential is found to be strongest (0.76 V) near Arg45 and is primarily due to the sum effects of Lys80, Arg45, Arg269 and Lys46. As mentioned above, neutralization of either Lys80, Arg45 or Arg269 impairs binding in our simulations in agreement with the experiments.[129] However, the effects of Arg269 and Lys46 can be partially subdued by Glu299 and Asp274, with which they form salt bridges. Our simulations also reveal partial closure of the cytoplasmic side upon substrate binding, consistent with the conformational changes required for the rocker-switch model. Substrate binding results in bending of the cytoplasmic halves of two helices, H5 and H11, which face each other across the lumen (Figure 10.6B). Invariably, in all of our binding simulations, substrate binding results in a reduction of the distances between the cytoplasmic half of these helices. The bending starts just below the plane of the substrate binding site and propagates to the cytoplasmic tips of the helices, at which point it exceeds 10 Å. No such bending is observed in the apo state protein, clearly indicating that substrate binding triggers the observed protein conformational changes. We note that although the observed bending is not sufficient to fully seal the cytoplasmic side within the time scale of our simulations, it likely represents the initial step in the formation of the occluded state, in which both sides of GlpT are closed. Hence our results not only support the rocker-switch model, but also capture the initial steps involved in the mechanism.

10.6 Membrane Potential-driven Nucleotide Exchange in ADP/ATP Carrier

In eukaryotic organisms, ATP is produced inside the mitochondria from ADP and inorganic phosphate (P_i). Maintaining a normal ATP concentration in the cytoplasm, therefore, requires continuous import of ADP into the mitochondria and export of ATP to the cytosol. The exchange of ADP and ATP across the inner mitochondrial membrane is achieved by the ADP/ATP carrier (AAC). When mitochondria are actively respiring, ADP and ATP molecules are exchanged on a strict 1-to-1 stoichiometry by AAC.[144] In each transport cycle, therefore, one negative charge is extruded from the mitochondrial matrix to the cytosol.[145] Mechanistically, it has been proposed that AAC undergoes transitions between two functional states involving large conformational changes (Figure 10.7A): the cytosolic-open state (c-state), in which ADP can bind from the cytoplasm and the matrix-open state (m-state), in which ADP is released to and ATP is taken up from the mitochondrial matrix.[146–150] Specific substrate binding to one state is believed to trigger the transition of AAC to the other state.

Despite considerable biochemical and functional characterization of AAC,[151–159] we have only just begun to understand the mechanism of this nucleotide transporter, primarily due to lack of sufficient structural information. The only available AAC structure is that of the c-state in complex with an inhibitor carboxyatractyloside (CATR), solved using X-ray crystallography at a resolution of 2.2 Å.[150] As shown in Figure 10.7B, six transmembrane helices (H1–H6) form the AAC basin, with a cone-shaped depression accessible from the cytosolic side. In the matrix half, three highly conserved proline residues introduce sharp helical kinks, bending the six helices towards the center, thus completely closing the lumen to the matrix side. The crystal structure of AAC solved a number of long-standing questions regarding the mechanism of the

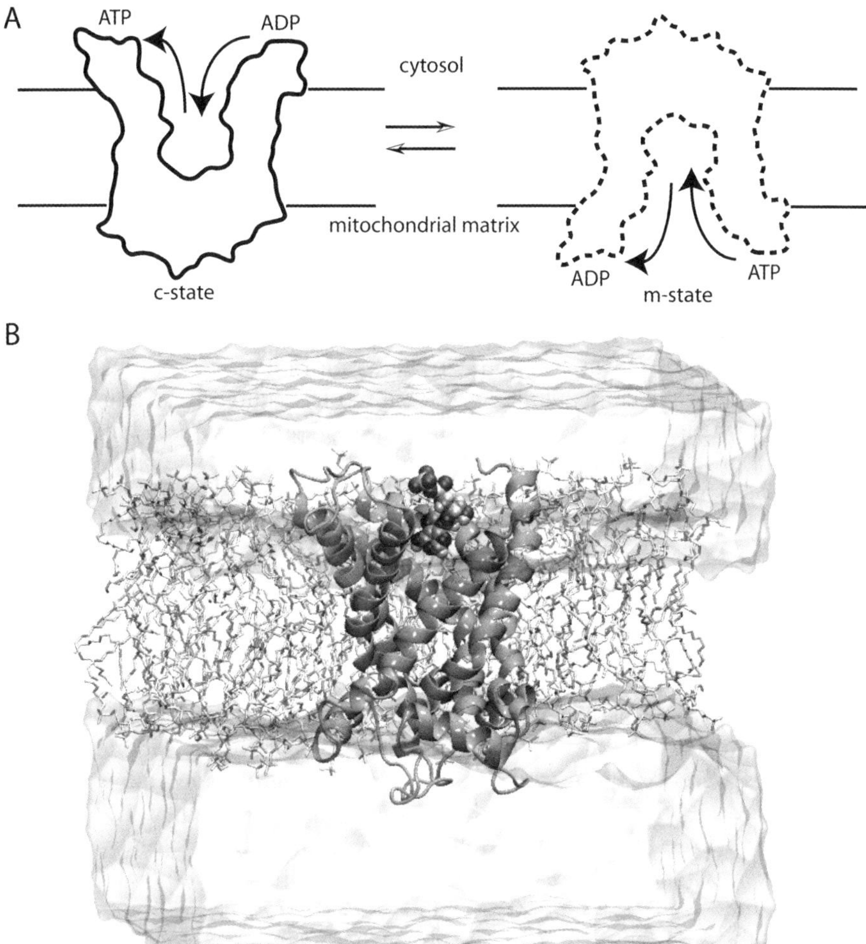

Figure 10.7 ADP–ATP exchange mediated by AAC. (A) Schematic representation of nucleotide translocation through AAC. ADP binds to the c-state AAC from the cytosol and is released to the mitochondrial matrix upon conversion of the protein to the m-state. ATP binds to the m-state AAC from the mitochondrial matrix and is released to the cytosol. (B) Side view of the AAC simulation system. In order to probe the nucleotide binding process, an ADP molecule is added to the cytosolic side of the protein at the beginning of the simulation. In three independent binding simulations, the phosphate groups of ADP point toward the cytosolic side, *i.e.* away from the protein. In another simulation, the phosphate groups of ADP point toward the matrix side, *i.e.* towards the protein (not shown).

carrier. Most importantly, it provided strong support for substrate transport through AAC monomers, ruling out a long-standing suspicion that nucleotides can only cross the membrane through a pathway provided by a dimeric form of AAC.[150] However, since the crystal structure is in complex with the inhibitor

CATR and not the physiological substrate ADP, it is not clear where the ADP binding site inside the AAC is located. Furthermore, little is known about the dynamic process of nucleotide translocation through AAC, or the conformational changes that are involved in this process (Figure 10.7A).

In order to probe the process of ADP binding to the c-state AAC, we performed four independent nucleotide binding simulations, in which an ADP molecule is initially placed on the cytosolic side of AAC (Figure 10.7B).[7] During these equilibrium simulations, ADP molecules are found to move spontaneously towards the bottom of the AAC lumen over a distance of more than 20 Å, penetrating deeply into the lumen within only a few nanoseconds. Regardless of their initial orientations, all AAC-bound ADP molecules converge into similar final orientations, in which phosphate groups are pointing down towards the matrix half and inserted into the binding site (Figure 10.8A). To our knowledge, these results are the first to report complete binding of a ligand to a protein described using unbiased simulations.[7] The observed rapid, spontaneous binding of ADP to AAC demonstrates a strong attraction between the nucleotide and the protein. The time-averaged electrostatic potential of the system calculated from an ADP-free simulation (apo-AAC) reveals a strong positive potential of $\sim 1.4\,V$ at the bottom of the AAC vestibule (Figure 10.8B). Such a strong potential, which has not been observed in any other membrane transport proteins, might be important for the carrier to recruit ADP against a strong transmembrane potential ($-0.2\,V$) across the inner mitochondrial membrane.[7]

In order to identify the ADP binding site, two of the nucleotide binding simulations were extended to 0.1 ms each.[7] These simulations allowed us to identify the ADP binding site buried deeply inside the AAC lumen. As shown in Figure 10.8C, both simulations resulted in the same binding site for the ADP phosphate groups, which form simultaneous, multiple salt bridges with four basic residues, Lys22, Arg79, Arg279 and Arg235. Two major binding modes are found for the adenine ring, in one of which a stacking interaction is formed between ADP and the aromatic residue Tyr186 (Figure 10.8C). The ADP binding site revealed by our simulations is supported by conservation of residues Lys22, Arg79, Arg279, Tyr186 and Arg235 among AACs of different species[144,150] and experimental results indicating that mutation of these residues severely impairs the transport activity of AAC.[150,160,161]

The binding of ADP has been suggested to perturb the salt bridge network at the bottom of AAC lumen, which in turn triggers large conformational changes of the carrier.[150,162] Our simulations provided a detailed picture of molecular events that initiate such a process.[7] As shown in Figure 10.8D–F, the phosphate groups of ADP dramatically alter the locations of binding site residues and interfere with the original salt bridge network at the bottom of AAC lumen. Such nucleotide binding induced conformational changes have been suspected[150,162] to represent the first step of the carrier's transition to the m-state. In order to probe further structural changes that might be involved in the activation of the carrier, ADP was forced to penetrate deeper into AAC beyond its binding pocket by two independent steered MD (SMD) simulations.[7]

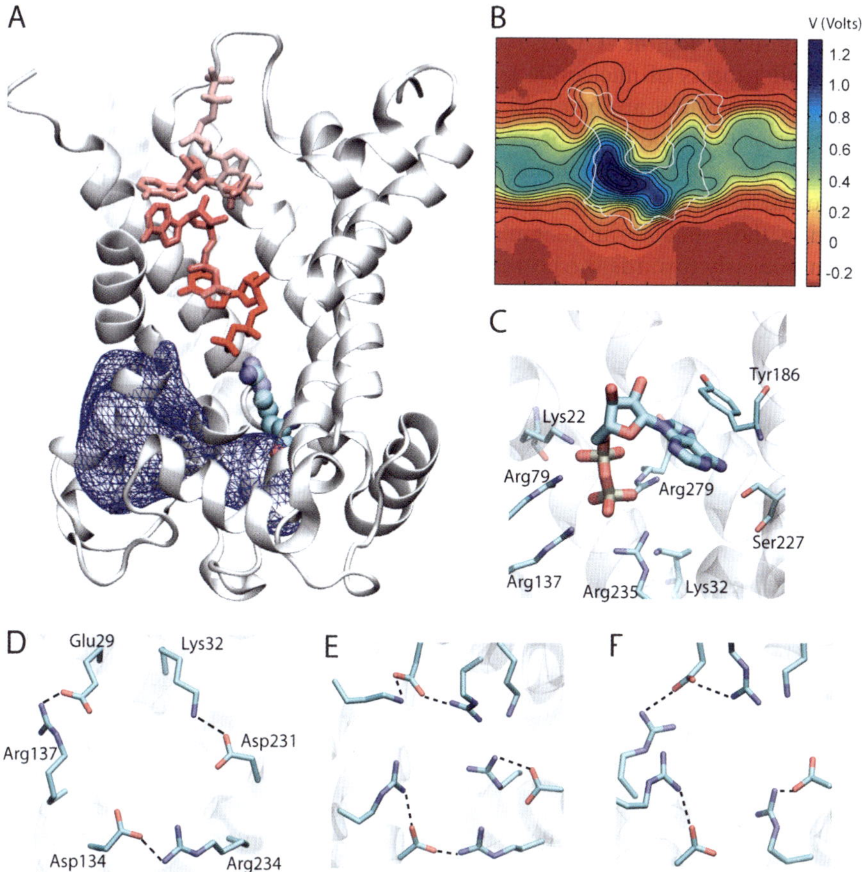

Figure 10.8　Spontaneous binding and induced translocation of ADP revealed by MD simulations. (A) Spontaneous ADP binding revealed by nucleotide binding simulations. ADP molecules at $t = 0$, 0.7, 3.2 and 100.0 ns are colored from light to dark red, respectively. The 1.0 V electrostatic potential isosurface is shown in a blue mesh. (B) Electrostatic potential map of AAC. A strong (~ 1.4 V) positive potential is located in the lumen of the protein and provides the driving force for ADP binding. (C) ADP binding pocket revealed by simulations at $t = 100$ ns. (D–F) Conformational changes of AAC during induced ADP translocation. The highly conserved salt bridge ring in AAC (D) undergoes significant rearrangement during two SMD simulations (E, F) in which ADP is induced to cross the membrane.

Although these simulations are clearly not describing the actual transport process in its entirety, they can provide information on certain aspects of protein conformational changes that likely accompany the nucleotide translocation. The results reveal that ADP translocation involves further rearrangement of the salt bridge network (Figure 8D). During this process, 'interhelical' salt bridges connecting the odd-numbered helices in the apo-AAC are replaced by either 'intrahelical' salt bridges or salt bridges between even-numbered and odd-numbered helices (Figure 10.8E and F). This rearrangement 'unlocks' the three odd-numbered helices, allowing the protein to expand in the matrix half, thus allowing the translocation of ADP.[7]

AAC belongs to the mitochondrial carrier family (MCF), which includes most of solute carriers in the inner mitochondrial membrane.[146–150] The distinct electrostatic features of AAC and their direct role in substrate binding and translocation prompted us to examine other MCF members in order to determine whether these features are shared by the entire MCF family. We calculated total net charges for all known 34 yeast *Saccharomyces cerevisiae* MCF members.[7] On average, a net charge of $+15e$ is found for 31 MCF members, whose substrates are either anionic or zwitterionic. Repeating the calculation for 1066 yeast membrane proteins results in an average net charge of essentially zero ($+0.3e$) and a net charge density distribution (see Section 10.2.2) distinct from that of MCF members. These results indicate that a large positive net charge is not an attribute shared by membrane proteins in general, but a distinct feature special to the MCF family.[7] As the inner mitochondrial membrane includes a significant fraction of negative lipids, the surface component of the positive charges carried by MCF members might enhance or even mediate partitioning the protein into the mitochondrial membrane. More important, however, is the strong luminal component of the positive electrostatic potential, which exhibits a profound effect on the dynamics of substrate recruitment and translocation in AAC, an attribute likely shared by other members of the MCF.[7]

10.7 Mechanically Driven Transport Across the Outer Membrane

Outer membrane (OM) transporters defy the conventional model of transport presented in previous sections for other transporters, in which energy is used locally to effect substrate transport. Because the OM of Gram-negative bacteria is porous, it is unable to maintain a membrane potential or to contain chemical energy such as ATP. Nonetheless, there are a number of substrates outside the cell that need to be actively imported. These substrates are often too large (>600 Da) to diffuse through porins and, furthermore, exist at vanishingly low concentrations.[163,164] To resolve this difficulty, bacteria have evolved unique multi-membrane systems that can couple energy generated at the cytoplasmic membrane (CM) to transport at the OM. One example in particular, TonB-dependent transporters (TBDTs), require interaction with a CM-bound protein, TonB, which spans the periplasmic space between membranes, to transport their substrates (see Figure 10.9A).[165–168] TonB further interacts with two additional CM proteins, ExbB and ExbD, which utilize the CM inner membrane proton motive force as a source of energy, transmitted to TonB.[169–171] This part of the process is still poorly understood, largely because there are almost no structural data for the CM components.[172,173]

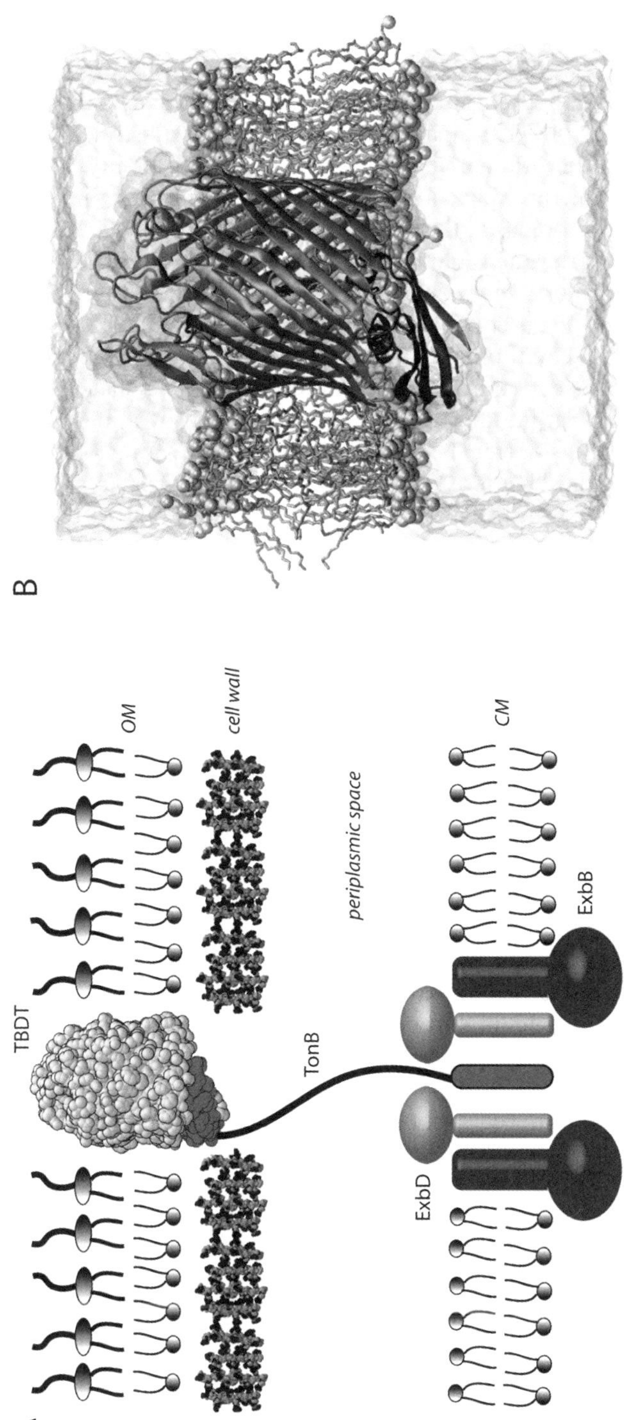

Figure 10.9 Active transport at the outer membrane. (A) Schematic of TonB-dependent transport. The TBDT, C-terminal domain of TonB and cell wall are structurally known;[192,206] ExbB and ExbD are not. The placement of ExbB and ExbD in the cytoplasm or periplasm is accurate, although the ratio of each to TonB is not (the stoichiometry is estimated to be 1 TonB : 2 ExbD : 7 ExbB[207]). The periplasmic width is not to scale. (B) System used for SMD simulations. The β-barrel structure of BtuB is illustrated in dark gray, with the C-terminal domain of TonB in black. The membrane and water are also shown.

In contrast, there is a significant amount of structural data for the OM counterpart, *i.e.* TBDTs, in both free and substrate-bound forms.[174–181] Known substrates of TBDTs are restricted to various iron complexes and cobalamins (*e.g.* vitamin B_{12}), although there might be many others as yet undiscovered.[182] Each TBDT is highly specific for a single substrate, binding it on the extracellular side with nanomolar or sub-nanomolar affinity.[184] Despite this specificity, the structures have a number of features in common. In particular, they all have two domains, the C-terminal domain forming a 22-stranded β-barrel (typical of OM proteins) and the N-terminal domain folding back into the barrel.[168,172] The interior domain completely occludes the barrel and thus is sometimes referred to as a 'cork' or 'plug' domain; alternatively, it has also been called the 'luminal' domain, a name which implies less about the mechanisms involved (see Figure 10.10A). Three extracellular-facing loops in the luminal domain form the substrate binding site, aided by a number of the extracellular loops of the barrel. The TonB interaction site, a region at the N-terminus known as the 'Ton-box', is found on the periplasmic side of the TBDT.[165,184–187] In the transporter's empty state, the Ton-box typically resides packed into the barrel; however, upon substrate binding, it undergoes a transition to a disordered state, presumably signaling occupancy to a waiting TonB in the periplasm.[188–190] The specific nature of the TonB–Ton-box interaction was resolved in two structures, one of the iron siderophore transporter FhuA in complex with the C-terminal domain of TonB and the other of the cobalamin transporter BtuB also in complex with the C-terminal domain of TonB, revealing a β-sheet interaction with three strands contributed by TonB and one strand by the Ton-box.[191,192] However, as the transporter is still completely closed in these structures, both the substrate translocation pathway and how TonB induces opening of the TBDT remain unknown.

Multiple models for the interaction of TonB and the TBDT and also for TBDT opening have been put forward. In one, referred to as the 'shuttle model', TonB is energized by its interactions with ExbB and ExbD; it then leaves the CM completely to cross the periplasm and impart the energy to the TBDT.[193–195] Alternatively, a 'pulling model' suggests that TonB remains anchored in the CM and induces transport through a mechanical interaction with the TBDT.[172,192,196] Recent experiments, in which GFP was covalently linked to the N-terminus of TonB in the cytoplasm but did not inhibit function, indicate that TonB spans the entire periplasm when interacting with TBDTs, in support of the pulling model.[190] Different hypotheses for the conformational response of the TBDT to interaction with TonB, whether mechanical or otherwise, have also been made, including conformational change of the luminal domain within the barrel, unfolding of the luminal domain, removal as a rigid 'plug' or, more likely, some combination of all three.[172,183,197] MD simulations of FhuA on the 10 ns time scale did not display large fluctuations of the luminal domain, but did illustrate many water molecules 'lubricating' the interface between the luminal domain and the barrel.[198] More recently, experiments have shown that residues in the luminal domain become exposed to solvent during transport, but residues inside the barrel do not.[197] Both

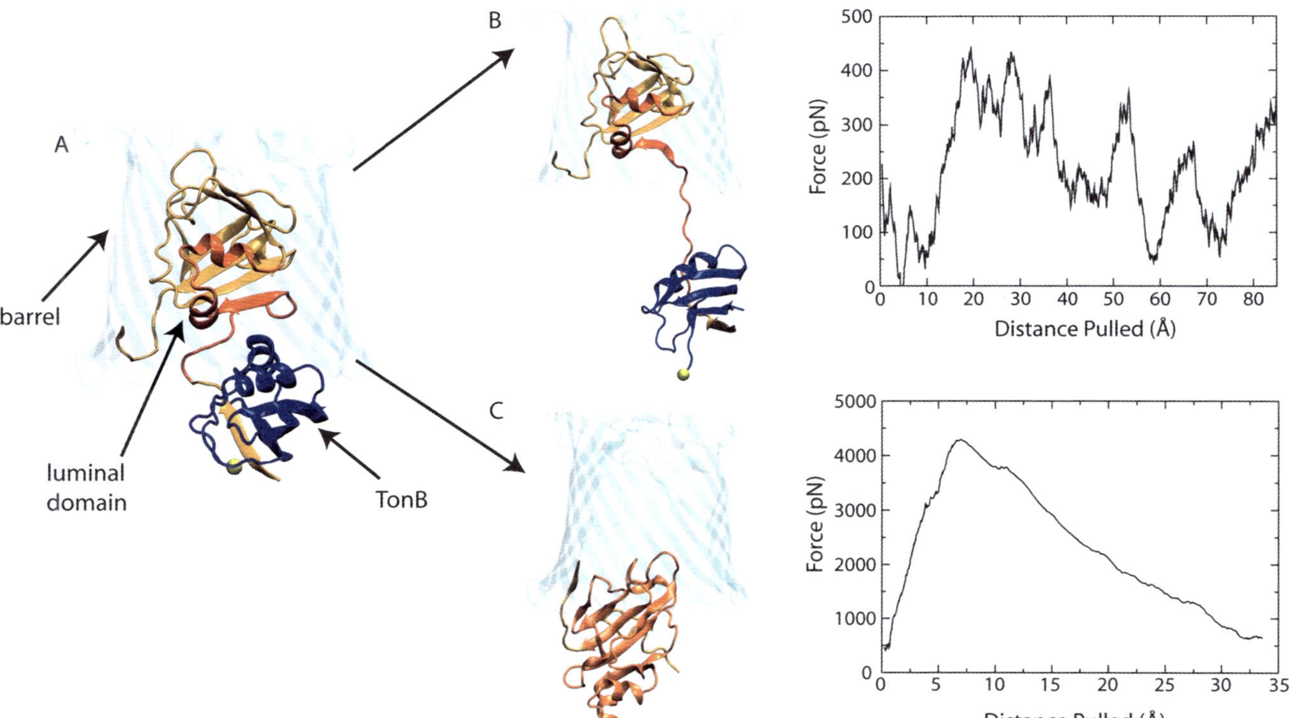

Figure 10.10 Two mechanisms for removal of the luminal domain. The barrel of BtuB is shown in a transparent, light blue representation with the front face cut away to expose the interior luminal domain in orange. (A) Crystallographic state. TonB is shown in blue on the periplasmic side. (B) Pulling on TonB. The yellow sphere of TonB represents the point of force application. The unfolded portion of the luminal domain is indicated in dark orange. The force required as a function of distance pulled is also shown. (C) Unplugging of the luminal domain. Here, force was applied to the center-of-mass of the luminal domain, causing it to come out as a singular unit, but with a significantly greater difficulty compared with unfolding in (B), evidenced by the much larger force required.

results suggest that the luminal domain leaves the barrel, at least partially, upon activation of the TBDT.

We performed a number of equilibrium MD and SMD simulations to investigate further the interaction between TonB and the TBDT BtuB; the simulation system is shown in Figure 10.9A.[196] We focused on the proposed mechanical mode of activation in which force is transduced by TonB to BtuB. In the simulations, the N-terminus of the portion of TonB co-crystallized with the transporter (see Figure 10.10B) was pulled at a constant velocity away from BtuB, towards the CM. Three pulling speeds, 2.5, 5 and 10 Å ns^{-1}, were used in order to confirm the results are reproducible and velocity independent. It was found that the connection between BtuB and TonB, consisting of 10–15 hydrogen bonds, is strong enough to transmit force throughout all simulations. This connection transmits forces of up to almost 1000 pN to the luminal domain of BtuB. The strength of the connection can be, at least partly, ascribed to the arrangement of the β-sheet between TonB and BtuB, parallel to the direction of pulling. This arrangement requires shearing of the β-strands, breaking all hydrogen bonds simultaneously, in order to disrupt the connection between the two proteins. This is in contrast to a perpendicular arrangement of the β-sheet, where the strands can be 'unzipped' by pulling, *i.e.* breaking hydrogen bonds sequentially. The orientation and stability of the complex is also aided by a salt bridge between a pair of conserved residues, Arg158 in TonB and Asp6 in BtuB, which persists during simulated pulling at 2.5 Å ns^{-1}, although not completely at faster speeds.

In order to effect transport, pulling on TonB must induce a conformational change in BtuB's luminal domain. The luminal domain, composed of around 110 residues, has a core β-sheet oriented parallel to the plane of the membrane and thus perpendicular to our pulling direction. Indeed, during simulated pulling, the luminal domain begins to unfold, the first strand of the β sheet coming unzipped (see Figure 10.10B).[196] Unfolding of the luminal domain is supported by cysteine cross-linking studies, which demonstrated that a disulfide bridge formed between the N-terminal region of the luminal domain (specifically residue 27 in FhuA) and the barrel prevents transport but one 50–80 residues downstream does not.[199–201] At 2.5 Å ns^{-1}, forces measured are less than 450 pN, shown in Figure 10.10B, comparable to simulated unfolding of similar proteins.[202–204] We continued unfolding up to nearly 100 Å at the slowest speed and to over 200 Å at the fastest. At a pulling distance of 215 Å, the opening formed in the barrel is large enough to accommodate the substrate, cyanocobalamin. As the width of the periplasm is between 180 and 250 Å in *E. coli*, 215 Å of pulling is not out of the question.[205] However, it is likely that at slower pulling speeds, unfolding will be accompanied by additional conformational change, making such a long pulling distance unnecessary.

We also examined an alternative to unfolding, namely removal of the luminal domain as a single unit.[196] In this simulation, TonB was removed from the system and, instead, force was applied directly to the center-of-mass of the luminal domain. We found that, at the same pulling speed (2.5 Å ns^{-1}), the force required to remove the luminal domain intact is 10 times that needed for unfolding (see Figure 10.10C). Hence unfolding, at least initially, is the more likely response of the luminal domain to an applied force. Further supporting the requirement of conformational change in the core β-sheet, when two adjacent β-strands are disulfide bonded, transport is severely reduced.[201] Nonetheless, further research is still needed to determine the precise conformational changes that the luminal domain undergoes during transport. How the interactions of ExbB, ExbD and TonB can generate a mechanical force at the CM is even less clear, although additional structural data will likely provide great insight into such mechanisms.

10.8 Conclusion

In this chapter, we have presented the results of large-scale MD simulations performed on a number of active membrane transporters investigating the mechanism(s) of energy transduction in this important family of membrane proteins. Active transporters are molecular machines specialized in coupling of diverse sources of free energy in the cell to efficient transport of molecules across the membrane, often against their electrochemical gradient. Common energy sources driving the transport cycle in membrane transporters include ATP binding and hydrolysis, electric membrane potential, proton motive force and electrochemical gradients of other ionic species. In order to harvest these sources of energy, and, more importantly, to couple them to substrate transport, transporter proteins need to interact intimately and specifically with more than one, often several, species and to be able to respond to their binding and stepwise translocation in a highly coordinated manner. As such, a detailed understanding of the transport cycle requires an atomic resolution description of the dynamics of the transporter and its interaction with various bound species.

The simulations described in this chapter demonstrate several successful applications of atomic MD simulations in membrane transporters. Although describing the entire transport cycle is still beyond the scope of such simulations, we have shown that they can successfully capture key *steps* involved in the transport cycle of a mechanistically diverse group of membrane transporters, *e.g.* the coupling and sequence of binding of the substrate and ions, the molecular nature of gates and the mechanism by which ATP hydrolysis results in domain conformational changes.

A key attribute of the simulations presented in this chapter is their rather extended time scale. Successful characterization of many of the molecular events reported in this chapter relied on simulations of at least several tens of nanoseconds of the entire transporter within its natural environment. Our ability to perform readily such long simulations continues to increase drastically the scope and success of molecular simulation in capturing more relevant biological events. The simulation studies of membrane transporters presented in this chapter are only a few examples of a fast-growing number of successful applications of extended simulations in describing biologically relevant phenomena.

Acknowledgements

This work was supported by grants from the NIH (R01-GM086749, R01-GM067887, R01-GM079800 and P41-RR05969). The authors acknowledge computer time at TeraGrid resources (grant number MCA06N060), specially on the BigRed cluster at Indiana University and the Abe cluster at NCSA, and also computer time from the DoD High Performance Computing Modernization Program at the Arctic Region Supercomputing Center, University of Alaska at Fairbanks. The authors also acknowledge Dr Zenmei Ohkubo's assistance in the preparation of some of the figures.

References

1. A. MacKerell Jr, D. Bashford, M. Bellott, R. L. Dunbrack Jr, J. Evanseck, M. J. Field, S. Fischer, J. Gao, H. Guo, S. Ha, D. Joseph, L. Kuchnir, K. Kuczera, F. T. K. Lau, C. Mattos, S. Michnick, T. Ngo, D. T. Nguyen, B. Prodhom, I. W. E. Reiher, B. Roux, M. Schlenkrich, J. Smith, R. Stote, J. Straub, M. Watanabe, J. Wiorkiewicz-Kuczera, D. Yin and M. Karplus, *J. Phys. Chem. B*, 1998, **102**, 3586–3616.
2. T. Darden, D. York and L. Pedersen, *J. Chem. Phys.*, 1993, **98**, 10089–10092.
3. M. Nelson, W. Humphrey, A. Gursoy, A. Dalke, L. Kalé, R. D. Skeel and K. Schulten, *Int. J. Supercomput. Appl. High Perform. Comput.*, 1996, **10**, 251–268.
4. L. Kalé, R. Skeel, M. Bhandarkar, R. Brunner, A. Gursoy, N. Krawetz, J. Phillips, A. Shinozaki, K. Varadarajan and K. Schulten, *J. Comput. Phys.*, 1999, **151**, 283–312.
5. J. C. Phillips, R. Braun, W. Wang, J. Gumbart, E. Tajkhorshid, E. Villa, C. Chipot, R. D. Skeel, L. Kale and K. Schulten, *J. Comput. Chem.*, 2005, **26**, 1781–1802.
6. B. Isralewitz, M. Gao and K. Schulten, *Curr. Opin. Struct. Biol.*, 2001, **11**, 224–230.
7. Y. Wang and E. Tajkhorshid, *Proc. Natl. Acad. Sci. USA*, 2008, **105**, 9598–9603.
8. A. Aksimentiev and K. Schulten, *Biophys. J.*, 2005, **88**, 3745–3761.
9. I. B. HollandS. P. C. ColeK. KuchlerC. F. Higgins, *ABC Proteins: from Bacteria to Man,* Academic Press, London, 2003.
10. M. Dean and T. Annilo, *Annu. Rev. Genomics Hum. Genet.*, 2005, **6**, 123–142.
11. B. Stieger, Y. Meier and P. J. Meier, *Pflug. Arch. Eur. J. Physiol.*, 2007, **453**, 611–620.

12. T. P. Burris, P. I. Eacho and G. Cao, *Mol. Genet. Metab.*, 2002, **77**, 13–20.
13. J. F. Oram, *Trends Mol. Med.*, 2002, **8**, 168–173.
14. K. Takahashi, Y. Kimura, K. Nagata, A. Yamamoto, M. Matsuo and K. Ueda, *Med. Mol. Morphol.*, 2005, **38**, 2–12.
15. G. Schmitz, G. Liebisch and T. Langmann, *FEBS Lett.*, 2006, **580**, 5597–5610.
16. A. Zarubica, D. Trompier and G. Chimini, *Pflug. Arch. Eur. J. Physiol.*, 2007, **453**, 569–579.
17. M. Fontenay, S. Cathelin, M. Amiot, E. Gyan and E. Solary, *Oncogene*, 2006, **25**, 4757–4767.
18. I. Napier, P. Ponka and D. R. Richardson, *Blood*, 2005, **105**, 1867–1874.
19. C. Pondarre, D. R. Campagna, B. Antiochos, L. Sikorski, H. Mulhern and M. D. Fleming, *Blood*, 2007, **109**, 3567–3569.
20. C. Schölz and R. Tampé, *J. Bioenerg. Biomembr.*, 2005, **37**, 509–515.
21. M. Herget and R. Tampé, *Pflug. Arch. Eur. J. Physiol.*, 2007, **453**, 591–600.
22. A. A. Aleksandrov, L. A. Aleksandrov and J. R. Riordan, *Pflug. Arch. Eur. J. Physiol.*, 2007, **453**, 693–702.
23. D. C. Gadsby, P. Vergani and L. Csanady, *Nature*, 2006, **440**, 477–483.
24. F. C. Lam, R. Liu, P. Lu, A. B. Shapiro, J.-M. Renoir, F. J. Sharom and P. B. Reiner, *J. Neurochem.*, 2001, **76**, 1121–1128.
25. J. T. Mack, V. Beljanskia, K. D. Tew and D. M. Townsend, *Biomed. Pharmacother.*, 2006, **60**, 587–592.
26. D. Kuhnke, G. Jedlitschky, M. Grube, M. Krohn, M. Jucker, I. Mosyagin, I. Cascorbi, L. C. Walker, H. K. Kroemer, R. W. Warzok and S. Vogelgesang, *Brain Pathol.*, 2007, **17**, 347–353.
27. W. S. Kim, C. S. Weickert and B. Garner, *J. Neurochem.*, 2008, **104**, 1145–1166.
28. G. D. Kolovou, D. P. Mikhailidis, K. K. Anagnostopoulou, S. S. Daskalopoulou and D. V. Cokkinos, *Curr. Med. Chem.*, 2006, **13**, 771–782.
29. J. Berger and J. Gartner, *Biochim. Biophys. Acta*, 2006, **1763**, 1721–1732.
30. S. Kemp and R. J. Wanders, *Mol. Genet. Metab.*, 2007, **90**, 268–276.
31. G. D. Leonard, T. Fojo and S. E. Bates, *Oncologist*, 2003, **8**, 411–424.
32. E. Bakos and L. Homolya, *Pflug. Arch. Eur. J. Physiol.*, 2006, **453**, 621–641.
33. R. G. Deeley, C. Westlake and S. P. C. Cole, *Physiol. Rev.*, 2006, **86**, 849–899.
34. B. Sarkadi, L. Homolya, G. Szakács and A. Váradi, *Physiol. Rev.*, 2006, **86**, 1179–1236.
35. W. T. Self, A. M. Grundenb, A. Hasonac and K. T. Shanmugam, *Res. Microbiol.*, 2001, **152**, 311–321.
36. K. Hantke, *Curr. Opin. Microbiol.*, 2005, **8**, 196–202.
37. S.-V. Albers, M. G. L. Elferink, R. L. Charlebois, C. W. Sensen, A. J. M. Driessen and W. N. Konings, *J. Bacteriol.*, 1999, **181**, 4285–4291.
38. E. E. Fetsch and A. L. Davidson, *Front. Biosci.*, 2003, **8**, 652–660.
39. A. H. F. Hosie and P. S. Poole, *Res. Microbiol.*, 2001, **152**, 259–270.

40. E. Webb, K. Claas and D. Downs, *J. Biol. Chem.*, 1998, **273**, 8946–8950.
41. W. Köster, *Res. Microbiol.*, 2001, **152**, 291–301.
42. I. B. Holland, L. Schmitt and J. Young, *Mol. Membr. Biol.*, 2005, **22**, 29–39.
43. W. T. Doerrler, *Mol. Microbiol.*, 2006, **60**, 542–552.
44. A. L. Davidson and J. Chen, *Annu. Rev. Biochem.*, 2004, **73**, 241–268.
45. C. Oswald, I. B. Holland and L. Schmitt, *Naunyn-Schmiedeberg's Arch. Pharmacol.*, 2006, **372**, 385–399.
46. J. Chen, G. Lu, J. Lin, A. L. Davidson and F. A. Quiocho, *Mol. Cell*, 2003, **12**, 651–661.
47. G. Lu, J. M. Westbrooks, A. L. Davidson and J. Chen, *Proc. Natl. Acad. Sci. USA*, 2005, **102**, 17969–17974.
48. R. J. P. Dawson and K. P. Locher, *Nature*, 2006, **443**, 180–185.
49. H. W. Pinkett, A. T. Lee, P. Lum, K. P. Locher and D. C. Rees, *Science*, 2007, **315**, 373–377.
50. K. Hollenstein, D. C. Frei and K. P. Locher, *Nature*, 2007, **446**, 213–216.
51. R. J. P. Dawson and K. P. Locher, *FEBS Lett.*, 2007, **581**, 935–938.
52. R. N. Hvorup, B. A. Goetz, M. Niederer, K. Hollenstein, E. Perozo and K. P. Locher, *Science*, 2007, **317**, 1387–1390.
53. M. L. Oldham, D. Khare, F. A. Quiocho, A. L. Davidson and J. Chen, *Nature*, 2007, **450**, 515–521.
54. A. Ward, C. L. Reyes, J. Yu, C. B. Roth and G. Chang, *Proc. Natl. Acad. Sci. USA*, 2007, **104**, 19005–19010.
55. S. Gerber, M. Comellas-Bigler, B. A. Goetz and K. P. Locher, *Science*, 2008, **321**, 246–250.
56. N. S. Kadaba, J. T. Kaiser, E. Johnson, A. Lee and D. C. Rees, *Science*, 2008, **321**, 250–253.
57. R. J. P. Dawson, K. Hollenstein and K. P. Locher, *Mol. Microbiol.*, 2007, **65**, 250–257.
58. K. Hollenstein, R. J. P. Dawson and K. P. Locher, *Curr. Opin. Struct. Biol.*, 2007, **17**, 412–418.
59. A. L. Davidson, E. Dassa, C. Orelle and J. Chen, *Microbiol. Mol. Biol. Rev.*, 2008, **72**, 317–364.
60. E. O. Oloo and D. P. Tieleman, *J. Biol. Chem.*, 2004, **279**, 45013–45019.
61. J. D. Campbell and M. S. P. Sansom, *FEBS Lett.*, 2005, **579**, 4193–4199.
62. E. O. Oloo, E. Y. Fung and D. P. Tieleman, *J. Biol. Chem.*, 2006, **281**, 28397–28407.
63. A. Ivetac, J. D. Campbell and M. S. P. Sansom, *Biochemistry*, 2007, **46**, 2767–2778.
64. J. Sonne, C. Kandt, G. H. Peters, F. Y. Hansen, M. O. Jansen and D. P. Tieleman, *Biophys. J.*, 2007, **92**, 2727–2734.
65. P. M. Jones and A. M. George, *J. Biol. Chem.*, 2007, **282**, 22793–22803.
66. P. M. Jones and A. M. George, *Proteins Struct. Funct. Bioinf.*, 2008, **75**, 387–396.

67. J. Weng, J. Ma, K. Fan and W. Wang, *Biophys. J.*, 2008, **94**, 612–621.
68. P.-C. Wen and E. Tajkhorshid, *Biophys. J.*, 2008, **95**, 5100–5110.
69. G. Tombline, L. A. Bartholomew, G. A. Tyndall, K. Gimi, I. L. Urbatsch and A. E. Senior, *J. Biol. Chem.*, 2004, **279**, 46518–46526.
70. K. Nikaido and G. F.-L. Ames, *J. Biol. Chem.*, 1999, **274**, 26727–26735.
71. N. H. Chen, M. E. Reith and M. W. Quick, *Pflug. Arch. Eur. J. Physiol.*, 2004, **447**, 519–531.
72. C. Grewer and T. Rauen, *J. Membr. Biol.*, 2005, **203**, 1–20.
73. R. Dingledine, K. Borges, D. Bowie and S. F. Traynelis, *Pharmacol. Rev.*, 1999, **51**, 7–61.
74. J. D. Clements, *Trends Neurosci.*, 1996, **19**, 163–171.
75. D. E. Bergles, J. S. Diamond and C. E. Jahr, *Curr. Opin. Neurobiol.*, 1999, **9**, 293–298.
76. D. J. Slotboom, W. N. Konings and J. S. Lolkema, *Microbiol. Mol. Biol. Rev.*, 1999, **63**, 293–307.
77. N. C. Danbolt, *Prog. Neurobiol.*, 2001, **65**, 1–105.
78. P. F. Behrens, P. Franz, B. Woodman, K. S. Lindenberg and G. B. Landwehrmeyer, *Brain*, 2002, **125**, 1908–1922.
79. J. H. Yi and A. S. Hazell, *Neurochem. Int.*, 2006, **48**, 394–403.
80. J. Dunlop and K. Marquis, *Drug Discov. Today Ther. Strategies*, 2006, **3**, 533–537.
81. N. Zerangue and M. P. Kavanaugh, *Nature*, 1996, **383**, 634–637.
82. L. M. Levy, O. Warr and D. Attwell, *J. Neurosci.*, 1998, **18**, 9620–9628.
83. G. Pines and B. I. Kanner, *Biochemistry*, 1990, **29**, 11209–11214.
84. M. P. Kavanaugh, A. Bendahan, N. Zerangue, Y. Zhang and B. I. Kanner, *J. Biol. Chem.*, 1997, **272**, 1703–1707.
85. D. Yernool, O. Boudker, Y. Jin and E. Gouaux, *Nature*, 2004, **431**, 811–818.
86. D. J. Slotboom, I. Sobczak, W. N. Konings and J. S. Lolkema, *Proc. Natl. Acad. Sci. USA*, 1999, **96**, 14282–14287.
87. M. Grunewald and B. I. Kanner, *J. Biol. Chem.*, 2000, **275**, 9684–9689.
88. R. P. Seal and S. G. Amara, *Neuron*, 1998, **21**, 1487–1498.
89. L. Borre, M. P. Kavanaugh and B. I. Kanner, *J. Biol. Chem.*, 2002, **277**, 13501–13507.
90. B. H. Leighton, R. P. Seal, K. Shimamoto and S. G. Amara, *J. Biol. Chem.*, 2002, **277**, 29847–29855.
91. R. M. Ryan and R. J. Vandenberg, *J. Biol. Chem.*, 2002, **277**, 13494–13500.
92. O. Boudker, R. M. Ryan, D. Yernool, K. Shimamoto and E. Gouaux, *Nature*, 2007, **445**, 387–393.
93. C. Grewer, P. Balani, C. Weidenfeller, T. Bartusel, Z. Tao and T. Tauen, *Biochemistry*, 2005, **44**, 11913–11923.
94. M. Grunewald and B. I. Kanner, *J. Biol. Chem.*, 1995, **270**, 17017–17024.
95. M. Grunewald, A. Bendahan and B. I. Kanner, *Neuron*, 1998, **21**, 623–632.
96. D. J. Slotboom, J. S. Lolkema and W. N. Konings, *J. Biol. Chem.*, 1996, **271**, 31317–31321.

97. D. J. Slotboom, W. N. Konings and J. S. Lolkema, *J. Biol. Chem.*, 2001, **276**, 10775–10781.

98. R. P. Seal, B. H. Leighton and S. G. Amara, *Neuron*, 2000, **25**, 695–706.

99. R. Zarbiv, M. Grunewald, M. P. Kavanaugh and B. I. Kanner, *J. Biol. Chem.*, 1998, **273**, 14231–14237.

100. H. P. Koch and P. H. Larsson, *J. Neurosci.*, 2005, **25**, 1730–1736.

101. P. H. Larsson, A. V. Tzingounis, H. P. Koch and M. P. Kavanaugh, *Proc. Natl. Acad. Sci. USA*, 2004, **101**, 3951–3956.

102. B. H. Leighton, R. P. Seal, S. D. Watts, M. O. Skyba and S. G. Amara, *J. Biol. Chem.*, 2006, **281**, 29788–29796.

103. C. Mim, Z. Tao and C. Grewer, *Biochemistry*, 2007, **46**, 9007–9018.

104. B. H. Leighton, R. P. Seal, K. Shimamoto and S. G. Amara, *J. Biol. Chem.*, 2002, **277**, 29847–29855.

105. Z. Tao, Z. Zhang and C. Grewer, *J. Biol. Chem.*, 2006, **281**, 10263–10272.

106. Z. Zhang, Z. Tao, A. Gameiro, S. Barcelona, S. Braams, T. Rauen and C. Grewer, *Proc. Natl. Acad. Sci. USA*, 2007, **104**, 18025–18030.

107. O. Haugeto, K. Ullensvang, L. M. Levy, F. A. Chaudhry, T. Honore, M. Mielsen, K. P. Lehre and N. C. Danbolt, *J. Biol. Chem.*, 1996, **271**, 27715–27722.

108. Y. Kanai, M. Stelzner, S. Nussberger, S. Khawaja, S. C. Hebert, C. P. Smith and M. A. Hediger, *J. Biol. Chem.*, 1994, **269**, 20599–20606.

109. B. I. Kanner and L. Borre, *Biochim. Biophys. Acta*, 2002, **1555**, 92–95.

110. Y. Zhang and B. Kanner, *Proc. Natl. Acad. Sci. USA*, 1999, **96**, 1710–1715.

111. A. Bendahan, A. Armon, N. Madani, M. P. Kavanaugh and B. I. Kanner, *J. Biol. Chem.*, 2000, **275**, 37436–37442.

112. S. G. Amara and A. C. K. Fontana, *Neurochem. Int.*, 2002, **41**, 313–318.

113. N. Watzke, T. Rauen, E. Bamberg and C. Grewer, *J. Gen. Physiol.*, 2000, **11**, 609–621.

114. C. Grewer, N. Watzke, M. Wiessner and T. Rauen, *Proc. Natl. Acad. Sci. USA*, 2000, **97**, 9706–9711.

115. C. Grewer, N. Watzke, T. Rauen and A. Bicho, *J. Biol. Chem.*, 2003, **278**, 2585–2592.

116. N. Rosental, A. Bendahan and B. I. Kanner, *J. Biol. Chem.*, 2006, **281**, 27905–27915.

117. Z. J. Huang and E. Tajkhorshid, *Biophys. J.*, 2008, **95**, 2292–2300.

118. S. Qu and B. I. Kanner, *Proc. Natl. Acad. Sci. USA*, 2008, **283**, 26391–26400.

119. I. H. Shrivastava, J. Jiang, S. G. Amara and I. Bahar, *J. Biol. Chem.*, 2008, **283**, 28680–28690.

120. D. E. Bergles, A. V. Tzingounis and C. E. Jahr, *J. Neurosci.*, 2002, **22**, 10153–10162.

121. Y. Huang, M. J. Lemieux, J. Song, M. Auer and D.-N. Wang, *Science*, 2003, **301**, 616–620.

122. S. B. Levy, *J. Appl. Microbiol.*, 2002, **92**, 65S–71S.

123. M. J. Lemieux, Y. Huang and D. Wang, *Res. Microbiol.*, 2004, **155**, 623–629.

124. L. Bartoloni, M. Wattenhofer, J. Kudoh, A. Berry, K. Shibuya, K. Kawasaki, J. Wang, S. Asakawa, I. Talior, B. Bonne-Tamir, C. Rossier, J. Michaud, E. R. B. McCabe, S. Minoshima, N. Shimizu, H. S. Scott and S. E. Antonarakis, *Genomics*, 2000, **70**, 190–200.

125. M. J. Lemieux, Y. Huang and D. N. Wang, *J. Electron Microsc.*, 2005, **54**, i43–i46.

126. M. Saidijam, G. Benedetti, Q. Ren, Z. Xu, C. J. Hoyle, S. L. Palmer, A. Ward, K. E. Bettaney, S. Gerda, J. Meuller, S. Morrison, M. K. Pos, P. Butaye, K. Walravens, K. Langton, R. B. Herbert, R. A. Skurray, I. T. Paulsen, J. O'Reilly, N. G. Rutherford, M. H. Brown, R. M. Bill and P. J. F. Henderson, *Curr. Drug Targets*, 2006, **7**, 793–811.

127. C. F. Higgins, *Nature*, 2007, **446**, 749–757.

128. C. J. Law, Q. Yang, C. Soudant, P. C. Maloney and D.-N. Wang, *Biochemistry*, 2007, **46**, 12190–12197.

129. C. J. Law, J. Almqvist, A. Bernstein, R. M. Goetz, Y. Huang, C. Soudant, A. Laaksonen, S. Hovmöller and D.-N. Wang, *J. Mol. Biol.*, 2008, **378**, 828–839.

130. S. S. Pao, I. T. Paulsen and M. H. Saier Jr, *Microbiol. Mol. Biol. Rev.1998, 62, 1–34.*

131. M. H. Saier Jr, *Microbiol. Mol. Biol. Rev.*, 2000, **64**, 354–411.

132. J. Abramson, I. Smirnova, V. Kasho, G. Verner, H. R. Kaback and S. Iwata, *Science*, 2003, **301**, 610–615.

133. O. Mirza, L. Guan, G. Verner, S. Iwata and H. Kaback, *EMBO J.*, 2006, **25**, 1177–1183.

134. Y. Yin, X. He, P. Szewczyk, T. Nguyen and G. Chang, *Science*, 2006, **312**, 741–744.

135. M. H. Saier Jr, *Mol. Microbiol.2000, 35, 699–710.*

136. J. Abramson, H. R. Kaback and S. Iwata, *Curr. Opin. Struct. Biol.*, 2004, **14**, 413–419.

137. M. J. Lemieux, *Mol. Membr. Biol.*, 2007, **24**, 333–341.

138. M. J. Lemieux, J. Song, M. J. Kim, Y. Huang, A. Villa, M. Auer, X.-D. Li and D.-N. Wang, *Protein Sci.*, 2003, **12**, 2748–2756.

139. M. Auer, M. J. Kim, M. J. Lemieux, A. Villa, J. Song, X.-D. Li and D.-N. Wang, *Biochemistry*, 2001, **40**, 6628–6635.

140. M.-C. Fann and P. C. Maloney, *J. Membr. Biol.*, 1998, **164**, 187–195.

141. M. J. Lemieux, Y. Huang and D. Wang, *Curr. Opin. Struct. Biol.*, 2004, **14**, 405–412.

142. S. Hayashi, J. P. Koch and E. C. C. Lin, *J. Biol. Chem.*, 1964, **239**, 3098–3105.

143. B. J. Calnan, B. Tidor, S. Biancalana, D. Hudson and A. D. Frankel, *Science*, 1991, **252**, 1167–1171.

144. H. Nury, C. Dahout-Gonzalez, V. Trezeguet, G. J. Lauquin, G. Brandolin and E. Pebay-Peyroula, *Annu. Rev. Biochem.*, 2006, **75**, 713–741.

145. T. Gropp, N. Brustovetsky, M. Klingenberg, V. Müller, K. Fendler and E. Bamberg, *Biophys. J.*, 1999, **77**, 714–726.
146. A. Bruni, S. Luciani and A. R. Contessa, *Nature*, 1964, **201**, 1219–1220.
147. E. D. Duee and P. Vignais, *Biochim. Biophys. Acta*, 1965, **107**, 184–188.
148. E. Pfaff, M. Klingenberg and H. W. Heldt, *Biochim. Biophys. Acta*, 1965, **104**, 312–315.
149. M. Klingenberg, *Arch. Biochem. Biophys.*, 1989, **270**, 1–14.
150. E. Pebay-Peyroula, C. Dahout-Gonzalez, R. Kahn, V. Trezeguet, G. J. Lauquin and G. Brandolin, *Nature*, 2003, **426**, 39–44.
151. S. D. Dyall, S. C. Agius, C. D. M. Lousa, V. Trezeguet and K. Tokatlidis, *J. Biol. Chem.*, 2003, **278**, 26757–26764.
152. Y. Kihira, A. Iwahashi, E. Majima, H. Terada and Y. Shinohara, *Biochemistry*, 2004, **43**, 15204–15209.
153. E. Pebay-Peyroula and G. Brandolin, *Curr. Opin. Struct. Biol.*, 2004, **14**, 420–425.
154. M. Klingenberg, *Biochemistry*, 2005, **44**, 8563–8570.
155. C. Dahout-Gonzalez, C. Ramus, E. P. Dassa, A. Dianoux and G. Brandolin, *Biochemistry*, 2005, **44**, 16310–16320.
156. H. Nury, C. Dahout-Gonzalez, V. Trezeguet, G. J. Lauquin, G. Brandolin and E. Pebay-Peyroula, *FEBS Lett.*, 2005, **579**, 6031–6036.
157. V. Postis, C. D. M. Lousa, B. Arnou, G. J. Lauquin and V. Trezeguet, *Biochemistry*, 2005, **44**, 14732–14740.
158. A. J. Robinson and E. R. Kunji, *Proc. Natl. Acad. Sci. USA*, 2006, **103**, 2617–2622.
159. C. Dahout-Gonzalez, H. Nury, V. Trezeguet, G. J. Lauquin, E. Pebay-Peyroula and G. Brandolin, *Physiology*, 2006, **21**, 242–249.
160. V. Müller, D. Heidkämper, D. R. Nelson and M. Klingenberg, *Biochemistry*, 1997, **36**, 16008–16018.
161. V. Müller, G. Basset, D. R. Nelson and M. Klingenberg, *Biochemistry*, 1996, **35**, 16132–16143.
162. E. R. Kunji and A. J. Robinson, *Biochim. Biophys. Acta*, 2006, **1757**, 1237–1248.
163. H. Nikaido, *Science*, 1994, **264**, 382–388.
164. H. Nikaido, *J. Biol. Chem.*, 1994, **269**, 3905–3908.
165. R. J. Kadner, *Mol. Microbiol.*, 1990, **4**, 2027–2033.
166. K. Postle and R. J. Kadner, *Mol. Microbiol.*, 2003, **49**, 869–882.
167. J. D. Faraldo-Gómez and M. S. P. Sansom, *Nat. Rev. Mol. Cell Biol.*, 2003, **4**, 105–116.
168. A. D. Ferguson and J. Deisenhofer, *Cell*, 2004, **116**, 15–24.
169. C. Bradbeer, *J. Bacteriol.*, 1993, **175**, 3146–3150.
170. P. I. Higgs, P. S. Myers and K. Postle, *J. Bacteriol.*, 1998, **180**, 6031–6038.
171. R. A. Larsen, M. G. Thomas and K. Postle, *Mol. Microbiol.*, 1999, **31**, 1809–1824.
172. D. P. Chimento, R. J. Kadner and M. C. Wiener, *Proteins Struct. Funct. Bioinf.*, 2005, **59**, 240–251.

173. A. Garcia-Herrero, R. S. Peacock, S. P. Howard and H. J. Vogel, *Mol. Microbiol.*, 2007, **66**, 872–889.
174. A. D. Ferguson, E. Hofmann, J. W. Coulton, K. Diederichs and W. Welte, *Science*, 1998, **282**, 2215–2220.
175. K. P. Locher, B. Rees, R. Koebnik, A. Mitschler, L. Moulinier, J. P. Rosenbusch and D. Moras, *Cell*, 1998, **95**, 771–778.
176. S. K. Buchanan, B. S. Smith, L. Venkatramani, D. Xia, L. Esser, M. Palnitkar, R. Chakraborty, D. van der Helm and J. Deisenhofer, *Nat. Struct. Biol.*, 1999, **6**, 56–63.
177. A. D. Ferguson, R. Chakraborty, B. S. Smith, L. Esser, D. van der Helm and J. Deisenhofer, *Science*, 2002, **295**, 1715–1719.
178. W. W. Yue, S. Grizot and S. K. Buchanan, *J. Mol. Biol.*, 2003, **332**, 353–368.
179. D. P. Chimento, A. K. Mohanty, R. J. Kadner and M. C. Wiener, *Nat. Struct. Biol.*, 2003, **10**, 394–401.
180. D. Cobessi, H. Celia, N. Folschweiller, I. J. Schalk, M. A. Abdallah and F. Pattus, *J. Mol. Biol.*, 2005, **347**, 121–134.
181. D. Cobessi, H. Celia and F. Pattus, *J. Mol. Biol.*, 2005, **352**, 893–904.
182. K. Schauer, D. A. Rodionov and H. de Reuse, *Trends Biochem. Sci.*, 2008, **33**, 330–338.
183. M. C. Wiener, *Curr. Opin. Struct. Biol.*, 2005, **15**, 394–400.
184. M. D. Lundrigan and R. J. Kadner, *J. Biol. Chem.*, 1986, **261**, 10797–10801.
185. E. Schramm, J. Mende, V. Braun and R. M. Kamp, *J. Bacteriol.*, 1987, **169**, 3350–3357.
186. N. Cadieux and R. J. Kadner, *Proc. Natl. Acad. Sci. USA*, 1999, **96**, 10673–10678.
187. N. Cadieux, C. Bradbeer and R. J. Kadner, *J. Bacteriol.*, 2000, **182**, 5954–5961.
188. H. J. Merianos, N. Cadieux, C. H. Lin, R. J. Kadner and D. S. Cafiso, *Nat. Struct. Biol.*, 2000, **7**, 205–209.
189. S. M. Lukasik, K. W. D. Ho and D. S. Cafiso, *J. Mol. Biol.*, 2007, **370**, 807–811.
190. W. A. Kaserer, X. Jiang, Q. Xiao, D. C. Scott, M. Bauler, D. Copeland, S. M. C. Newton and P. E. Klebba, *J. Bacteriol.*, 2008, **190**, 4001–4016.
191. P. D. Pawelek, N. Croteau, C. Ng-Thow-Hing, C. M. Khursigara, N. Moiseeva, M. Allaire and J. W. Coulton, *Science*, 2006, **312**, 1399–1402.
192. D. D. Shultis, M. D. Purdy, C. N. Banchs and M. C. Wiener, *Science*, 2006, **312**, 1396–1399.
193. T. E. Letain and K. Postle, *Mol. Microbiol.*, 1997, **24**, 271–283.
194. R. A. Larsen and K. Postle, *J. Bacteriol.*, 2003, **49**, 211–218.
195. K. Postle and R. A. Larsen, *Biometals*, 2007, **20**, 453–465.
196. J. Gumbart, M. C. Wiener and E. Tajkhorshid, *Biophys. J.*, 2007, **93**, 496–504.
197. L. Ma, W. Kaserer, R. Annamalai, D. C. Scott, B. Jin, X. Jiang, Q. Xiao, H. Maymani, L. M. Massis, L. C. Ferreira, S. M. C. Newton and P. E. Klebba, *J. Biol. Chem.*, 2007, **282**, 397–406.

198. J. D. Faraldo-Gómez, G. R. Smith and M. S. P. Sansom, *Biophys. J.*, 2003, **85**, 1406–1420.

199. F. Endrib, M. Braun, H. Killmann and V. Braun, *J. Bacteriol.*, 2003, **185**, 4683–4692.

200. H. A. Eisenhauer, S. Shames, P. D. Pawelek and J. W. Coulton, *J. Biol. Chem.*, 2005, **280**, 30574–30580.

201. R. Chakraborty, E. Storey and D. van der Helm, *Biometals*, 2007, **20**, 263–274.

202. M. Gao, D. Craig, O. Lequin, I. D. Campbell, V. Vogel and K. Schulten, *Proc. Natl. Acad. Sci. USA*, 2003, **100**, 14784–14789.

203. E. H. Lee, M. Gao, N. Pinotsis, M. Wilmanns and K. Schulten, *Structure*, 2006, **14**, 497–509.

204. M. Sotomayor and K. Schulten, *Science*, 2007, **316**, 1144–1148.

205. V. R. F. Matias, A. Al-Amoudi, J. Dubocher and T. J. Beveridge, *J. Bacteriol.*, 2003, **185**, 6112–6118.

206. S. O. Meroueh, K. Z. Bencze, D. Hesek, M. Lee, J. F. Fisher, T. L. Stemmler and S. Mobashery, *Proc. Natl. Acad. Sci. USA*, 2006, **103**, 4404–4409.

207. P. I. Higgs, R. A. Larsen and K. Postle, *Mol. Microbiol.*, 2002, **44**, 271–281.

Molecular Dynamics Studies of the Interactions Between Carbon Nanotubes and Biomembranes

E. JAYNE WALLACE AND MARK S. P. SANSOM

Department of Biochemistry, University of Oxford, Oxford, OX1 3QU, UK

11.1 Introduction

Carbon nanotubes (CNTs) are hollow, cylindrical tubes that are composed of seamlessly closed rolled-up graphene sheets. In recent years, CNTs have attracted intense interest owing to their unique properties. These properties include structural robustness, high electrical and thermal conductivity and lengths that far exceed their diameters. In addition, many research groups have demonstrated the propensity of CNTs to functional modification. Together these distinct properties have led to many potential biological applications.

Within the field of nanomedicine, the use of CNTs as drug delivery vehicles is being investigated. In fact, it has been shown that CNTs are able to cross the cellular membrane to deliver cargoes to a range of cell types *in vitro*.[1–5] Moreover, it may be possible to impart tissue selectivity on these potential drug carriers if the CNT has a specific functional modification. There is also interest in the fundamental properties of CNTs as nanopipes.[6] For example, studies have demonstrated the possibility of using CNTs as 'nanosyringes' to inject fluids across cell membranes.

RSC Biomolecular Sciences No. 20
Molecular Simulations and Biomembranes: From Biophysics to Function
Edited by Mark S.P. Sansom and Philip C. Biggin
© Royal Society of Chemistry 2010
Published by the Royal Society of Chemistry, www.rsc.org

These and other potential applications of CNTs in nanomedicine and bionanotechnology require an improved understanding of how CNTs interact with biomembranes. In order to exploit the unique properties of CNTs to their full biological potential, it is important to understand the fundamental interplay between CNTs and cell membranes and their components.

For example, greater understanding of the cellular uptake pathway of CNTs is required. There is experimental evidence that suggests that CNTs can be internalized into cells *via* endocytosis.[2,3] However, other evidence implies that CNTs undergo spontaneous insertion and diffusion across the cell membrane.[1,4,5] Understanding the mechanism(s) of insertion into a membrane and of stabilization of an inserted, transmembrane orientation will be a key aspect of the successful exploitation of CNTs as nanosyringes.

In addition to exploiting the interactions of CNTs with biomembranes, we need to avoid interactions which are potentially detrimental to cells. Therefore, the cytotoxicity of CNTs is another area of research that requires deeper understanding. Studies have shown that particle size, surface chemistry and solubility are all involved in the toxicity of nanoparticles.[7]

Computer simulations play a key role in exploring the nature of CNTs and their interactions. A number of recent reviews have explored the role of computer modelling in bionanotechnology in general[8] and in simulations of the interactions of water with CNTs.[9] In this chapter, we focus on how simulations and related studies have been used to inform our understanding of the interactions of CNTs with membranes and their components and of the potential properties of CNTs when acting as nanopores through biological membranes.

11.1.1 Carbon Nanotube Structure

CNTs are categorized as either single-walled carbon nanotubes (SWNTs) or multi-walled carbon nanotubes (MWNTs). SWNTs are composed of a single graphene cylinder, whereas MWNTs are composed of concentrically nested graphene cylinders. The way in which a graphene sheet is rolled up to form the cylindrical structure of a CNT is referred to as CNT 'chirality' (Figure 11.1). 'Armchair' and 'zigzag' are two special CNT chiralities. Unlike ordinary materials, CNTs exhibit metallic or semiconducting behaviour depending upon their chirality.

CNTs can grow as long as several millimetres,[10] but ultrashort CNTs with average lengths of the order of 7.5 nm have been obtained.[11] The diameters of SWNTs range from 0.4 to 3 nm,[12,13] whereas MWNTs have diameters as large as 100 nm.

CNTs have a highly organized and near ideal sp^2-bonded carbon structure, with delocalized π-electrons along the tube surface. The π-electrons in CNTs are highly polarizable and easily influenced by their environment.[10] It is this feature that provides CNTs with their unique strength and conductivity.

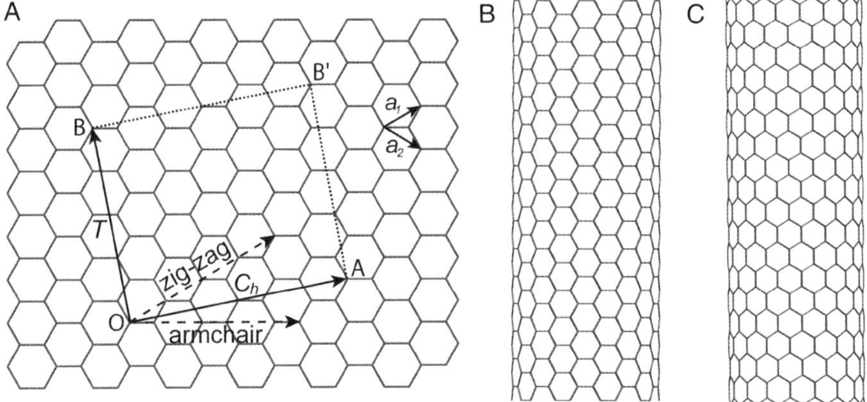

Figure 11.1 (A) A graphene sheet with lattice vectors a_1 and a_2. A CNT can be formed if the sheet is rolled up such that points O and A and B and B' are connected. The chiral vector $C_h = \overrightarrow{OA} = 4a_1 + 2a_2$ of the (4,2) CNT is shown. The tube axis is perpendicular to C_h. The minimum translational period is given by the vector $T = \overrightarrow{OB}$. The vectors C_h and T form a rectangle which is the unit cell of the CNT. Structure of an armchair (B) and zigzag (C) CNT.

11.1.2 Experimental Techniques for Studying CNTs in a Biological Environment

Although the mechanical and electrical properties of CNTs can be measured explicitly in experiments,[14] experimental investigation of CNT interactions with biological materials remains at an early stage. The difficulty of investigating the interplay between nanoscale systems and biology is due partially to a lack of imaging tools: the nanoscale is too small for light microscopy and too large for X-ray crystallography; it is too heterogeneous for NMR and too 'wet' for electron microscopy.[8] Furthermore, manipulation of CNTs in biological environments is impeded by their lack of solubility.

However, individual SWNTs have been visualized in cells through a novel technique called low-loss energy-filtered transmission electron microscopy in combination with confocal microscopy.[15] This technique made it possible to determine directly the distribution of SWNTs in both stained and unstained human cells. It also permitted visualization of CNTs entering the cytoplasm, localizing within the cell nucleus and causing cell mortality in a dose-dependent manner.

11.2 Molecular Dynamics Simulations

Although there is promise in being able to monitor experimentally the interplay between individual CNTs and biological molecules, it is currently more convenient to employ computational approaches. For example, quantum mechanical simulations, classical molecular dynamics and Monte Carlo simulations have all been used to study CNTs in biological environments. In this chapter, we focus on molecular dynamics simulations that provide atomistic/molecular level descriptions of CNT interactions.

11.2.1 Methodology

Molecular dynamics (MD) is a simulation technique that describes the time evolution of a simulation system. The simulation system is composed of, *e.g.*, a carbon nanotube, lipids and water and their interactions are described by a suitable molecular mechanics forcefield. The forcefield refers to the functional form and parameter set that describes the potential energy of the particles within the system. In order to generate the time evolution, *i.e.* trajectory, of the simulation system, Newton's laws of motion are integrated. This process yields successive particle positions and velocities as a function of time.

11.2.2 Parameterization of CNT Models

In recent years, both atomistic and coarse-grained models have been developed to study CNTs. Below we describe the different models in detail.

11.2.2.1 Atomistic Models

In the majority of classical MD simulations, the carbon atoms that comprise CNTs are modelled as neutral atoms with pairwise additive Lennard-Jones (LJ) potentials.[16–25] Hence both the electronic properties of CNTs and the effect of CNT polarization on its interaction with the environment are ignored. The carbon atoms are typically modelled as sp^2-hybridized carbons, *e.g.* graphite or benzene atoms.

Density functional theory (DFT) calculations on open-ended finite-length SWNTs suggest that at the non-saturated end, C–C triple bonds are formed for armchair CNTs.[26,27] This causes a finite charge and a significant dipole moment at the tube ends. However, C–C triple bonding is not observed at the tube ends of zigzag CNTs. This work suggests that CNT chirality may influence electrostatics of the tube. In light of this, several groups have modelled partial charges on the CNT atoms and found that they affect the translocation of molecules across the tubes.[28–32]

In order to model accurately the intermolecular forces between CNTs and their environment, the delocalized π-electrons need to be considered. These electrons render CNTs highly polarizable.[8] Polarizable CNT models need to be computationally very efficient, because in MD simulations, the forces acting on atoms are updated frequently. Hence, to date, only very short tube sections have been simulated.[33,34]

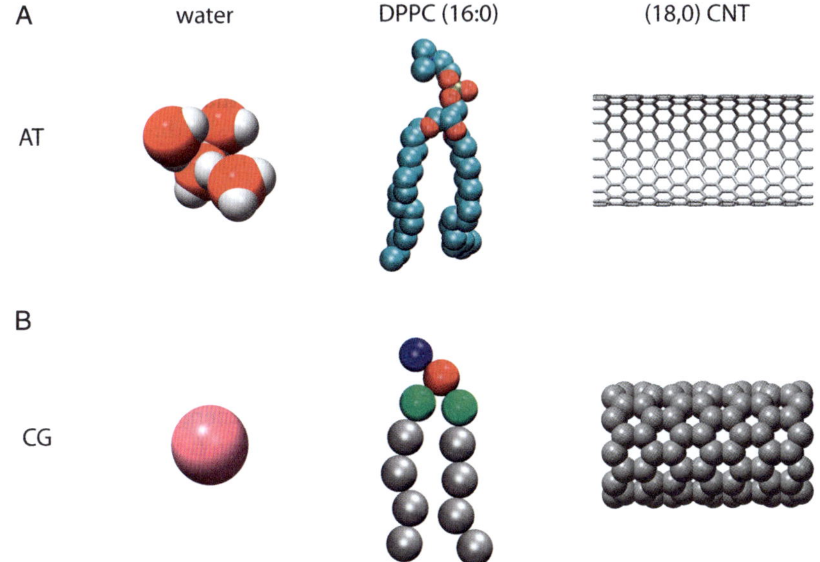

Figure 11.2 Atomistic (A) and coarse-grained (B) representations of water, the lipid dipalmitoylphosphatidylcholine (DPPC) and a CNT. In the AT model, atoms are coloured using the CPK convention. In the CG model, the particles are coloured as follows: pink = polar particle; blue/red = positive/negative charged particle; green = mixed polar/non-polar particle; and grey = hydrophobic particle. In the CG model, all particles are of the same size.

11.2.2.2 *Coarse-grained Models*

In recent years, advances have been made in the use of coarse-grained (CG) models of CNTs in a biological environment. CG models[35–39] reduce the overall system size by grouping together several atoms into a single particle (Figure 11.2). Additionally, the models use simpler and softer potentials that further increase computing efficiency. CG models lead to an ~100-fold increase in speed with respect to atomistic simulations, therefore allowing the simulation of larger length and longer time scales.

The CG approach has been successfully implemented to simulate lipid self-assembly into planar bilayers and vesicles.[35,40,41] The same CG methodology has been extended to allow simulation of both the self-assembly of proteins and detergents into mixed micelles and of the self-assembly of membrane proteins with lipid bilayers.[42,43]

The CG protocol has recently been applied to the simulation of CNTs.[44–50] This method can be particularly useful for simulating CNTs since these molecules are often very large. Figure 11.2 shows the CG structure of a CNT implemented in the authors' laboratory. This CG structure reflects the underlying atomistic CNT structure. Other groups have simulated CG nanopores built from individual rings of CG particles.[47–50]

11.3 CNT Interactions with Lipids and Related Molecules

We now go on to discuss MD studies that focus on CNT interactions with lipids and related molecules such as detergents. Understanding such interactions is of importance with respect to modelling the interactions of CNTs with biomembranes, but also with respect to understanding mechanisms of solubilization and delivery of CNTs to cells.

The application of CNTs in biological systems is critically related to their solubility in aqueous environments. As-synthesized CNTs are insoluble in most common solvents, including water.[51,52] The individual tubes form ropes or bundles that contain up to 100 CNTs packed in a hexagonal lattice with an inter-tube distance comparable to the inter-sheet distance in graphite.[53,54] Solubilization of CNTs has been facilitated by covalently attaching polar or charged groups to CNT surfaces. Usually the covalent modification involves esterification or amidation of acid-oxidized CNTs and side-wall attachment of functional groups.[55–59] Unfortunately, however, covalent modification can alter the inherent properties of CNTs. Therefore, attention has turned towards the non-covalent adsorption of lipids and detergents, polymers and other biomolecules on the surface of CNTs, thereby preserving the extended π networks of the tubes.

In this section, we focus on MD studies of the non-covalent adsorption of lipids and detergents on CNTs. We focus on lipids and detergents since not only are the interactions between these amphiphilic molecules and CNTs of interest from a solubility perspective, but also there is interest in the nature of interactions of CNTs with human cells. However, in order to interpret fully data on cell–CNT interactions, it is important that we understand how CNTs interact with cellular components, such as membranes. This in turn requires a more detailed characterization of the interactions of CNTs with the lipid molecules which make up cell membranes and with detergent molecules that may be used to solubilize both CNTs and membrane proteins.

Several studies have shown that dissolution of CNTs can be achieved *via* non-covalent adsorption of detergents or lipids on the CNT surface.[60–66] To date, the mechanism by which these amphiphilic molecules solubilize CNTs is poorly understood. Currently, there are three commonly used models for lipid/detergent adsorption on CNTs (Figure 11.3): CNT encapsulation within a cylindrical micelle, hemimicellar adsorption of detergents on CNTs and random adsorption of molecules on the tube surface.

MD simulations have been employed to study the adsorption profile of amphiphiles on CNTs. The advantage of simulations is that they allow both temporal and spatial resolution of the adsorption dynamics. O'Connell *et al.*[62] performed both experimental and MD simulation studies to examine the adsorption mode of detergents on CNTs. The detergent they studied was sodium dodecyl sulfate (SDS), a commonly used detergent for CNT solubilization. They found that SDS encapsulates CNTs in a cylindrical micelle. Their AT MD simulations were 0.4 ns in length and the CNT was modelled as a smooth cylinder.

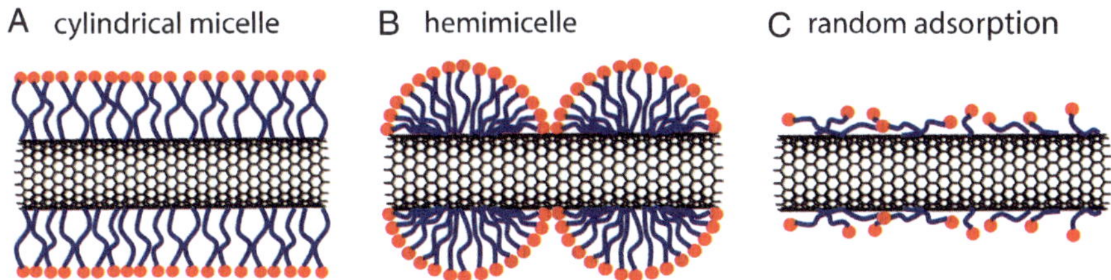

Figure 11.3 Possible mechanisms of adsorption of amphiphilic molecules on CNTs. (A) The cylindrical micelle, (B) hemimicelle and (C) random adsorption models. Red spheres represent the hydrophilic headgroups and the blue lines represent the hydrophobic tails. Illustration adapted from Wallace and Sansom.[46]

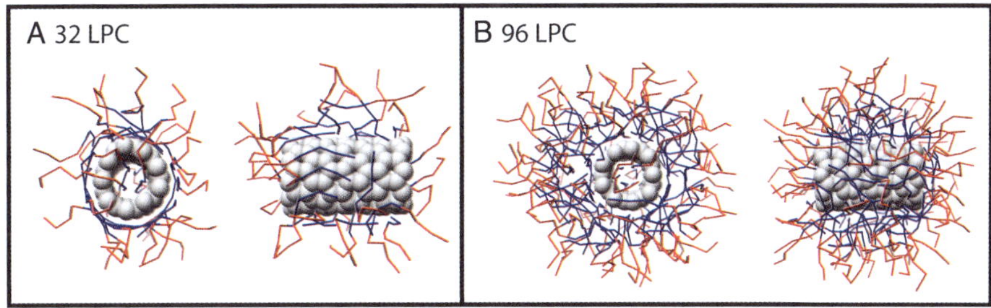

Figure 11.4 End and side snapshots of equilibrated LPC–CNT complexes, at low (A) and high (B) LPC/CNT ratios. The same colour scheme is used as in Figure 11.3. Adapted from Wallace and Sansom.[44]

Later, Qiao and Ke[67] performed relatively brief (24 ns) AT MD simulations of lysophospholipids adsorbing on CNTs in aqueous solution. Lysophospholipids (single-tailed phospholipids) are signalling molecules that occur naturally in cell membranes and hence are biocompatible. Recent experimental studies have shown that lysophospholipids provide superior solubility for CNTs[63,65,68] in comparison with the solubility provided by other detergents. The study by Qiao and Ke[67] focused on the adsorption mode of lysophosphatidylcholine (LPC, 18-carbon chain) on CNTs. It was found that at low LPC/CNT ratios, the detergent is predominantly aligned along the CNT axis.

In order to reach longer time scales than are routinely accessible *via* AT simulation, we performed ≥ 400 ns CG MD simulations of the self-assembly of LPC with a CG CNT. Our CNT was based on an atomistic (18,0) CNT, the same tube as that simulated by Qiao and Ke. In our CG simulations we did not observe LPC alignment along the CNT axis. Instead, we found that for a low LPC/CNT ratio, detergent molecules wrap around the tube (Figure 11.4A), with the wrapping angle dependent on tube chirality. However, for a high LPC/CNT ratio, the LPC molecules adopt a more micelle-like conformation on the tube surface (Figure 11.4B).

It is likely that the discrepancy between our CG study and the AT study by Qiao and Ke may reflect the more limited time scale accessible by AT simulations. Thus, in the AT simulations LPC aligns with the CNT axis after 24 ns, with the detergent headgroups lying adjacent on the tube surface. This conformation of LPC exposes large areas of the hydrophobic LPC tails to the aqueous environment. However, in the CG simulations the LPC tails wrap around the tube at low detergent concentration, with the detergent headgroups scattered evenly over the complex surface. This conformation of LPC reduces hydrophobic exposure of the LPC tails. It is anticipated that if an all-atom simulation was extended to CG time scales, then LPC would eventually redistribute over the tube surface so that hydrophobic exposure of the detergent tails is reduced.

Interestingly, we find that when LPC adsorbs on a long CNT at high detergent concentration, clear striations occur in the adsorption profile (see Figure 11.5B and C). The distance between striation peaks is ~ 40 Å, comparable to the spacing of LPC striations observed *via* TEM experiments (Figure 11.5A). Our results suggest that a hemimicelle model for adsorption may be preferred for a high detergent concentration.

We have also used CG MD to compare the adsorption mechanisms of the bilayer-forming lipid dipalmitoylphosphatidylcholine (DPPC) and of the detergents dihexanoylphosphatidylcholine (DHPC) and LPC on a CNT. All of these molecules have the same phosphatidylcholine headgroup and so differ only in the number and length of their hydrocarbon tails. We compared the behaviour of the two detergents with that of the bilayer-forming lipid, as a lipid-coated CNT may be less potentially disruptive of cell membranes.

We found that DPPC, DHPC and LPC all wrap around a CNT at a low amphiphile/CNT ratio, thereby minimizing contact between water and the hydrophobic CNT and amphiphile tails. Hence the random adsorption model shown in Figure 11.3 most accurately describes this process. Upon increasing the number of amphiphiles, a transition in adsorption is observed: DPPC encapsulates the CNT within a cylindrical micelle, whereas both DHPC and LPC adsorb onto CNTs in hemimicelles (Figure 11.5B and C).

The smooth configuration that DPPC adopts when adsorbed on a CNT reflects the packing preference of DPPC, namely, DPPC preferentially forms planar bilayers in an aqueous environment. Alternatively, both DHPC and LPC preferentially form micelles in aqueous solution, giving rise to the striated configuration when they are in the CNT–amphiphile complex. Our finding that DPPC forms a cylindrical micelle around a CNT suggests that lipids may be used to solubilize CNTs. This is supported by recent experiments on (functionalized) CNTs which suggest that solubilization by lipids may indeed be possible.[69]

Several studies have reported that there is an optimal detergent concentration for CNT dispersion.[63,64,66] This suggests that there may indeed be concentration-dependent amphiphile adsorption mechanisms, as found in our MD studies. Hence it is of interest to perform further simulation studies that probe the efficacies of solubilizing CNTs *via* the various adsorption mechanisms

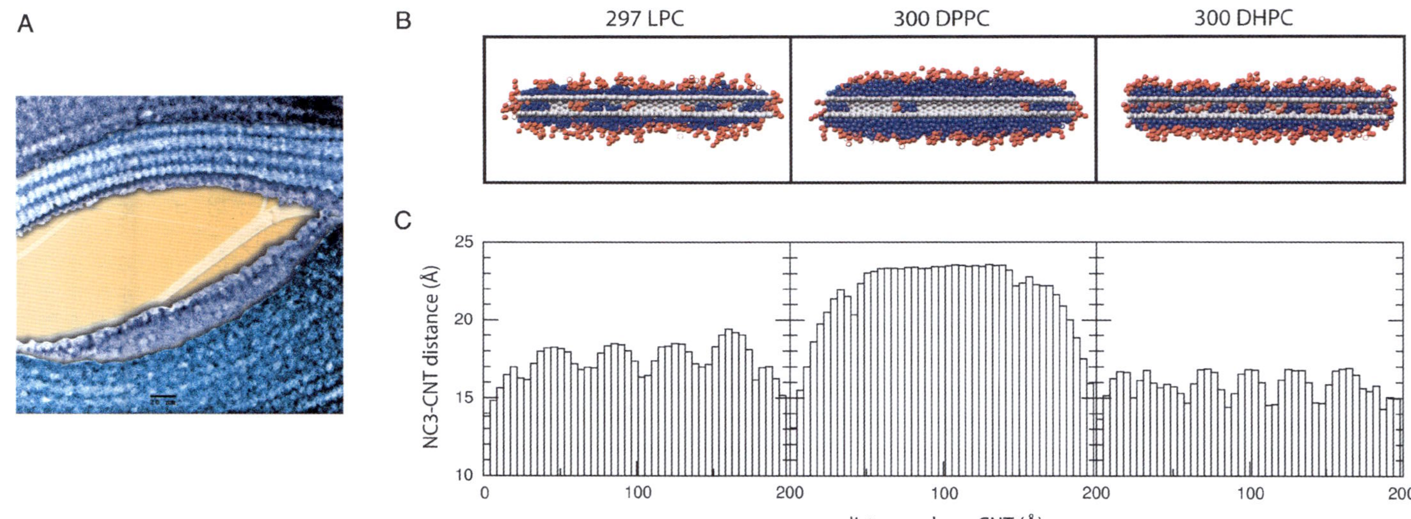

Figure 11.5 Adsorption profiles of amphiphiles on CNTs. (A) TEM image of CNT/LPC complexes.[65] Scale bar: 20 nm. (B) and (C) CG simulation results[46] at high amphiphile/CNT ratio. (B) Cross-sections through equilibrated amphiphile/CNT complexes. The CNT has a diameter of 14 Å and a length of 200 Å. (C) Average radial distance of the terminal amphiphile headgroup particle (NC3) from the CNT surface as a function of distance along the CNT. The data are averaged over a 200 ns period from 1.4 to 1.6 μs.

found here. Moreover, it would be useful to convert equilibrated CG amphiphile–CNT complexes into AT representations in order to examine the interactions in more detail. Such a multiscale simulation approach has been applied previously to lipid bilayers.[70,71]

11.4   Interaction of CNTs with Lipid Bilayers

If CNTs are to be exploited in both bionanotechnology and nanomedicine, then it is important to understand the interaction between CNTs and biological membranes. Surprisingly, there have been relatively few simulation studies that address this interaction mechanism. Klein and co-workers performed CG MD simulations of nanopores inserting into lipid bilayers.[48,72] Their generic nanopores are models for nanosyringes that may be based on CNTs. The nanopores are either purely hydrophobic tubes or hydrophobic tubes with hydrophilic sites at the ends of the tube. In the CG MD simulations, Klein and co-workers observed that the hydrophilic nanosyringe inserts spontaneously into the lipid bilayer (Figure 11.6). Interestingly, the reorientation of the nanosyringe such that it adopts a transmembrane orientation involves the assistance of chaperone lipids. These lipids enable one of the hydrophilic ends of the nanosyringe to cross the hydrophobic core of the membrane. In subsequent studies by Klein and co-workers,[47,49] it was found that the length of the nanosyringe plays an important role in the stability of the transmembrane structure. If the hydrophobic region of the tube is longer or shorter than that of the lipid bilayer, a swelling or compression of the bilayer is observed such that a meniscus forms around the tube. However, it is found that strong ordering in the membrane plane occurs regardless of the nanotube length. The addition of hydrophilic caps to the tube favourably stabilizes the transmembrane structure. Klein and co-workers also found that longer CNTs tilt to maximize the contact between the hydrophobic tube and the hydrophobic core of the membrane.

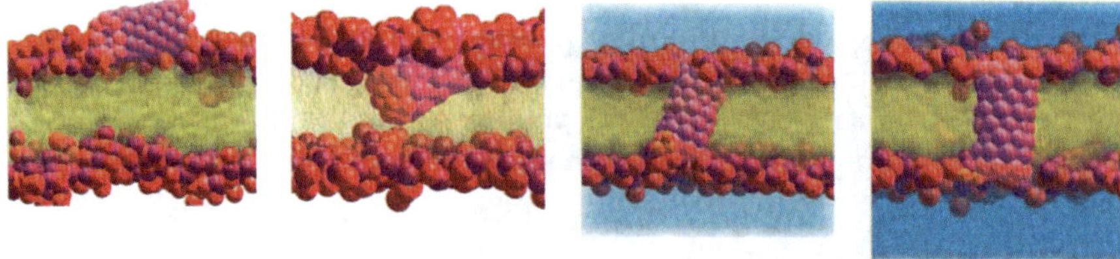

Figure 11.6   Snapshots showing the evolution of the insertion and rotation of a coarse-grained hydrophobic nanopore with hydrophilic caps into a lipid bilayer. The snapshots are coloured as follows: blue = water, pale purple = hydrophobic nanopores sites, orange = hydrophilic caps, yellow = lipid tails, dark purple = lipid phosphate unit and red = lipid choline unit. Image adapted from Lopez *et al.*[47]

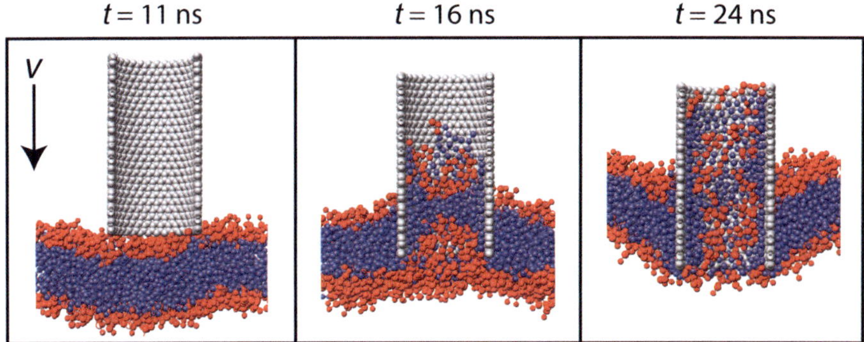

Figure 11.7 Snapshots of a CNT of diameter $d = 50$ Å and length $l = 100$ Å as it is pulled through a DPPC bilayer at pulling velocity $v = 5$ Å ns^{-1}. Each snapshot shows a cross-section through the bilayer and CNT. The same colour scheme is used as in Figure 11.3. Image adapted from Wallace and Sansom.[45]

Simulations performed in the authors' laboratory have addressed CNT entry into and exit from a lipid bilayer.[45] In these CG simulations, steered MD was employed to permit the penetration of DPPC lipid bilayers by hydrophobic SWNTs. This work was inspired by experimental studies that successfully used atomic force microscopy (AFM) to deliver molecules to cells by pushing a cargo-loaded CNT though a cell membrane, suggesting the use of CNTs as nanoinjectors. In the CG MD simulations, Wallace and Sansom observed that lipids are extracted from a bilayer during CNT penetration and reside on both the inner and outer tube surfaces.[45] The lipids that interact with the CNT interior wall spread out and therefore may 'block' the tube (Figure 11.7). Therefore, if CNTs are to be exploited as delivery systems, it may be more feasible for cargo to be carried outside the tube, since lipid blocking of the CNT interior could prevent delivery of internally carried cargo. Encouragingly, the authors did not observe an apparent effect on bilayer integrity after CNT penetration, with the bilayer able to self-seal.

Shi *et al.*[50] also addressed CNT entry into and exit from a lipid bilayer. However, unlike the tubes that we studied,[45] their CNTs were effectively functionalized, having energetically favourable interactions with lipid headgroups. Shi *et al.* found that CNTs with small radii are able to pierce through a lipid bilayer, whereas larger MWNTs having no internal cavity tend to cross a bilayer *via* a wrapping mechanism.

To date, the study of CNTs interacting with lipid bilayers is in its infancy. However, the research mentioned above highlights the rich interplay between CNTs and bilayers and suggests the need for further characterization of the fundamental nature of CNT–lipid interactions, including the influence of, *e.g.*, CNT functionalization and cargo on CNT insertion into and exit from a lipid bilayer.

11.5   CNTs as Nanopores

The size of a CNT is comparable to the size of a biological ion channel (typically of internal radius ~ 1–$5\,\text{Å}$ and length $\sim 30\,\text{Å}$).[73] Hence studies on water–electrolyte and CNT interactions have attracted more attention because of their potential applications in bionanotechnology.[6,8] In particular, there is considerable interest in CNTs as 'nanosyringes' spanning membranes, forming biomimetic pores capable of drug delivery or of selective transport of ions and water in biosensor devices.

In this section, we explore simulation studies of CNTs as model nanopores, which have focused on the transport properties of water and ions and more recently of charged biopolymers, *e.g.* nucleic acids, through such pores. It will be seen that as such simulations become more realistic, it will be important to integrate them with studies of the interactions of CNTs with the lipidic components of membranes, as described in the previous sections.

A number of different types of simulation system have been used to explore the transport properties of CNT nanopores. These range from isolated CNTs in water, through CNTs in a bilayer-mimicking 'slab' (which may be modelled by a water-excluding low dielectric region, a graphite sheet or a 'membrane' formed by alkane molecules), to CNTs embedded in a phospholipid bilayer (Figure 11.8). Whereas the simpler systems (*e.g.* CNTs in a slab) allow more extensive and detailed studies of *e.g.* water permeation, the more complex systems (*e.g.* a CNT in a lipid bilayer) may capture important effects of the membrane environment on the functional behaviour of a nanopore.

A further model which has been explored is a 'membrane' which is formed by a parallel array of hexagonally packed CNTs. Although this does not directly mimic a CNT-containing biomembrane, such a system is clearly related to one of the CNTs in a bilayer. Recent *experimental* studies have addressed water and

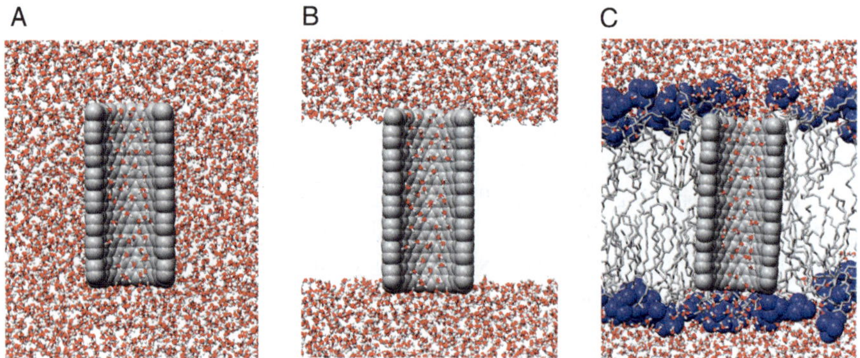

Figure 11.8   Snapshots of an (18,0) CNT nanopore in water (A), in a bilayer-mimicking 'slab' (B) and in a lipid bilayer (C). The hydrophobic CNT and lipid tails are shown in grey and the hydrophilic lipid headgroups in blue.

ion transport through a CNT-containing membrane formed by deposition of silicon nitride on a densely packed 'forest' of vertically aligned CNTs and have shown that decoration of the mouths of the nanotubes with charged groups can lead to selective ion transport properties.[74]

11.5.1 Transport of Water and Ions Through CNT Nanopores

There have been many simulation studies of the structural and transport properties of water in CNTs, and these have been reviewed recently.[9] In pioneering early studies, Hummer *et al.*[19] demonstrated that an SWNT immersed in water could undergo transitions between an empty and full state (*i.e.* a transition between liquid and vapour within the nanotube) and suggested that water occupancy, and hence transport, could be tuneable by, *e.g.*, local changes in nanopore polarity.[75] Beckstein and co-workers[76–78] observed similar liquid–vapour oscillation in simplified models of biomimetic nanopores (Figure 11.9) and also suggested the possibility of 'gating' of transport through such pores by changes in pore size, polarity and/or flexibility.

Subsequent studies of water in CNT nanopores have built upon these initial observations to reveal some of the complexities of simulated behaviour of these systems. Membranes formed of hexagonally packed parallel CNTs have been

Figure 11.9 A simplified model of a biomimetic nanopores, showing the van der Waals surface of a pore model (blue) embedded in a membrane-mimetic slab (brown). Image provided by Oliver Beckstein.

used to simulate directly the osmotic flow of water through nanopores.[20] These studies confirmed a high flow rate of water through CNT nanopores and suggested that the limiting factors in such flow included water entry/exit at the mouths of the pores.

A number of studies have modelled the membrane environment of a single CNT nanopore by a 'slab'-type model, either a simple low-dielectric region from which water molecules are excluded or by embedding the CNT in one or more graphite-like sheets that mimic a membrane. Comparing a (16,16) CNT with other model nanotubes, it was shown that the fast hydrodynamic flow of water through CNT nanopores was a result of the smooth water/nanotube interface and that making the wall of the nanotubes more polar or rougher resulted in a slower rate of water flow.[29]

Some interesting studies of a CNT nanopore embedded in a graphite sheet revealed some potential sensitivities to the nature of the membrane model. Studies with a single graphite sheet and a charged particle in the plane of the sheet just external to the CNT[79] revealed a marked dependence of the flow of water on the distance (δ) between the charge and the CNT wall. Thus, for a separation of $\delta = 0\,\text{Å}$ the nanopore did not support water flow, whereas for $\delta > 0.85\,\text{Å}$ the external charge had little effect on the water flow seen through a pristine CNT nanopore. Studies with a pristine CNT embedded within a 'membrane' formed by two parallel graphite sheets[22] showed an up to two-fold increase in the water flux as the vertical separation of the graphite sheets ξ was increased from 0 to $10\,\text{Å}$. Hence what lies outside the nanopore may have an important effect on its transport properties, arguing for more realistic membrane models to be used.

More recently, there have been a number of simulations of CNT nanopores in more realistic lipid bilayer and related membrane models. For example, Garate *et al.*[80] simulated SWNTs ranging from (5,5) to (11,11) in POPC bilayers and showed a switch from single-file to classical diffusion of water through the nanopores. Another recent study explored some of the complexities that might be introduced by functionalization of CNT pores embedded in a DMPC bilayer.[81] In particular, the authors compared single-walled with double-walled CNT nanopores and suggested that shielding of the inner CNT from lipid-induced distortions was the cause of observed two-fold slower water transport in the double-walled CNT nanopore.

Recent work by Mashl and Jakobsson addressed the phase and transport properties of water confined inside a hydrophobic nanopore.[82] They found that water undergoes a series of distinct transitions as the diameter of the pore is varied.

Notwithstanding the complexities of water within CNT nanopores, there have been a number of studies of ions. In an early study of a (16,16) CNT nanopore embedded in a low-dielectric slab with $1.85\,\text{M}$ KCl solution on either side of the slab, Mashl and co-workers[30] showed that for a pristine CNT very few (less than one at a time) ions entered the nanopore. However, if the two mouths of the nanotubes were decorated with charges, then ions entered the pore in a selective fashion. In a wide-ranging review of the uses of modelling and simulation for bionanotechnology, Schulten and co-workers[8] emphasized the importance of inclusion of polarizability in the CNT model when exploring energetics of ions within a nanopore. For an isolated CNT nanopore, the energetics of interaction with a K^+ ion are significantly altered, especially at entry/exit of the nanopore, by the presence/absence of polarizability from the forcefield.[33,83] In an earlier study of a membrane formed by an array of CNTs, Schulten's group demonstrated by comparing pristine and decorated CNTs that modification of the CNT could modulate the dipolar ordering of waters within the nanopores.[32] Recently, Song and Corry studied ion transport through narrow hydrophobic pores in membranes comprised entirely of SWNTs.[84] They found that Na^+, K^+ and Cl^- faced different free energy barriers when entering the hydrophobic pores, explaining the intrinsic ion selectivity of narrow CNTs. Experimental data on both water and ion transport through CNTs are sparse.

In summary, simulation studies reveal the sensitivity of water and ion transport not only to the simulation parameters, but also more importantly to the structure and dynamics (*i.e.* size, polarity and environment) of the nanopore. This suggests that further simulation studies are needed before we can fully understand the behaviour of water within CNT nanopores in biological and biomimetic membranes.

11.5.2 Nanopores as Nanosyringes

A major motivation for studying CNTs as nanopores is to explore their potential as nanosyringes. To this end, there have been studies of the translocation of single-stranded RNA molecules through CNTs – the latter either in a membrane formed by a hexagonal array of CNTs[85] or more recently as single CNT nanopores in a lipid (DMPC) bilayer.[86] In the former study, it was revealed that the exit of RNA from the nanopores was controlled by hydrophobic interactions between the bases and walls of the CNT. In the latter bilayer simulations, pristine CNTs were compared with CNT nanopores whose mouths were decorated with hydroxyl groups. It was found that the decoration of the mouth of the nanopore prevented its occlusion by 'creep' of lipid molecules at the rims of the mouth. Furthermore, the lipid headgroups interacted with the translocating RNA molecule. The authors concluded that interactions of suitable decorations of the CNT and of membrane components might be exploited to control the transport properties of a nanopore. This encourages and justifies the continued development of more accurate (*e.g.* multi-scale) and more realistic (in terms of, *e.g.*, CNT derivatization and lipid bilayer composition) simulations of transport through CNT nanopores.

Klein and colleagues have used CG simulations to study the insertion of simplified models of nanopores into lipid bilayers (Figure 11.6).[47–49] These studies have emphasized the role of polar derivatization of the mouths of the nanopores in stabilizing a bilayer spanning orientation. It will be of interest in future studies to see how more detailed model of CNT derivatization may be explored in terms of nanopore localization and stabilization in lipid bilayer membranes.

11.6 Conclusion

Potential applications of CNTs in nanomedicine and bionanotechnology are an increasingly important field of research. However, in order for this field to develop fully it is imperative that we obtain a deeper understanding of how CNTs interact with model biomembranes and their cellular counterparts. In recent years, significant progress has been made using molecular modelling methods to explore the fundamental interplay between CNTs and biological systems. In particular, molecular dynamics has been implemented to study the interactions between CNTs and lipid bilayers and the transport properties of water in CNT nanopores. Moreover, there appears to be an encouraging shift towards more realistic models of membranes in nanopore studies. However, significant challenges still remain to be solved. As mentioned above, the electronic systems of π-electrons in CNTs are highly polarizable and so are readily influenced by their environment. Therefore, polarizable CNT models are needed to model more accurately the intermolecular forces between CNTs and their environment. However, polarizable CNT models are computationally demanding. Hence to date, only very short tube sections have been simulated.[33,34] In addition, there is a need to adopt a multi-scale modelling approach[87] in order to combine greater biological realism with more robust and accurate calculations. Nevertheless, so far molecular modelling has contributed significantly to our understanding of how CNTs interact with a biological environment and this field is expected to continue to grow at a rapid pace.

References

1. A. Bianco, J. Hoebeke, S. Godefroy, O. Chaloin, D. Pantarotto, J. Briand, S. Muller, M. Prato and C. D. Partidos, *J. Am. Chem. Soc.*, 2005, **127**, 58.
2. P. Cherukuri, S. M. Bachilo, S. H. Litovsky and R. B. Weisman, *J. Am. Chem. Soc.*, 2004, **126**, 15638.
3. N. W. S. Kam, T. C. Jessop, P. A. Wender and H. J. Dai, *J. Am. Chem. Soc.*, 2004, **126**, 6850.
4. Q. Lu, J. M. Moore, G. Huang, A. S. Mount, A. M. Rao, L. L. Larcom and P. C. Ke, *Nano Lett.*, 2004, **4**, 2473.
5. D. Pantarotto, J. Briand, M. Prato and A. Bianco, *Chem. Commun.*, 2004, **16**.
6. M. Whitby and N. Quirke, *Nat. Nanotechnol.*, 2007, **2**, 87.
7. G. Oberdorster, E. Oberdorster and J. Oberdorster, *Environ. Health Perspect.*, 2005, **113**, 823.
8. D. Y. Lu, A. Aksimentiev, A. Y. Shih, E. Cruz-Chu, P. L. Freddolino, A. Arkipov and K. Schulten, *Phys. Biol.*, 2006, **3**, S40.
9. A. Alexiadis and S. Kassinos, *Chem. Rev.*, 2008, **108**, 5014.

10. M. S. Dresselhaus, G. Dresselhaus and P. Avouris, *Carbon Nanotubes: Synthesis, Structure, Properties and Applications*, Springer, Berlin, 2001.
11. X. Sun, S. Zaric, D. Daranciang, K. Welsher, Y. Lu, X. Li and H. Dai, *J. Am. Chem. Soc.*, 2008, **130**, 6551.
12. N. Wang, Z. K. Tang, G. D. Li and J. S. Chen, *Nature*, 2000, **408**, 50.
13. Q. H. Yang, S. Bai, J. L. Sauvajol and J. B. Bai, *Adv. Mater.*, 2003, **15**, 792.
14. Z. L. Wang, P. Poncharal and W. A. de Heer, *J. Phys. Chem. Solids*, 2000, **61**, 1025.
15. A. E. Porter, M. Gass, K. Muller, J. N. Skepper, P. A. Midgley and M. Welland, *Nat. Nanotechnol.*, 2007, **2**, 713.
16. R. J. Mashl, S. Joseph, N. R. Aluru and E. Jakobsson, *Nano Lett.*, 2003, **3**, 589.
17. T. Nanok, N. Artrith, P. Pantu, P. A. Bopp and J. Limtrakul, *J. Phys. Chem. A*, 2009, **113**, 2103.
18. Y. D. Zhu, M. J. Wei, Q. Shao, L. H. Lu, X. H. Lu and W. F. Shen, *J. Phys. Chem. C*, 2009, **113**, 882.
19. G. Hummer, J. C. Rasaiah and J. P. Noworyta, *Nature*, 2001, **414**, 188.
20. A. Kalra, S. Garde and G. Hummer, *Proc. Natl. Acad. Sci. USA*, 2003, **100**, 10175.
21. G. Cicero, J. C. Grossman, E. Schwegler, F. Gygi and G. Galli, *J. Am. Chem. Soc.*, 2008, **130**, 1871.
22. X. Gong, J. Li, H. Zhang, R. Wan, H. Lu, S. Wang and H. Fang, *Phys. Rev. Lett.*, 2008, **101**, 257801.
23. Y. Kang, Y. C. Liu, Q. Wang, J. W. Shen, T. Wu and W. J. Guan, *Biomaterials*, 2009, **30**, 2807.
24. J. H. Walther, R. Jaffe, T. Halicioglu and P. Koumoutsakos, *J. Phys. Chem. B*, 2001, **105**, 9980.
25. J. H. Walther, R. L. Jaffe, E. M. Kotsalis, T. Werder, T. Halicioglu and P. Koumoutsakos, *Carbon*, 2004, **42**, 1185.
26. C. W. Chen and M. H. Lee, *Nanotechnology*, 2004, **15**, 480.
27. S. M. Hou, Z. Y. Shen, X. Y. Zhao and Z. Q. Xue, *Chem. Phys. Lett.*, 2003, **373**, 308.
28. B. D. Huang, Y. Y. Xia, M. W. Zhao, F. Li, X. D. Liu, Y. J. Ji and C. Song, *J. Chem. Phys.*, 2005, **122**, 84708.
29. S. Joseph and N. R. Aluru, *Nano Lett.*, 2008, **8**, 452.
30. S. Joseph, R. J. Mashl, E. Jakobsson and N. R. Aluru, *Nano Lett.*, 2003, **3**, 1399.
31. C. Y. Won, S. Joseph and N. R. Aluru, *J. Chem. Phys.*, 2006, **125**, 114701.
32. F. Zhu and K. Schulten, *Biophys. J.*, 2003, **85**, 236.
33. D. Y. Lu, Y. Li, U. Ravaioli and K. Schulten, *J. Phys. Chem. B*, 2005, **109**, 11461.
34. D. Y. Lu, Y. Li, S. V. Rotkin, U. Ravaioli and K. Schulten, *Nano Lett.*, 2004, **4**, 2383.
35. S. J. Marrink, A. H. de Vries and A. E. Mark, *J. Phys. Chem. B*, 2004, **108**, 750.

36. J. C. Shelley, M. Y. Shelley, R. C. Reeder, S. Bandyopadhyay and M. L. Klein, *J. Phys. Chem. B*, 2001, **105**, 4464.
37. B. Smit, A. J. Hilbers, K. Esselink, L. A. M. Rupert, N. M. Van Os and G. Schlijper, *Nature*, 1990, **348**, 624.
38. V. Tozzini, *Curr. Opin. Struct. Biol.*, 2005, **15**, 144.
39. L. Whitehead, C. M. Edge and J. W. Essex, *J. Comput. Chem.*, 2001, **22**, 1622.
40. S. J. Marrink and A. E. Mark, *J. Am. Chem. Soc.*, 2003, **125**, 15233.
41. H. J. Risselada and S. J. Marrink, *Phys. Chem. Chem. Phys.*, 2009, **11**, 2056.
42. P. J. Bond, J. Holyoake, A. Ivetac, S. Khalid and M. S. P. Sansom, *J. Struct. Biol.*, 2007, **157**, 593.
43. P. J. Bond and M. S. P. Sansom, *J. Am. Chem. Soc.*, 2006, **128**, 2697.
44. E. J. Wallace and M. S. P. Sansom, *Nano Lett.*, 2007, **7**, 1923.
45. E. J. Wallace and M. S. P. Sansom, *Nano Lett.*, 2008, **8**, 2751.
46. E. J. Wallace and M. S. P. Sansom, *Nanotechnology*, 2009, **20**, 045101.
47. C. F. Lopez, S. O. Nielsen, B. Ensing, P. B. Moore and M. L. Klein, *Biophys. J.*, 2005, **88**, 3083.
48. C. F. Lopez, S. O. Nielsen, P. B. Moore and M. L. Klein, *Proc. Natl. Acad. Sci. USA*, 2004, **101**, 4431.
49. S. O. Nielsen, B. Ensing, V. Ortiz, P. B. Moore and M. L. Klein, *Biophys. J.*, 2005, **88**, 3822.
50. X. H. Shi, Y. Kong and H. J. Gao, *Acta Mech. Sin.*, 2008, **24**, 161.
51. R. S. Ruoff, D. S. Tse, R. Malhotra and D. C. Lorents, *J. Phys. Chem.*, 1993, **97**, 3379.
52. W. A. Scrivens and J. M. Tour, *J. Chem. Soc., Chem. Commun.*, 1993, **1207**.
53. E. Nativ-Roth, R. Shvartzman-Cohen, C. Bounioux, M. Florent, D. S. Zhang, I. Szleifer and R. Yerushalmi-Rozen, *Macromolecules*, 2007, **40**, 3676.
54. A. Thess, R. Lee, P. Nikolaev, H. J. Dai, P. Petit, J. Robert, C. H. Xu, Y. H. Lee, S. G. Kim, A. G. Rinzler, D. T. Colbert, G. E. Scuseria, D. Tomanek, J. E. Fischer and R. E. Smalley, *Science*, 1996, **273**, 483.
55. J. L. Bahr, E. T. Mickelson, M. J. Bronikowski, R. E. Smalley and J. M. Tour, *Chem. Commun.*, 2001, **193**.
56. S. Banerjee and S. S. Wong, *J. Am. Chem. Soc.*, 2002, **124**, 8940.
57. F. Pompeo and D. E. Resasco, *Nano Lett.*, 2002, **2**, 369.
58. M. Sano, A. Kamino, J. Okamura and S. Shinkai, *Langmuir*, 2001, **17**, 5125.
59. Y. Sun, S. R. Wilson and D. I. Schuster, *J. Am. Chem. Soc.*, 2001, **123**, 5348.
60. M. F. Islam, E. Rojas, D. M. Bergey, A. T. Johnson and A. G. Yodh, *Nano Lett.*, 2003, **3**, 269.
61. V. C. Moore, M. S. Strano, E. H. Haroz, R. H. Hauge, R. E. Smalley, J. Schmidt and Y. Talmon, *Nano Lett.*, 2003, **3**, 1379.
62. M. J. O'Connell, S. H. Bachilo, C. B. Huffman, V. C. Moore, M. S. Strano, E. H. Haroz, K. L. Rialon, P. J. Boul, W. H. Noon, C. Kittrell, J. Ma, R. H. Hauge, R. B. Weisman and R. E. Smalley, *Science*, 2002, **297**, 593.

63. C. Richard, F. Balavoine, P. Schultz, T. W. Ebbesen and C. Mioskowski, *Science*, 2003, **300**, 775.
64. H. Wang, W. Zhou, D. L. Ho, K. I. Winey, J. E. Fischer, C. J. Glinka and E. K. Hobbie, *Nano Lett.*, 2004, **4**, 1789.
65. Y. Wu, J. S. Hudson, Q. Lu, J. M. Moore, A. S. Mount, A. M. Rao, E. Alexov and P. C. Ke, *J. Phys. Chem. B*, 2006, **110**, 2475.
66. K. Yurekli, C. A. Mitchell and R. Krishnamoorti, *J. Am. Chem. Soc.*, 2004, **126**, 9902.
67. R. Qiao and P. C. Ke, *J. Am. Chem. Soc.*, 2006, **128**, 13656.
68. P. C. Ke, *Phys. Chem. Chem. Phys.*, 2007, **9**, 439.
69. N. C. Toledo, M. R. R. de Planque, S. A. Contera, N. Grobert and J. F. Ryan, *Jpn. J. Appl. Phys.*, 2007, **46**, 2799.
70. R. Chang, G. S. Ayton and G. A. Voth, *J. Chem. Phys.*, 2005, **122**, 244716.
71. A. P. Lyubartsev, *Eur. Biophys. J.*, 2005, **35**, 53.
72. C. F. Lopez, S. O. Nielsen, B. Ensing, P. B. Moore and M. L. Klein, *Biophys. J.*, 2005, **88**, 3083.
73. M. S. P. Sansom and M. S. P. Biggin, *Nature*, 2001, **414**, 156.
74. F. Fornasiero, H. G. Park, J. K. Holt, M. Stadermann, C. P. Grigoropoulos, A. Noy and O. Bakajin, *Proc. Natl. Acad. Sci. USA*, 2008, **105**, 17250.
75. S. Sriraman, I. G. Kevrekidis and G. Hummer, *Phys. Rev. Lett.*, 2005, **95**, 130603.
76. O. Beckstein, P. C. Biggin and M. S. P. Sansom, *J. Phys. Chem. B*, 2001, **105**, 12902.
77. O. Beckstein and M. S. P. Sansom, *Proc. Natl. Acad. Sci. USA*, 2003, **100**, 7063.
78. O. Beckstein and M. S. P. Sansom, *Phys. Biol.*, 2004, **1**, 42.
79. J. Y. Li, X. J. Gong, H. J. Lu, D. Li, H. P. Fang and R. H. Zhou, *Proc. Natl. Acad. Sci. USA*, 2007, **104**, 3687.
80. J. A. Garate, N. J. English and J. M. D. MacElroy, *Mol. Simul.*, 2009, **35**, 3.
81. B. Liu, X. Y. Li, B. L. Li, B. Q. Xu and Y. L. Zhao, *Nano Lett.*, 2009, **9**, 1386.
82. R. J. Mashl, S. Joseph, N. Alarn and E. Jakobsson, *Nanotech 2003*, 2003, **1**, 152.
83. D. Y. Lu, Y. Li, U. Ravaioli and K. Schulten, *Phys. Rev. Lett.*, 2005, **95**, 246801.
84. C. Song and B. Corry, *J. Phys. Chem. B*, 2009, **113**, 7642.
85. I. C. Yeh and G. Hummer, *Proc. Natl. Acad. Sci. USA*, 2004, **101**, 12177.
86. U. Zimmerli and P. Koumoutsakos, *Biophys. J.*, 2008, **94**, 2546.
87. G. S. Ayton and G. A. Voth, *Curr. Opin. Struct. Biol.*, 2008, **19**, 138.

Subject Index

Page references to *figures* and *tables* are shown in *italics*.